W9-DEI-823

MODERN WAR STUDIES

Theodore A. Wilson
General Editor

Raymond Callahan

J. Garry Clifford

Jacob W. Kipp

Allan R. Millett

Carol Reardon

Dennis Showalter

David R. Stone

Series Editors

U.S. Army Doctrine

U.S. Army Doctrine

From the American Revolution to the
War on Terror

WALTER E. KRETCHIK

UNIVERSITY PRESS OF KANSAS

Published by the University Press of Kansas (Lawrence, Kansas 66045), which was organized by the Kansas Board of Regents and is operated and funded by Emporia State University, Fort Hays State University, Kansas State University, Pittsburg State University, the University of Kansas, and Wichita State University

© 2011 by the University Press of Kansas

Library of Congress Cataloging-in-Publication Data

Kretchik, Walter E. (Walter Edward), 1954–
    U.S. Army doctrine : from the American Revolution to the War on Terror / Walter E. Kretchik.
        p.   cm. — (Modern war studies)
    Includes bibliographical references and index.
    ISBN 978-0-7006-1806-4 (cloth : alk. paper)
1. Military doctrine—United States—History.
2. Strategic culture—United States—History.
3. United States. Army—History. I. Title.
    UA23.K778 2011
    355.020973—dc22

                                                2011010714

British Library Cataloguing-in-Publication Data is available.

Printed in the United States of America
10   9   8   7   6   5   4   3   2   1

To Pamela

To be at the head of a strong column of troops, in the execution of some task that requires brain, is the highest pleasure of war—a grim one and terrible, but which leaves on the mind and memory the strongest mark; to detect the weak point of an enemy's line; to break through with vehemence and thus lead to victory; or to discover some key-point and hold it with tenacity; or to do some other distinct act which is afterward recognized as the real cause of success. These all become matters that are never forgotten.

—William Tecumseh Sherman,
Memoirs of Gen. William T. Sherman

CONTENTS

ACKNOWLEDGMENTS

I am pleased to acknowledge the many friends and colleagues who offered advice, assistance, and encouragement during my work on this project. My academic mentors Ted Wilson and Roger Spiller reinvigorated languishing intellectual energy and provided the inspiration to make this book possible. My thanks to the reviewers, Brian McAllister Linn, Roger Spiller, and Jonathan M. House, for their comments and suggestions that improved the manuscript. I am indebted to the series editors and staff at the University Press of Kansas and Michael Briggs, Jennifer A. Dropkin, Susan K. Schott, and Amy Sherman, in particular. Thanks also to Erin Girard for transcribing oral histories and to those who agreed to be interviewed. Joseph G. D. "Geoff" Babb, A. J. Bacevich, Larry T. Balsamo, Charles H. Baumann, Robert F. Baumann, Robert Berlin, Jerry Brown, Edwin Burgess, Michael Burke, Richard E. Cavazos, Carmen J. Cavezza, Alexander "Sandy" Cochran, William C. Combs, Tony Cucolo, Robert Epstein, David M. Fishback, John T. Fishel, Christopher Gable, George W. Gawrych, Greg Heritage, Michael Jallo, Virginia Jelatis, Virginia Leonard, Alan Lowe, Jim McDonough, Allan Millet, Larry Moores, Jerry D. Morelock, Tony R. "Randy" Mullis, Tim Nenninger, William G. Pierce, Jeffrey C. Prater, William Glenn Robertson, Douglas P. Scalard, Peter J. Schifferle, James Schneider, Dwayne Wagner, David S. Weisman, James Willbanks, Larry T. Yates, and the exceptionally professional librarians and staff at the Combined Arms Research Library (CARL), Fort Leavenworth, Kansas, made many contributions that may not be readily apparent to them, but I am indebted nonetheless. I greatly appreciate the candor and professionalism of Scott W. Palmer, a true friend and colleague who proofread the manuscript and offered perceptions that significantly strengthened this work. I appreciate the expedient efforts of Heather Moore of the U.S. Senate Historical Office, Jim Parker at Double Delta, and the staff members of Art Resource, the Granger Collection, the U.S. Army Heritage and Education Center, and the National Archives in providing documents and images. To the conference attendees at the various Society of Military History panels over the years, many thanks for the feedback on the papers I presented regarding this topic.

I thank Western Illinois University for a travel stipend that helped to fund my research. To my colleagues and students in the Department of

History, your kind words of encouragement over the years were much appreciated. Thanks go out to my colleagues and the many graduate and undergraduate students at Bilkent University in Ankara, Turkey, as well as those at the U.S. Army Command and General Staff College.

At the more personal level, my wife, Pamela J. Kontowicz, helped in more ways than I can count. She was always there for me and sustained me during many difficult days. I would be lost without her. I am also indebted to my parents, Walter V. Kretchik and Martha L. (Selvaggi) Kretchik, for their encouragement and the many personal sacrifices they made on my behalf.

LIST OF ABBREVIATIONS

AAF	Army Air Forces
AEF	Army Expeditionary Forces
AFG	American Forces in Germany
AMAG	American Mission to Aid Greece
AO	area of operations
AOE	Army of Excellence
ARFOR	Army Forces
ARTEP	Army Training and Evaluation Program
ARVN	Army of the Republic of Vietnam
ASCOPE	area, structures, capabilities, organizations, people, and events
ASPIA	American State Papers Indian Affairs
BCTP	Battle Command Training Program
CAC	Combined Arms Center
CALL	Center for Army Lessons Learned, Fort Leavenworth, Kansas
CARL	Combined Arms Research Library, United States Army Command and General Staff College, Fort Leavenworth, Kansas
CENTCOM	U.S. Central Command
CFLCC	Combined Forces Land Component Command
CGSC	Command and General Staff College
CINC	Combatant Commander
CJTF-7	Combined/Joint Task Force 7
COIN	counterinsurgency
CONARC	Continental Army Command
CSI	Combat Studies Institute
CTCs	Combat Training Centers
DMH	Department of Military History
DoD	Department of Defense
DTIC	Defense Technical Information Center
DUCO	Department of Unified and Combined Operations
EM	cyber/electromagnetic activities
EUCOM	European Command
FID	foreign internal defense

FLOT	forward line of own troops
FM	Field Manual
FSCL	fire support coordination line
FSR	*Field Service Regulations*
GPO	Government Printing Office
IFOR	Implementation Force
IIA	inform and influence activities
JCS	joint chiefs of staff
JSEAD	joint suppression of enemy air defenses
JSOTF	Joint Special Operations Task Force
JTF	Joint Task Force
JUSMAG	Joint United States Military Advisory Group
JUSMAPG	Joint United States Military Advisory and Planning Group, Greece
LIC	low-intensity conflict
LRRPS	long-range reconnaissance patrols
MACV	Military Assistance Command Vietnam
MEF	Marine Expeditionary Force
MLRS	Multiple Launch Rocket System
MNC-I	Multinational Corps–Iraq
MNF-I	Multinational Force–Iraq
MOMAR	Modern Mobile Army
MSTC-I	Multinational Security Transition Command–Iraq
MOOTW	military operations other than war
NARA	National Archives and Records Administration
NATO	North Atlantic Treaty Organization
NDA	National Defense Act
ODSS	offense, defense, stability, and support
OOTW	operations other than war
ORHA	Office of Reconstruction and Humanitarian Assistance
OTCs	Officer Training Camps
PROVN	Program for the Pacification and Long-Term Development of South Vietnam
ROAD	Reorganization Objectives Army Division
ROE	rules of engagement
ROTC	Reserve Officers Training Corps
SAMS	School of Advanced Military Studies
SOF	special operations forces

SPACECOM	U.S. Space Command
STANAG	standard agreements
SQT	Skill Qualification Test
TRADOC	Training and Doctrine Command
TRANSCOM	U.S. Transportation Command
TRICAP	triple capability
UAR	United Arab Republic
UN	United Nations
UNMIH	UN Mission in Haiti
UNPROFOR	UN Protection Force
UNSCR	UN Security Council Resolution
USAREUR	U.S. Army Europe
USARPAC	U.S. Army Pacific
USSOCOM	U.S. Special Operations Command
USSOUTHCOM	U.S. Southern Command

CHRONOLOGY OF U.S. ARMY KEYSTONE DOCTRINAL MANUALS

See the text for published changes or updates to various manuals.

1779   *Regulations for the Order and Discipline of the Troops of the United States, Part 1*

1812   *Regulations for the Field Exercise, Maneuvers, and Conduct of the Infantry of the United States Drawn Up and Adapted to the Organization of the Militia and Regular Troops*

1813   *Regulations to be Received and Observed for the Discipline of Infantry in the Army of the United States*

1815   *Infantry Tactics*

1855   *Rifle and Light Infantry Tactics for the Exercise and Maneuvers of troops when acting as light infantry or riflemen*

1862   *Infantry Tactics*

1867   *Infantry Tactics*

1891   *Infantry Drill Regulations*

1905   *Field Service Regulations*

1910   *Field Service Regulations*

1913   *Field Service Regulations*

1914   *Field Service Regulations*

1923   *Field Service Regulations*

1939   Tentative *Field Service Regulations* Field Manual 100-5 *Operations*

1941   *Field Service Regulations* Field Manual 100-5 *Operations*

1944   *Field Service Regulations* Field Manual 100-5 *Operations*

1949   *Field Service Regulations* Field Manual 100-5 *Operations*

1954   *Field Service Regulations* Field Manual 100-5 *Operations*

1962   *Field Service Regulations* Field Manual 100-5 *Operations*

1968   Field Manual 100-5 *Operations of Army Forces in the Field*

1976   Field Manual 100-5 *Operations*

1982   Field Manual 100-5 *Operations*

1986   Field Manual 100-5 *Operations*

1993   Field Manual 100-5 *Operations*

2001   Field Manual 3-0 *Operations*

2008   Field Manual 3-0 *Operations*

INTRODUCTION

U.S. ARMY DOCTRINE IN
HISTORICAL PERSPECTIVE

It is only analytical that these attempts at theory can be
called advances in the realm of truth; synthetically, in
the rules and regulations they offer, they are absolutely
useless. They aim at fixed values; but in war everything
is uncertain, and calculations have to be made with
variable quantities.—Carl von Clausewitz in *On War*

Shall I be understood as saying that there are no such
things as tactical rules, and that no theory of tactics can
be useful? What military man of intelligence would be
guilty of such an absurdity?—Antoine Henri de Jomini
in *The Art of War*

Contemporaries in life, the nineteenth-century Prussian master Carl von Clausewitz and the Swiss theorist Antoine Henri
de Jomini both wrote about war. While they agreed in many ways, they
differed significantly in others. For Clausewitz, war is a dynamic process
beyond absolute control and enduring principles. An army operates in the
realm of chaos. Its leaders must cope with the intrinsic pandemonium
and unique nature of each conflict. To Jomini, methods change in war, but
principles endure; the fundamentals of war are simple enough for those
with the genius to comprehend it.

Despite their differences, both men have nonetheless shaped the perceptions and approaches of U.S. Army leaders toward armed conflict.
Clausewitz has held sufficient sway to warrant a cenotaph at the U.S.
Army War College. Many discussions of war within the halls of the service's Command and General Staff College and the School of Advanced
Military Studies consider him. Jomini's precepts are also present within
the Army's educational institutions, for one cannot deliberate the nuances of campaign design without considering him.

Carl von Clausewitz (Bildarchiv Preussischer Kulturbesitz/Art Resource, New York)

Clausewitz and Jomini's ideas about war illustrate a quandary that has bedeviled army leaders for centuries: How does one reconcile the polar absolutes of chaos and order to achieve national objectives in peace and war? In attempting to answer this important question, the Army has published doctrine. Doctrinal manuals seek to impose a system of equipment, training, organization, and procedure upon the Army, to create a common understanding of individual and unit actions to be undertaken when necessary, and to produce a cohesive force capable of succeeding despite the inherent mayhem of military conflict.

Antoine Henri de Jomini (Bildarchiv Preussischer Kulturbesitz/Art Resource, New York)

Given doctrine's authority in shaping the Army's preparation and behavior during hostilities and in peacetime, it is not surprising that scholars have produced an immense body of literature dedicated to explaining doctrine's purpose and effect. While A. J. Bacevich, Andrew J. Birtle, Christopher C. S. Cheng, Robert A. Doughty, Kenneth Finlayson, Paul H. Herbert, L. D. Holder, Jonathan House, William O. Odom, John L. Romjue, and John P. Rose have focused on army doctrinal developments primarily within the twentieth century, Birtle has produced a two-volume series on counterinsurgency doctrine from 1860 to 1941 and 1942 to 1976. Stephen Ambrose, Harry P. Ball, T. R. Brereton, Archer Jones, Brian McAllister Linn, David R. Mets, Timothy Nenninger, Carol A. Reardon, Roger Spiller, William B. Skelton, and Harold R. Winton explore institutional nuances that have influenced doctrine. Focusing on tactical application in peace and war are Donald E. Graves, Paddy Griffith, Earl J. Hess, Perry D. Jamieson, Andrew F. Krepinevich Jr., Jay Luvaas, Grady McWhitney, Allan Millett, Steven T. Ross, Robert H. Scales, Ingo Trauschweizer, and Robert M. Utley. Doctrinal implications regarding an "American way of war" and service culture are addressed by John Grenier, Adrian R. Lewis, Linn, and Russell Weigley. Robert M. Citino, R. Clayton Newell, and Michael D. Krause examine the Army's search for an operational-level doctrine during the twentieth century. Creating doctrine through an examination of intellectual process rests with Noel K. Evans, Herbert, and Romjue.[1]

Noticeably absent from the existing literature is an overview of U.S. Army doctrine as a singular body of evolving work. Although valuable studies devoted to army doctrine have been produced, none traces the progression of the service's doctrine by means of the Army's "keystone" publications, which I define as the dominant manuals that have shaped army operations for over two centuries. This book attempts to address that deficiency. By examining primary and secondary material, it scrutinizes the development of keystone army doctrine from 1779 through the early twenty-first century. The book establishes what constituted the service's dominant manual during a particular era and its historical lineage. It seeks to ascertain the fundamental characteristics of army doctrine and to judge its impact in preparing the service to accomplish its missions in both domestic and foreign venues.[2]

This book is not a typical work about doctrinal manuals, although many audiences will find it instructive. It is not intended to consider every army publication nor does it focus upon every detail within each manual, although some receive more coverage than others. The subject of this book is the historical development of army doctrine traceable from the first tactical drill manual, the 1779 *Regulations for the Order and Discipline of the Troops of the United States, Part 1.* The book's intellectual thread begins with that manual, carries on through a series of successive nineteenth-century drill manuals, and continues through the twentieth century into the early twenty-first century via the *Field Service Regulations,* Field Manual 100-5 *Operations,* and Field Manual 3-0 *Operations.* It explains how doctrine reflected army policymakers' understanding of strategy (the search for solutions to military problems through logic), operations (the planning and execution of missions), and tactics (the arrangement and employment of forces relative to each other to accomplish tasks). It further examines how and why doctrine advanced particular approaches toward regulating the chaos of conflict while offering assessments of doctrine's "real-world" application. By exploring these matters, this book seeks to answer a straightforward but salient question: What is the relationship between doctrine and the U.S. Army?

Doctrine is a term that has abundant interpretations. The word appeared in 1382 when John Wycliffe used it in a religious sense to define the teaching of a body of instruction. In 1848, I. E. Holmes mentioned doctrine as it pertained to President James Monroe's speech to Congress in 1823. The use of doctrine in a military sense appeared only in the twentieth century, although its first use is murky. Prior to the twentieth century, few officers or enlisted men would have thought of the tactical manuals they consulted as doctrine. Fewer still would have fully appreciated its intellectual value in directing an army.[3]

For my purposes, doctrine is considered to be a subcategory of military literature, the body of published work that describes the nonfictitious war or nonwar activities of the armed forces. As a subcategory, doctrine is distinguished by two characteristics. The first is approval by an authority, typically the government. The second is that the approving authority mandates its use by all the armed forces or by a particular service. As an approved and prescribed publication, doctrine stands juxtaposed to "informal practice," which evolves from custom, tradition, and experience passed on through assorted writings, circulated materials, and conversation. Historically, in peace and war, the Army concurrently adhered to both doctrine and informal practice, as this study elucidates.[4]

My approach includes the term "regulating." "Regulating" is a variant of "regulate," which is to bring something under control. "Regulating" is used to describe the process of producing doctrine in an effort to control the innate chaos of the battlefield and to guide those who plan and execute military operations. I also use the terms "conventional" and "unconventional" to refer to types of forces and forms of warfare. "Conventional" indicates armies prepared to operate for extended periods of time through arranged formations designed to attack, defend, or withdraw by use of fire and maneuver. Conversely, "unconventional" refers to specific units smaller than armies such as guerrillas, insurgents, or special-operations forces. Unconventional forces are designed to operate independently or behind enemy lines while avoiding orthodox methods and, more often than not, conducting ambushes and raids of limited duration.[5]

As written guidance reflecting how service leaders understood the theory and practice of warfare at a given moment in time, doctrinal manuals vary in both content and scope. Their contents are not consistent from one publication to the next. In some cases, manuals consider how to employ a force in peace or war operationally or tactically, while others describe those procedures plus the administration and sustainment of a field army or any number of subordinate units contained within it. While overtly operational or tactical, some doctrinal publications include strategic matters, such as how the service achieves national-security objectives. The primary focus of this book concerns how the service leadership perceived the conduct of military operations, for those ideas are generally consistent from one manual to the next. Less attention is given to the administration and sustainment of the force. However, where such matters are deemed a significant shift in the service's doctrinal views in my judgment, they are illuminated.

Tracing the evolution of doctrine and the ideas behind it is a thorny proposition. Cultural artifacts in the form of printed official and nonofficial manuals exist, but the authors are often unnamed. Judging a particular author's influence can be problematic, especially when more than one writer created the manual. Frequently, the prepublication dialogue within the service that ultimately framed a manual's contents is absent or incomplete. In the case of late-eighteenth- and nineteenth-century U.S. Army doctrine, the dominant manuals were initially narrow in scope, for the Army was an infantry-dominated force. From the early-twentieth century onward, doctrine broadened in content to consider combined-arms warfare and other aspects of conflict, including multiservice and multinational operations. Still, sources such as diaries, government documents, letters, memoirs, and oral histories help to illuminate the motives and thinking behind certain manuals. The actions taken by the Army in peace and war also shed additional light upon the subject. To understand the genesis and subsequent development of doctrine, however, one must first gain an appreciation for what came before it.

Origins of American Army Doctrine

Although the American Army is the rebellious offspring of the British Empire, its doctrinal origins are arguably Dutch. In the early 1500s, armies were dynastic, conventional, and drilled. They typically arrayed themselves according to the Spanish *tercio*, a 3,000-man or larger infantry-based formation consisting of three columns (*colonellas*), each headed by a "colonel." By mid-century, the block-style formations consisted of arquebusiers and pikemen arranged in ten ranks to dispatch opposing infantry and cavalry forces. Spanish armies were so successful in battle and rapine that the Hapsburgs and others copied their tactics.

Military pressure by Roman Catholic Spain eventually led the Protestant leader Maurice of Nassau, prince of Orange, to seek out ways to counter Spanish tactical power. Through the study of military literature of the time and antediluvian Roman warfare publications in translation, Maurice produced a tactical system that employed an arrangement of three lines preceded by skirmishers. This linear design utilized the skirmishers to disrupt opposing attacking forces before they engaged the first two defensive lines of infantry battalions, each constituting upwards of 500 pikemen and arquebusiers. The third rank was held in reserve. The "Dutch system," as it became known, provided for excellent defense and proved more maneuverable than the Spanish *tercio*.

While acting as the Dutch civil authority from 1585–1625, Maurice of Nassau produced an illustrated drill manual and directed its use by the army. The Dutch system was thus based upon written doctrine. Under Gustavus Adolfus the Swedes later adopted, modified, and improved Maurice's methods. English mercenaries serving in Europe's Wars of Religion (1562–1598) mastered the existing linear tactics while serving in the Dutch and Swedish armies, among others.[6]

Between 1607 and 1778, North American English colonists turned British-American subjects prepared for and conducted tactical army operations in the absence of doctrine despite waging intermittent wars against various security threats, conventional and unconventional, foreign and indigenous. In the early 1600s, mercenaries such as Lion Gardiner, John Mason, George Percy, John Smith, Miles Standish, and John Underhill created an informal tactical approach to war suitable for North America. Neither military literature nor doctrine, the informal approach blended concepts garnered from European military writing with the linear tactical experience the mercenaries had acquired by taking part in Europe's religious wars. By means of word-of-mouth instruction and repetition through locally devised tactics, techniques, and procedures, English mercenaries gradually converted security-minded farmers and artisans into "trained bands"—militia companies capable of waging offensive tactical warfare against Indians and employing blockhouses for local defense.[7]

Although colonial forces followed informal practice in the sense that they lacked doctrine, militia companies trained periodically to master the basic conventional motions and commands necessary to load firearms, move into linear ranks, and give fire as a tactical unit. However, conventionally trained militias often fought unconventionally because mercenary instructors rejected European conventional warfare tactics due to unsuitability in defeating the Indians who "skulked" through the woods using hit-and-run methods. Typically, militias learned enough conventional procedures to organize, train, march, and give fire against Indians. Their tactics, however, were the basis of unconventional "extirpative" war characterized by "razing and destroying enemy villages and fields; killing enemy women and children; raiding settlements for captives; intimidating and brutalizing enemy noncombatants; and assassinating enemy leaders." The Virginia Powhatan (or Tidewater) Wars (1622–1632 and 1644–1646), the New England Wars (1637), wars against the Hudson Valley Indians (1642 and 1663), raids in Massachusetts (1671), and later wars such as King Philip's War (1675–1678), the Tuscarora and Yamassee Indian Wars (1712 and 1719), Pontiac's War (1763), and Lord Dunmore's War (1774) are representative of American informal tactics expressed through extirpative warfare.[8]

As the English moved away from coastal settlements and established isolated farms deeper within the backcountry, colonial governments formed specialized Ranger units to conduct strikes against marauding Indians. In July 1675, Captain Benjamin Church of Plymouth colony organized North America's first Ranger Company consisting of sixty Englishmen and 140 friendly Indians for this purpose. Additional Ranger units subsequently formed in most colonies. As with militia, Rangers were organized conventionally into companies. However, they fought unconventionally in campaigns from New France to Spanish Florida. Unlike settlement-based militia, which periodically drilled to stand in ranks and to give fire on command as a group, Rangers were primarily frontiersmen who traveled unencumbered far from settlements, fought in small factions like Indians (when necessary), and relied upon stealth and speed rather than marching. Notoriously independent, unit members chafed at discipline and, when fighting, shunned linear formations, preferring to exchange blows using muskets, hatchets, and knives.[9]

Cavalier and often uncontrollably vicious, Rangers never operated under English government-approved doctrine. In 1716, Thomas Church (son of Colonel Benjamin Church) published a *History of the Great Indian War*, which described Ranger practices. But this was a narrative account, not doctrine. Robert Rogers, the commander of His Majesty's Independent Company of American Rangers, published his *Standing Orders* of 1757. Yet, Rogers's work was also not doctrine in that it offered guidance for how his unit alone was to organize and to fight tactically. His methods were never officially approved by the Crown for universal adoption among all American Ranger forces, thus they were also informal.[10]

From Jamestown's founding until the 1680s, militias and Rangers followed informal practice to fight domestic, extirpative wars tactically. In 1689, however, colonial militias and Rangers began to augment Crown troops fighting in North America against the French and Spanish. For Americans, participation in international wars meant using their methods alongside British conventional doctrine. During the War of the Grand Alliance (or King William's War, 1689–1697), a group of Massachusetts troops joined with English forces conducting a sea movement, landed on hostile shores, and lay siege to Port Royal in Acadia. Later the same year, New England sent troops and ships to attack the French along the St. Lawrence River, while French and Indian raids destroyed English settlements in New York, New Hampshire, and Maine. The War of the Spanish Succession (or Queen Anne's War, 1701–1713), the War of Jenkin's Ear (1739–1743), and the War of the Austrian Succession (or King George's War, 1740–1748) employed South Carolina volunteers in three different assaults (1702, 1728, and 1740) on St. Augustine, Florida. American militia and Rangers were used extensively in those wars. They also fought alongside their British counterparts during the French and Indian War (or Seven Year's War, 1754–1763).[11]

American warfighting methods and British conventional tactics often clashed during execution. To colonials who were accustomed to fighting more informally than not, the British Army appeared so steadfast in their conformist practices that they were deemed "pigheaded, intransigent, and, with few notable exceptions, incapable of adapting their dogmatically held doctrines to the exigencies of war in America." Conversely, the British, who attempted to force militia units to fight conventionally along European lines, determined that "Americans [were] in general the dirtiest, the most contemptible, cowardly dogs that you can conceive." Yet, for some American officers, exposure to British doctrine generated envy, not denigration. During the French and Indian War, Virginia militia Colonel

George Washington came to see British Army discipline and the adoption of its doctrine as a means of attaining levels of effectiveness and prestige enjoyed by European armies.[12]

Prior to the French and Indian War, European military literature had proven of little value to colonial troops whose informal tactics differed radically from Europeans. European military authors had debated the value of linear tactics vice massive columns. The Chevalier de Folard, Maurice de Saxe, and Mesnil-Durand advocated columns of troops of various sizes to produce shock effect in breaking an opposing line. In 1748, the French government issued standard drill regulations to replace the system wherein individual regimental commanders trained their troops by whim. France's 1755 infantry doctrine was overtly linear but also included the use of columns, when appropriate. Still, the French Army failed miserably on the Continent during the French and Indian War. Under Fredrick the II, the Prussian tactical system with its oblique order came to dominate European warfare.[13]

By the 1760s, many Americans admired the Prussian system and its means of transforming raw recruits into disciplined soldiers. Consideration of Prussian doctrine reflected not only a shift in dependency upon the traditional British mother country-colonial relationship, but also an American desire to emulate the most current and successful foreign doctrine as a means for becoming the equal of European armies. Colonial Americans believed the British had underappreciated their contributions to the French and Indian War and the difficulty in turning apprentices, servants, and transients recruited from outside the militia system into effective soldiers.

For George Washington, his 1750s appeal for adopting British doctrine allowed the provincials to gain a modicum of respectability and legitimacy relative to their Crown counterparts. Yet, Americans in the 1760s were torn over accepting British or other European methods of warfare to prepare for future conflict versus continuing to rely on their informal practice. In 1777, Nathaniel Greene explained this intellectual discord in that experience was the best school of war for Americans, but if European ideas meshed with colonial military methods, then use them.[14]

Regardless of what Americans thought about foreign ideas and war, from the mid-1750s until the American Revolution eagerness increased among military-minded individuals to read European sources. Copious amounts of military literature arrived in the colonies, primarily British publications and translations and reprints ranging from ancient and contemporary war, conventional and partisan (*petite guerre*). Popular works included translations of Aelian and Flavus Vegitius Renatus, L. M. de Jeney's *Le Partisan, ou, L'Art de Faire la Petite-Guerre*, and Humphrey Bland's *A Treatise of Military Discipline* that reflected procedures used by the Duke of Marlborough. Used throughout the Southern colonies after 1755, it was replaced by Edward Harvey's *The Manual Exercise as ordered by His Majesty in 1764*, reprinted in Boston in 1774 and adopted unofficially by the Continental Army early in the American Revolution. Also imported was French Count Turpin de Crisse's *Essai sur l'Art de la Guerre* that was translated into English by Captain Joseph Otway in 1761. Crisse and de Saxe's translated *Reveries* both advocated using line infantry in broken terrain (such as North America), provided that they were trained as light infantry, troops more accustomed to fighting as individuals rather than in large and well-drilled formations. Also of note was the publication of "Reflections on War With the Savages of North America" by Henri Bouquet, contained within William Smith's *A Historical Account of the Expedition Against the Ohio Indians in the Year MDCCLXIV*. Smith's work served as a source of information for the Continental Army. In 1754, William Fawcett translated Frederick the Great's infantry *Instructions* and three years later his directives for Prussian cavalry, which proved popular in North America. Over forty military literature publications reached North America prior to the American Revolution. Even more arrived during the war.[15]

After the French and Indian War, foreign military literature and doctrinal manuals proliferated to such an extent that determining precisely what source influenced a particular militia or Ranger unit is problematic. In 1773, for example, Lord Dartmouth requested that colonial governments report on the readiness of militias. The colonial governors responded that training was neither universal nor enforced, noting it was not even possible to ascertain how many militias actually existed. This situation had developed despite the 1745 appearance of the Parliament-approved *A Plan for Establishing and Disciplining a National Militia in Great Britain, Ireland, and in all the British Dominions of America*. Whig apprehension over Britain's growing dependency upon foreign mercenaries had resulted in inquiry over substituting English militias, domestic and colonial, in foreign wars. Intended as a guide for future conflict with France in Europe or in North America, the *Plan* prompted Parliament to pass the Compromise Militia Act of 1757 (or the so-called "Norfolk Discipline") during the French and Indian War. The Norfolk Discipline advocated the organization of militias into two groups, a superior militia made up of propertied men and a subordinate militia consisting of commoners. It also included various instructions for employing the militias. The *Plan* was never formally institutionalized; the British government rejected the deployment of domestic militias in international wars, although American militias had been previously used in that manner within North America. Some Connecticut, Massachusetts, and Rhode Island militias adopted the *Plan* but it was never universally accepted within the colonies.[16]

Still, Americans followed European military developments with a keen interest. In 1766, following France's defeat in the French and Indian War, Comte du Guibert sought to reconcile an ongoing debate between those who advocated using heavy, concentrated formations armed with muskets and bayonets (*ordre profond*) and those who favored the more linear form of war that maximized firepower (*ordre mince*). Guibert's solution was the *ordre mixte*, a tactic later refined in 1772 within the *Essai General de Tactique*. Guibert argued that infantry possess sufficient flexibility for both line and light-infantry missions, as well as an ability to fight from either the column or the line formation depending upon the tactical situation. France failed to adopt these changes until 1791, although they generated some interest in North America, especially with George Washington who advocated light-infantry units instead of Rangers. Some colonies attempted to adopt procedures for their militia units, but enforcing such a standard was problematic because colonial governments lacked the capacity to inspect militias for compliance.[17]

By 1775, Americans had fought for over 160 years using informal practice methods without a bona fide doctrine or continuous British Army oversight. For Americans such as George Washington, Crown doctrine had proven effective in training troops for future war and in mobilizing, employing, and sustaining troops during campaigns and was thus worthy of emulation. Yet, others believed that European doctrine was absurd and should be rejected, given the realities of North American warfighting practice and terrain. For their part, New York's Superintendent of Indian Affairs William Johnson and Massachusetts Governor William Shirley had grown to appreciate American Rangers. The same was true of British Major General John Campbell Loudon, who authorized raising several Ranger companies of "Stout able men," deeming them "much better than their Provincial [militia] Troops." Units such as Gorham's Ranger Company and others proved themselves from 1747 through 1759, but many conventional officers, such as Washington, found unconventional forces to be disorderly and their methods contemptible.[18]

By the outbreak of the American Revolution, many communities throughout British North America had shifted defense concerns to only a small portion of the populace, thus fewer and fewer people were familiar with formal military procedures. Nonetheless, persons with and without military skill soon rallied to the revolt and the colonists had to "improvise an Army from the variegated militia systems within the colonies." Without an approved doctrine to guide it, the Continental Army and its predecessors such as the New England Army attempted to use existing American informal practice to organize and fight the current war. The results were disastrous.[19]

As Americans flocked to the rebellion, individuals with British Army doctrinal expertise made use of it within their regiments. Others perused British, French, and Prussian treatises hastily reprinted within the colonies. Some relied upon their militia training or learned what they could upon enlistment. Washington consulted several works, including Thomas Webb's *A Military Treatise on the Appointments of the Army*, written in 1759 for use by the British in North America during the French and Indian War. Although Washington personally endorsed *A Military Treatise*, the manual was not universally adopted.[20]

The depth to which the informal approach had imbedded itself in American tactical warfare showed when the Continental Army attempted to adopt it to fight conventionally to European norms. American colonialism had never allowed for a standardized military organization; the states now formed various units and assigned military rank according to their own preferences. Washington's forces were primarily infantry and lacked artillery, engineers, and cavalry initially; logistics procedures were noticeably unequal to the task of supplying an army. On 4 August 1775, Washington complained in writing to the president of Congress over the lack of organization.[21]

Most troubling, however, was the militia and the Army's lackluster performance in battle from 1775–1777. Reliance upon informal methods meant that most militia members were unskilled in facing conventionally trained opponents. Results in battle were mixed. In 1775, at Lexington, Massachusetts, the militia had formed into line, stood its ground, and gave fire with some effect. Still, as the force was too small and ill-disciplined to withstand the closely controlled attack of British regulars, it ran off. In the British retreat from Concord, however, American militiamen did not form into ranks to give fire but attacked in small groups from behind trees and fences, more analogous to informal practice or light-infantry methods, while the British withdrew according to their conventional doctrine. Later, an unsophisticated New England Army fought a conventional defense behind a redoubt at Breeds (Bunker) Hill, where it inflicted serious damage upon the linear British formations. American Army defeats in Canada and New York resulted not only from spotty leadership, poor training, and inadequate discipline but also the absence of an approved doctrine to provide overarching guidelines. In sum, Americans achieved mixed results in the early days of the rebellion due to an inability to execute orders and maneuvers cohesively stemming from a lack of standardized tactical procedures and reliance upon informal practice. Each militia unit and regiment reflected whatever drill manual the commander consulted.[22]

To be sure, the Continental Army was in doctrinal disarray from the highest levels of command downward to the rank and file. Given its ad hoc nature, Congress was in no position to produce standardized doctrine, for the government lacked the power to direct the states to cooperate and thus conform to one manual. Although Congress had created the Board of War on 12 June 1776, its three members and clerks were deficient in requisite military experience. Crafting doctrine was far beyond their capabilities. Washington was unable to devise a proper doctrine, for he was continually concerned with numerous command details, of which recruiting, logistics, and fighting pitched battles with the British Army were of more immediate interest.[23]

Yet, Washington and his officers were acutely aware that the Army required a standardized doctrine to regulate tactical warfare procedures. In May 1777, writing to Brigadier General Alexander McDougall, Washington remarked that he agreed with McDougall's observation of the "impropriety of that diversity in the modes of training our Regiments which had prevailed hitherto." He indicated his intent as commanding general "to digest and establish a regular system of discipline, manoeuvers, evolutions, regulations for guards &Ca. to be observed throughout the Army." In the meantime, Washington asked McDougall to "be particular attentive to having them [soldiers] instructed in the proper use of their feet, so as to enable them to perform the necessary movements in marching and forming, with ease, order, agility, and expedition." Still, by November 1777, Washington's command duties prevented him from finding the time to write a drill manual, although Continental officers considered a "New Establishment and regulations of the Army" that would abolish individual colonial distinctions among the force. Frustrated, Washington turned to a self-study approach and issued standing orders for Continental officers to spend their spare time reading the publications of foreign military authors for guidance. His officers evidently followed that direction both before and after it was issued, for captured American knapsacks often contained well-used copies of diverse military literature.[24]

Despite American informal practice proving unsuitable for fighting a conventional war, the Northern Army in New York under the command of Major General Horatio Gates proved its worth at Saratoga in 1777. Doctrinally, the American force was no more homogenous than Washington's Continentals. Yet, militia forces, augmented by regulars, succeeded in defeating British Major General John Burgoyne. Gate's victory had as much to do with vague British planning and poor execution as American leadership and army cohesion. But an American doctrinally dysfunctional force had defeated the British, some small comfort for Washington and his noncohesive army that had failed to spare the rebel capital of Philadelphia, Pennsylvania, the insult of British occupation in October 1777.[25]

During the winter of 1777–1778, despondency descended upon Washington's army, which lay encamped at Valley Forge, Pennsylvania. Officers trained their regiments using whatever military literature was available. Soldiers endured the routine drudgeries of camp life: drill, guard duty, foraging for food, and tossing wood into continuously burning fire pits. Informal practice had failed not only the Army but every American who supported the insurrection. Critics both uniformed and civilian sensed that the rebellion was in danger.[26]

CHAPTER ONE

MIMICS AT WAR: TACTICAL DRILL, 1778–1848

I think "Regulations for the Infantry of the United States" will be sufficient.—George Washington to Baron von Steuben, *The Writings of George Washington*

Since English colonization of North America commenced in the early 1600s, the colonists fought wars in the absence of doctrine. While capable of defeating the indigenous population in local engagements, the colonists' informal practice was less effective when facing well-drilled, conventional European armies. Having rebelled against Britain in 1775, the American Congress and army leadership needed a tactical doctrine capable of forming a cohesive army from among the diverse state militias.

For two years, informal battlefield practice contributed far more often to defeat than victory. By mid-1777, the Army had been tactically unable to bludgeon their British opponents into acquiescence. By October, the Army leadership's inability to regulate the Continental Army's performance in battle had contributed to the British occupation of Philadelphia, Pennsylvania, the rebel capital. While many Americans believed their rebellion was doomed, a troubled Major General George Washington withdrew a dispirited Continental Army to Valley Forge, Pennsylvania. As the winter intensified, a bloodied army licked its wounds and struggled to rebuild. Volunteers arrived at an encampment filled with hardship and despair. Despondency permeated not only Washington's camp but the thirteen states, as well. Civilians and soldiers alike who had supported the rebellion were growing increasingly pessimistic about their future and the coming campaign season of 1778.

Toward an American Army Doctrine

On 8 January 1778, Major General George Washington learned of the pending arrival of Frederich Wilhelm Ludolf Gerhard Augustin von Steuben, an experienced Prussian military officer turned opportunistic mercenary. This news broke at a time when seasoned officers capable of forging an army were in short supply. After writing a letter of welcome to Steuben, Washington admitted to Brigadier General George Weedon, "I can see clearly that instead of having the proper Officers to assist in arranging, training, and fitting the Troops for the field against the next campaign, that we shall be plunged into it as we were last year heels over head without availing ourselves of the advantages which might be derived from our present situation and prospects." Steuben's pending arrival notwithstanding, Washington envisioned the forthcoming campaign season to be no better than the previous.[1]

Steuben's advent at Valley Forge on 23 February 1778 provoked mixed reactions from the encampment's inhabitants. Already present was Thomas-Antoine, Chevalier de Mauduit du Plessis, a veteran of the Battle of Germantown and a respected drill instructor who, perhaps, saw a rival in Steuben. Some, including Colonel Timothy Pickering, distrusted yet another European professional soldier and a Prussian automaton in particular. Familiar with Prussian doctrine, Pickering believed it held no value for American military culture, claiming, "Tis the boast of some that their men are mere machines, but God forbid that my countrymen should ever be thus degraded." Congress had allowed Steuben access to Washington, so he listened as the Prussian pled his case in French with John Laurens, Washington's aide, acting as translator. During the course of the conversation, Steuben offered his assistance in any capacity.[2]

Washington contemplated Steuben's potential value. Given Prussian Army success on the European continent from 1756 to 1760, having an officer intimate with the inner workings of that army was indeed fortuitous. During early March 1778, in an attempt to rejuvenate an inadequate winter training regimen, Washington decided to appoint Steuben as the Army's inspector general. This choice spawned immediate complaints from American officers that foreigners were overly ambitious and that Steuben's appointment denigrated those who had served Washington from the beginning. Officers also voiced that their enlisted soldiers might balk at taking instruction from a foreign drillmaster (although the Frenchman Thomas-Antoine had in all probability provided some drill instruction advice). There was also uncertainty over how Steuben might handle an American army composed of partially trained officers and noncommissioned officers leading a multiethnic rabble of transient Whites (both native and newcomer), African Americans, and Indians. More troubling still was that Major General Thomas Conway already served as the inspector general. But Conway had complained to Horatio Gates that Washington proved to be a weak commander and should be replaced. When the grumblings reached Congress, Washington was irritated enough to seek Conway's reassignment. In the interim, Steuben acted in Conway's capacity, without rank. He began to train the Army for a campaign against British conventional forces within a few weeks.[3]

Through happenstance, Washington had acquired a competent drillmaster from the kingdom that many believed possessed the finest army in the world. On 19 March 1778, the commanding general issued orders directing the Army to obey the acting inspector general and to appoint subinspectors to assist him. Together, Washington and Steuben pursued a common purpose: to prepare the Continental Army at Valley Forge to fight a conventional war. Given that informal practice methods had been embedded within American military culture as early as Jamestown, the complexity of this undertaking was colossal. As it was, the Army was barely functional, incapable of even marching to the same cadence between regiments. In battle, soldiers could not shift from line to column formation or vice versa with any sense of order. Tactical diversity had proven to be a recipe for disaster when facing British Army precision on the battlefield. Now, Washington had a Prussian officer skillful enough to train the Army to fight conventionally along a common European standard.[4]

Steuben soon discovered that imposing Prussian or British doctrine upon the Army was feasible but unsuitable. American soldiers were far more individualistic than their European or Atlantic counterparts. The ranks constituted a cross section of Colonial America reflecting varying degrees of education and views of service. Collectively, the assembled multitude loathed physical punishment and pointless maneuvers; training required a delicate balance between Prussian discipline, British procedures, and American attitudes. Most troops were familiar with the 1764 British manual of arms, but Steuben nonetheless confiscated a potpourri of well-worn foreign military literature because "each Colonel exercised his regiment according to his own ideas or to those of any military author that might have fallen into his hands." In sum, Steuben had to sort out the accumulated chaos of over 160 years of informal American warfighting practice that had culminated in a hollow force lacking the homogeneity needed to win a war. To succeed, Steuben had to consider diversity while in search of unity. More important, he had to create a military system simple enough for a mob to understand its nuances and perform each task perfectly. And he had to do so in a matter of weeks during winter.[5]

George Washington and Baron von Steuben (The Granger Collection, New York)

Imposing a system of order on the Continental Army began with training the members to discard their individual regimental procedures so they could become a cohesive conventional force. With Washington's permission, Steuben created a model company composed of 50 Virginians and 100 additional men from the various states. On 19 March 1778, Steuben personally schooled the soldiers to stand at attention properly and march to 75 steps per minute at slow tempo and 120 steps for double time. Speaking through translators in German and French (enriched with English profanities), he taught the soldiers to advance in columns of four instead of single file, thereby making formations compact and able to bring more firepower to bear more quickly on the battlefield. Troops learned how to shift from column into line and back again, insight from Prussian, British, and Guibert's French Army doctrine. Drawing upon American skill with the musket, Steuben added bayonet training and included defensive procedures such as how to form a hollow square to defend against cavalry or other attacks. Colonel Alexander Scammell of New Hampshire remarked that it was a lark "to see a gentleman dismissed with a lieutenant general's commission from the Prussian monarch condescend, with a grace peculiar to himself, to take under his direction a squad of ten or twelve men in capacity of a drill sergeant." Steuben's passion earned the American's admiration and respect, for "he has undertaken the discipline of the Army and shows himself to be a perfect master of it, not only in grand maneuvers, but in every minutia." Slowly, a disciplined army formed by mimicking British and Prussian standards with a bit of French tactical procedure thrown in, all leavened by American sensitivities.[6]

After several days of drill, Washington notified his subordinate commanders that new regulations were forthcoming. To produce the commanding general's long-desired drill manual, Steuben turned to four officers to act as deputy inspectors general while he devised a suitable text. Unable to write in English, Steuben composed in German and at times in French, then passed his notes on for translation. The manual was "composed in good German, translated into bad French, put into good French by Captain Fleury, then translated into poor English by Captain Walker." Steuben wrote each lesson several days in advance of implementation while his deputies transcribed the translated documents and passed the copies along to the brigades and regiments.[7]

Even as the drill manual took form, Washington ordered all brigade and regimental commanders to comply with the new instructions, stating that "establishing a uniform system of useful maneuvers and regularity of discipline must be obvious; the deficiency of our Army in these respects must be equally so." In response, Continental Army officers and noncommissioned officers from brigade and regiment down through company level observed the model company drill twice each day with Steuben and his assistants explaining each lesson. By following the concept of "training the trainer," unit leaders learned firsthand what needed to be accomplished and then instructed their own troops in what they had been taught.[8]

Steuben's manual combined military philosophy with an individual- and unit-training system. The publication contained standardized procedures for an infantry-dominated force, for the Continental Army was overwhelmingly composed of infantrymen. The Army now had a standardized method for how to give, receive, and implement orders and to organize and maneuver units small and large. Basic training began with the individual soldier's manual of arms and individual drill movements before progressing to larger formations. Philosophical directives were included in an effort to tutor officers, regimental commanders, majors, adjutants, and quartermasters concerning their responsibilities. For example, instructions "for the Captain" prescribed that an officer "cannot be too careful of the company the state has committed to his charge," an appeal to duty and the obligations accompanying commissioning and command. In deference to the American soldier, officers were to attend to their health, discipline, arms, clothes, and other necessities so they could foster cohesion and respect. As American regiments were understrength, Steuben directed them to merge into two full battalions during training. A Continental regiment now contained two battalions, each consisting of 486 soldiers divided into 9 companies of 54 individuals each.[9]

By April 1778, the Continental Army had undergone a miraculous transformation. With Washington's monitoring and approval along the way, Steuben wrote the first training manual applicable for any conventional multicolonial American army operating in North America. He created the American Army training model: explanation followed by demonstration and then practical exercise. He also achieved the recognition he had long sought. In response to Steuben's efforts, Washington sent a glowing report to Congress on 30 April. On 5 May 1778, Congress responded by appointing Steuben as a major general with pay of $165 per month. American troops at Valley Forge now had a unified conventional standard and were soon ready to begin the summer campaign season.[10]

Given subsequent events at Monmouth Court House, New Jersey, on 28 June 1778, there is little doubt that Steuben's manual dramatically altered the performance of the Continental line. On a hot and humid day, the Army marched into battle, executed complex maneuvers with precision, and engaged a crack force under the command of Lieutenant General Sir Henry Clinton. Although the battle was a tactical draw, the American line had held. The war soon ended in the northern colonies.[11]

While Steuben's manual was a success and Washington had ordered the Army to use it, government approval was still forthcoming. Absent congressional authorization, the manual could not be considered doctrine, but it was well on its way toward that end. After taking a field command in 1778, Steuben continued to expand and revise his work well into early 1779 in an effort to create a publication that not only made sense to the Army but was also capable of securing congressional endorsement. Washington shared this goal. His position as commanding general allowed him to take an active role in reading the drafts and modifying them as he deemed necessary. His letter to Steuben of 26 February 1779 reflects that personal involvement and is partially reproduced here:

> my remarks on the first part, the Remainder shall follow as soon as other affairs of equal importance will permit. I very much approve the conciseness of the work, founded on your general principle of rejecting every thing superfluous; though perhaps it would not be amiss in a work of instruction, to be more minute and particular in some parts. One precaution is rendered necessary by your writing in a foreign tongue, which is to have the whole revised and prepared for the press by some person who will give it perspicuity and correctness of diction, without deviating from the appropriated terms and language of the Military Science. These points cannot be too closely attended to, in Regulations which are to receive the sanction of Congress and are designed for the general Government of the Army.

Later, on 11 March 1779, Washington again edited the manual before notifying Congress of his willingness to submit it for their "public sanction, that the regulations may be carried into execution as soon as possible."[12]

The First Doctrine and Its Outcome

On 29 March 1779, Congress approved Steuben's *Regulations for the Order and Discipline of the Troops of the United States, Part 1.* In authorizing its immediate use for the Army as a whole, Congress announced the birth of the first American Army doctrine. It also indirectly began the decline of informal practice in the sense that the Army was ordered to adopt certain standardized procedures, at least regarding conventional tactical warfare. Steuben's efforts had produced the first keystone manual of "invariable rules, for the order and discipline of the troops, especially for the purpose of introducing an uniformity in their formation and manoeuvres, and in the service of the camp." On 12 April 1779, Washington ordered the Army to comply immediately with the congressionally approved *Regulations*. At least 3,000 copies were printed with a blue cover (thus the colloquial term *Blue Book*). Although the manual was intended for use by all regiments, the war ended before that goal was attained.[13]

The government-approved 1779 *Regulations* enabled the military leadership to regulate the Army's behavior tactically and systematically. It also enabled the Continentals to fight effectively alongside the conventionally trained French Army. Although the doctrine had not discussed the nuances of coalition warfare, the American procedures were familiar enough to the skilled European professional eye. When the Yorktown campaign concluded in the fall of 1782, Commanding General Jean Baptiste Donatien de Vimeur, Count of Rochambeau, commented to Washington: "You must have formed an alliance with the King of Prussia; these troops are Prussians." Baron Ludwig von Closen, a Prussian officer under French employ, further observed, "It is incredible that soldiers composed of men of every age, even children of fifteen, of whites and blacks, almost naked, unpaid, and rather poorly fed, can march so well and withstand fire so steadfastly." Even so, the congressional order that the *Regulations* be used by all the troops of the United States was problematic, for the manual discussed infantry units alone. It did not address artillery, engineers, and other service specialties such as cavalry. Enforcing its contents was also troublesome, for although copies were widely distributed, not all American Regulars were trained under the *Regulations* owing to a variety of reasons such as continued personnel turnover. Various state militia units received copies, but training was spotty and highly dependent upon the whims of local government officials and commanders.[14]

In approving the *Regulations,* Congress had legitimatized the conventional form of war that the Army had fought previously in Massachusetts, parts of New York, New Jersey, and Eastern Pennsylvania in its engagements against conventionally trained British forces. The 1779 *Regulations*'s conventional guidelines proved effective not only at Monmouth Court House but in many of the Southern campaigns, as well. On the frontier, however, portions of the Army continued to employ informal methods. British-provoked Cherokee raids in North and South Carolina included the burning of settlements, scalping, and the butchering of men, women, and children. Regulars, militia, and Rangers also fought within the frontier areas of the Wyoming Valley of Pennsylvania and elsewhere. Americans, British, and Indians battled into the early 1780s; actions at Fort Sackville under George Rogers Clark (1779) and the Battles of Blue Licks (1782) and Arkansas Post (1783) reflected the continuation of an informal tactical approach to war much as it had been fought since the 1600s. A doctrine for American unconventional warfare was nowhere to be found.[15]

As the Revolution ended in March of 1783, Baron von Closen observed an army marching to the sound of the fife and drums whose "direction and the ease and precision of its movements really astonished us." The force moved as one within perfect columns, then shifted into ranks and wheeled to the left and right with ease. To be sure, nothing less was expected of the Prussians, regarded as Europe's finest troops. Von Closen, however, was not in Europe but in America. The soldiers that he praised were not the Prussians but the Continental Army, which had recently secured America's independence. Doctrine had not only accomplished its purpose of producing a cohesive army. It had assisted in the creation of a sovereign nation.[16]

Despite having become a doctrinally proficient conventional force, the Continentals were disbanded on 2 June 1784 (retaining eighty men to guard West Point, New York, and Fort Pitt, Pennsylvania) over Article of Confederation issues concerning a standing national army. The Army, however, soon rose again out of necessity along the frontier. Although threats of invasion from British Canada and Spanish colonies were very real, more immediate security considerations arose from the presence of British forts, mischief against American settlers, land speculation, and squatters clashing with Indian confederations.[17]

Adding to these considerations was the fact that the 1783 Treaty of Paris had accomplished little in bringing about peace. The British retained forts within the frontier pending American compensation for pre-Revolutionary debts. Moreover, Indian claims and rights had been disregarded, leading to continued disputes and violence within the frontier. In 1784, under the command of Revolutionary War veteran Lieutenant Colonel (later Brigadier General) Josiah Harmar, the First American Regiment was formed to address the dilemma posed by the frontier. An understrength 640-man regiment composed of eight infantry and two artillery companies was divided into small units and dispatched to outposts scattered about the Ohio Valley. Doctrinally, the Army trained under the congressionally approved *Regulations*, the manual written to fight the British Army conventionally during the recently concluded American Revolution.[18]

For six years, Harmar drilled his Regulars to conventional doctrinal standards. Yet, the frontier was far from conventional. The Regular's missions were more akin to peace operations and preemptive diplomacy of the twentieth century insofar as they sought to avert war stoked by British agents, Indians, and Americans. As a conventional tactical doctrine in the European tradition, the *Regulations* had not addressed peacekeeping duties or the role of military officers as diplomats. Still, as Captain Jonathan Heart noted, the situation on the frontier was indeed a delicate balance of diplomacy and bribes, coercion and compassion, skirmishing and intimidation, with Indians "constantly dying all around" from smallpox. Without a doctrine to guide their efforts, frontier army officers depended more upon their own ingenuity and informal practice in serving the nation's interests.[19]

In the late-1780s and into the 1790s, as peacekeeping gradually gave way to war, the *Regulations* were again put to the test. By 1789, Americans had moved down the Ohio River onto land acquired by the Miami Purchase. The resultant occupation of Indian hunting land by white settlers led to widespread security problems and frequent murders on both sides. Harmar reported to Secretary of War Henry Knox numerous incidents caused by Wabash Indians on the south side of the Ohio River and constant warfare in Kentucky since 1783. The situation was so volatile that Northwest Territory Governor Arthur St. Clair wrote to President Washington, beseeching the former commanding general to "give me orders you think proper. The people of Kentucky will not wait" in seeking retribution against marauding Indians.[20]

Protecting the settlers required the construction of Fort Washington near the present city of Cincinnati. Harmar moved his headquarters there in December 1789. Even so by spring 1790, the security situation had deteriorated. Robberies and murders among whites and Indians were an everyday occurrence. Government attempts to quell the fighting failed. On 7 June 1790, Harmar was ordered to extirpate marauding bands of treaty-breaking Indians, primarily among the Miami, Kickapoo, Shawnee, and Wea tribes.[21]

Harmar's force of 320 Army Regulars and 1,200 militia was culled from various locations including Pennsylvania, Virginia, and Kentucky, which formed at Fort Washington in September 1790. Learning to operate as a doctrinally cohesive unit was impossible, for the militia units, mounted and dismounted, trickled in over several weeks and trouble ensued over who should command. The Regulars attempted to drill the militia according to the *Regulations*, but the instructions were far too detailed for the assembled rabble to master in the few days available. On 26 September 1790, three battalions of raw militia departed camp. They were followed three days later by the regular Army.

Over the 175-mile advance, the dysfunctional force proved unwieldy, noisy, and slow. On 17 October, Harmar failed to trap the Miami and Shawnee Indians and their British and French traders. Kentucky Rangers, however, managed to burn the village of Kekionga and its food supplies. Over three additional days, army troops killed several Indians and burned six abandoned villages. Harmar then deviated from conventional regiment and company formations by dividing his force into smaller parties of 50-odd Regulars and 200 militiamen to seek out and annihilate Indians. On 19 October, one group skirmished at the Eel River. The militia bolted when pressed, while the accompanying Regulars were virtually annihilated while attempting to execute a bayonet charge by the book.[22]

Returning to the Indian reoccupied village of Kekionga on the morning of 22 October, Harmar's forces planned a complex attack involving converging forces to drive the Indians toward a nearby river and into a blocking force. The plan went awry when a stray shot warned the Indians, who then fled. The militia, sensing victory, broke ranks and charged as a mob while the Regulars advanced under doctrinal norms. Separated, the militia attempted to engage their adversaries in small groups within thick woods. This made a mockery of the Regular's linear formations and volley fire. The Indians closed with the two disjointed forces and, after a brisk cascade of musketry, settled the issue with hatchets and scalping knives at close quarters. When the dust settled, 50 Regulars and 100 militiamen lay dead. Total casualties from the various engagements numbered 183 killed, including several key officers. Indian losses were estimated at about 100 killed. The combined American Army of Regulars and militia had suffered defeat at the hands of an unconventional force led by Little Turtle, Blue Jacket, and Le Gris. Harmar's force returned to Fort Washington in disgrace.[23]

In the first major engagement under the *Regulations* since the Revolutionary War, American Regulars executed their conventional tactics as trained. But the militia, terrain, and foe proved inappropriate for their methods. Lack of interoperability with militia units proved detrimental; militias were recalcitrant in adopting the Army's drill system. In clinging to their informal practice, the militias were without sufficient discipline to execute a complicated and unrehearsed encirclement maneuver involving divided and converging forces. The outcome enraged Congress and sowed fear in the American populace. Harmar was accused of cowardice and court-martialed, only to be later exonerated. A second expedition to cow the bloodied but emboldened Indians occurred the following year.[24]

Major General Arthur St. Clair's subsequent campaign fared no better. After Harmar's debacle, the public became despondent as Indian raids became more frequent. Yet, the U.S. government's principal objective was to prevent an Indian war. In pursuit of this goal, Secretary of War Henry Knox notified St. Clair on 21 March 1791 that he was to negotiate a treaty and establish a garrison in Miami Indian territory to provide security for settlers. He was to engage in war only as a last resort. Accomplishing these tasks would require a force large enough to awe the Indians into signing a treaty or if necessary to defeat them in battle.[25]

In March 1791, Congress added a second regiment of Regulars to the existing one, and authorized 2,000 levies who would serve for six months. St. Clair was to raise local militia, if needed, and to recruit cavalry from Kentucky. Ultimately, St. Clair assembled 2,000 men consisting of 600 Regulars of the 1st and 2nd United States regiments and some artillery, 800 levies, and 600 militiamen. Cavalry was deemed too expensive and a logistical strain, although some accompanied the force. Although the reconstituted Regulars were partially trained to government standards, they proved hardly better than the militia and the levies. Many men had never been in the woods or fired a weapon in anger. The force had not maneuvered as an army, nor was it possible to train them to do so due to the lateness of the campaign season, political pressure to act quickly, and the general aversion of the troops to military service. St. Clair's force departed Fort Washington in September 1791 with families and camp followers in tow. For several weeks, the assembly cut its way through the wilderness averaging no more than 5 or 6 miles a day. The terrain, a baggage train, the presence of civilians, and the need to construct and secure several blockhouse forts en route slowed the advance considerably.[26]

On 3 November 1791, the Army reached the Wabash River, where scouts reported Indians nearby. Some troops briefly skirmished with a small war party. Due to the lateness of the day, St. Clair encamped for the night and, following doctrine, arranged his force in a defensive perimeter. He selected high ground approximately a half mile in circumference; constituting the outer perimeter were regular forces and some of the militia. Cavalry were arrayed outside of the camp to give early warning. Within the perimeter was the main body, which consisted of the 2nd Regiment and the artillery positioned within a conventional "hollow square" defense, two parallel lines that faced outward joined at the top and bottom at either end by frontier riflemen to form a rectangle about 350 yards long. Shallow creeks assisted in securing the flanks. The remaining militia was located about 500 yards from the main body and across the eastern branch of the Wabash. St. Clair's best troops, the 1st Regiment, were not present. He had dispatched them to bring up the supply wagons and track down some deserting militia members.

Although St. Clair intended to attack the following day, the Indians struck first. Just before sunrise, approximately 1,000 warriors representing ten tribes and allegedly led by Little Turtle, hit the militia east of the Wabash hard. The men broke ranks after one volley then fled in terror to the rear. As nearby Regulars attempted to line up and give fire, the Indians mixed with the panicked militia to run as one mob into the main body. Chaos ensued as soldiers and Indians fought for their lives. About 300 women and children ran about screaming as additional Indians encircled from both flanks. The perimeter ruptured as hundreds of soldiers dropped dead or fell from wounds. Officers and artillerymen, important targets to be sure, were identified by the Indians and gradually picked off. Only several bayonet charges by the 2nd Regiment staved off annihilation. Following several hours of combat, St. Clair extracted the survivors more by rout than order. Unnerved soldiers and camp followers ran off to Fort Jefferson 29 miles to the south. The dead and dying were left behind. Pursuing Indians dispatched the weak and slow. Later, St. Clair reported the fight "as warm and as unfortunate action as almost any that has been fought." This was something of an understatement given that 637 soldiers had been killed and 263 wounded, nearly half of his force.[27]

St. Clair's defeat remained the most impressive Indian victory over American soldiers until George Armstrong Custer's rout at Little Big Horn in 1876. Fortunately, few Indians were emboldened enough to raid the Ohio Valley, despite their victory. Gone, however, was the American

Army that had awed Prussian and French observers by defeating the British in the field but a few years before; unconventional Indian forces had handed out two severe beatings in a matter of months. The *Regulations,* with its emphasis on conventional tactics, was partially to blame. It was never intended for use on the frontier where unconventional, informal practice held sway and civilians accompanied combat forces. President Washington nonetheless blamed St. Clair for having been caught unawares and not putting men behind trees "in the Indian manner." These were remarkable words from someone who had himself placed little value in institutionalizing unconventional warfare. Following closely on the heels of Harmar's defeat, St. Clair's near annihilation at the Battle of the Wabash was as much a setback for doctrine as for the nation and the Army.[28]

In fairness to St. Clair, where Steuben had been given weeks and numerous assistants to prepare the Continental line to face the British Army, the American Army at the Wabash had been hastily formed and only partially trained. Moreover, for doctrine to have had any chance of success, the Regulars would have had to convince recalcitrant frontiersmen to master the *Regulations* and then practice the maneuvers jointly. The regular Army had failed in this regard, for the frontiersmen shunned convention in favor of informal methods. Attempting to establish authority over the militia commanders only provoked argument and obstruction. St. Clair was aware of the problems he faced, noting that punitive expeditions "bring out the worst of men," a circumstance that doctrine alone could not fix.[29]

Where the Continental line had used the up-to-date *Regulations* to win the American Revolution just over a decade before, the manual was now woefully obsolete. The combined effects of the Harmar and St. Clair debacles caused some to question the manual's merits. As Thomas Jefferson put it, "The federal council has yet to learn by experience, what experience has long taught us in Virginia, that rank and file fighting will not do against Indians." In February 1792, that lesson was made painfully aware to Lieutenant Colonel James Wilkinson when his 2nd Regiment detachment visited the St. Clair battlefield to bury the dead and retrieve the cannons. The burial party found every tree and shrub torn by musket fire and the bodies of 600 tomahawked, mutilated, scalped, and stripped men and women only slightly decomposed due to the early onset of winter. Inappropriate and poorly executed doctrine had contributed to their deaths.[30]

Despite these failures, it was the reorganization of the Army, not the replacement of its keystone doctrine, which became the government's focus. On 27 December 1792, rather than supplant the *Regulations*, President Washington approved Secretary of War Henry Knox's restructuring proposal, which exchanged the Army's regimental system for a combined-arms force, one that Knox and Steuben had envisioned in 1784. At that time, Knox and Steuben called for an army reorganized into eight "legions," each composed of one regular and seven militia companies capable of fighting independently in the wilds of North America. The Congress rejected the idea due to cost but after Harmar's defeat it authorized one additional conventional regiment. Now, in the wake of a second disaster, Congress approved a new force labeled the "Legion of the United States" to comprise 5,334 officers and men. Within the Legion were four sublegions, each one containing two infantry battalions, one rifle battalion, one artillery company, and a cavalry company. In many ways the Legion was more the result of Henry Bouquet's organizational innovation during the French and Indian War rather than the efforts of Steuben, Washington, or Knox. Still, Congress had approved a conventional force that, on paper at least, was organized to operate within the North American frontier and seemed capable of defeating the unconventional tactics employed by Indians.[31]

Given the national embarrassment and public terror generated by the Army's two stunning defeats, the government and the service could not afford another trouncing at the hands of unconventional forces. The task of rebuilding and leading the Army soon fell to Major General Anthony Wayne more by default than preference. Washington selected Wayne not so much for his command qualities but because he was acceptable to Congress and his fame was sufficient to stimulate recruiting. As the "Hero of Stony Point," Wayne met both political and military prerequisites. Along with his track record of getting results came a well-earned reputation for discipline, drink, a hot temper, and vanity.[32]

From June 1792 until April 1793, Wayne trained the new force at "Legionville," his camp near Pittsburgh, Pennsylvania. As a conventionally trained officer experienced with conventional troops, Wayne did not embrace extirpative war per se in crafting an army designed to defeat Indians. Rather than delve into military literature of the era that addressed unconventional forces, such as Grandmaison's *La Petite Guerre*, Wayne used the congressionally approved *Regulations* but modified the precepts. Recruits drilled to army standard until they could execute basic orders for marching and firing weapons. That accomplished, the bulk of the Army then adopted an open order of several feet between men instead of standing shoulder to shoulder, as doctrine prescribed. The men collectively learned to fire in volley at a concentration of Indians rather than at specific targets. Select riflemen and light infantry troops focused on killing individual opposing leaders. Wayne also embraced dragoons (mounted infantry) as an essential element of his force for their quick-strike capability. He also stressed the importance of flankers, a corps of observation, and a vanguard to give early warning and to protect the main body from attack. Tactical training did not involve raids and ambushes in small groups but rather mock battles where the Legion formed into two sides that maneuvered against each other under the commander's personal tutelage. The "Indians" wore appropriate attire and skulked from tree to tree while the Army formed into columns then wheeled and turned into ranks to give fire. Wayne retained but modified conventional doctrine to achieve an optimum mix of forces suitable for woodland combat.[33]

Following the completion of training, the Legion moved into the Ohio Valley in May 1793. There, it waited at Hobson's Choice while U.S. government negotiators attempted to forge a treaty with Blue Jacket. When talks failed in October 1793, Wayne advanced further into the Ohio Valley and spent December at Fort Recovery, the scene of St. Clair's defeat. Hostilities commenced in early 1794, with an unsuccessful Indian attack on the fort. On 27 July, Wayne's Legion, now bolstered by mounted Kentucky volunteers, advanced toward the Maumee River to confront Indians who had been promised support from the nearby British garrison at Fort Miamis.[34]

On the morning of 20 August 1794, with the enemy in close proximity, 900 men of the Legion of the United States advanced conventionally in columns with fixed bayonets into an area where trees lay blown down by a tornado years before. The Columbian Scouts, under the command of Captain Ephraim Kibbey, advanced well in front followed by an advanced guard of two companies commanded by Major William Price. The main body trailed with Lieutenant Colonel John F. Hamtramck and the 4th sublegion on the left wing and Brigadier General Robert Todd's brigade of mounted volunteers slightly to the rear. Brigadier General James Wilkinson's 1st sublegion moved on the right wing. Captain Jacob Kingsbury's 3d sublegion, Wayne's headquarters, pack horses, and sixteen howitzers were between wings with Lieutenant Colonel David Strong's 2d sublegion just behind; Brigadier General Thomas Barbee's mounted Kentuckians rode behind the entire formation. A light infantry screen surrounded the Legion. The Americans faced between 1,500 and 2,000 men constituted from Canadian militia under Captain Daniel McKillip and an Indian Confederacy of Wyandotte, Mingo, Delaware, Shawnee, Pottawatomie, Ottawa, and Chippewa, as well as a few Miami and some Mohawk, deployed in a scattered line about six men deep.[35]

From the Legion's perspective, the ensuing Battle of Fallen Timbers unfolded according to doctrine and training. After an early and brief rain shower, the scouts and advanced guard made contact with the Indians before being driven back to the main body. The Indians advanced furtively through the woods, fighting in groups by "treeing" themselves to fire from behind timber, and employing crescent tactics, thinning and extending their ranks to envelop both American flanks. To foil the maneuver, Wayne ordered the Legion to change formation from column into one or two lines, as the terrain allowed. Battalions and companies returned fire as the battle raged back and forth with portions of both sides advancing and retreating. Eventually, the Legion stretched for some two and one-half miles in a linear formation, its right flank anchored on the Maumee River flood plain. After fifteen minutes of hot fire and with casualties all around, Wayne ordered the Legion to advance in trail arms with bayonets. Ultimately, the force drove the Indians backward for 2 miles. At this point, the cavalry "fell upon and entirely routed them," while the British at Fort Miamis declined to intervene. Only after securing victory by fighting conventionally did the Legion break ranks to locate and burn Indian villages and stored food.[36]

Fallen Timbers is without question one of the most significant military encounters in American history. Given the Harmar and St. Clair catastrophes, Wayne had achieved an astonishing victory. With only 900 men, he had engaged and decisively defeated a force of 2,000 Canadian militia and their Indian allies. Triumph came at the cost of 33 men killed and 100 wounded. The battle had national repercussions. Fallen Timbers bolstered American public opinion, calmed anxiety on the frontier, and restored government confidence in the Army. The Indian cause was lost and peace ensued, for the moment. With the ensuing Treaty of Greenville (signed the following year), the Ohio Valley was again open for American settlement.

In terms of army doctrine, Fallen Timbers is significant for other reasons. Wayne had used conventional doctrine written for conventional warfare during the American Revolution. He did not use the *Regulations* verbatim, however, but modified them to suit Indian tactics. Despite the Legion's disbandment in 1796 and the Army's subsequent reduction in size to four regiments and a few dragoons, Wayne had demonstrated that conventional doctrine when appropriately implemented was fully capable of defeating unconventional forces. As George Washington had envisioned long before, conventional doctrine trumped the unconventional force in war, this time decisively. Much more than an example of "considerable discipline and efficiency in suppressing Native American resistance," Fallen Timbers established that a conventional doctrine when properly applied by effective army leadership was fully capable of defeating unconventional foes. This principle would carry forward into future army keystone doctrine.[37]

The *Regulations* was official army doctrine, as approved by Congress, although militia units often used whatever manual suited their needs. Still, a common drill continued to prove its worth into the early 1800s within the recently organized Indiana territory. Although Little Turtle and Blue Jacket became more cooperative with the U.S. government, other Indian leaders, such as Tecumseh, continued to resist as settlers violated agreements. In 1809, the Treaty of Fort Wayne ceded 3 million acres of Indian land, a treaty that Tecumseh and his brother Tenskwatawa (the Prophet) refused to acknowledge. By 1810, the Indians had formed a federation of tribes and the situation was precarious in the ceded territory. Indiana's territorial governor and statehood advocate William Henry Harrison was able to buy off several destitute tribes for an annuity, but talks between Harrison and Tecumseh broke down. The Army soon faced yet another Indian war.[38]

Harrison had been Wayne's aide-de-camp at Fallen Timbers. He was well acquainted with the *Regulations,* having seen firsthand what modified conventional doctrine was capable of accomplishing against unconventional forces. Still, the governor and commander in chief spent only a little time training his combined force of about 1,100 regular and militia infantry, dragoons, and mounted rifles consisting of one regular infantry regiment, two separate companies, and a militia so green that its members were deemed intimidating only when on horseback. After gaining President James Madison's approval, Harrison moved these troops toward Prophet's Town, the home of Tecumseh and his brother, an Indian spiritual leader.[39]

Harrison's advance used conventional doctrine. The main body moved in a column about a mile long while advanced guards and flank security provided early warning. Mounted elements forward of the main body were well-placed to react faster than infantry and provide additional early warning. The force eventually crossed the Wabash, arriving at the Vermillion River and the extreme boundary of the land cession without detection. Now, 60 miles from Prophet's Town, Harrison constructed a blockhouse to cache supplies.[40]

On 3 November 1811, the Army crossed into the ceded Indian Territory using tactics "similar to that used by General Wayne." The infantry were arrayed conventionally in two columns of files on either side of a road, with mounted riflemen and cavalry in front, at the rear, and on both flanks. Expecting an attack from the front, the Army changed formation into battle order during the last 4 miles of the march. A vanguard of mounted riflemen preceded the main body by 300 yards followed by a first line of short columns with the Regulars in the center and two companies of militia infantry and one mounted company on each flank. Immediately behind was the baggage train composed of cannon, cattle, hogs, and wagons. Short columns of cavalry formed a second line at a distance of 300 yards behind the baggage train. Three companies of militia in column constituted a reserve directly behind the second line. Along the way, scouts reported potential ambush sites. As a precaution, Harrison changed formations as often as three times in a mile and a half.[41]

The Army came to within a mile and a half from the objective, as parties of Indians gathered and moved about. After attempts to negotiate with the Indians through interpreters failed, Harrison halted the columns and ordered the men to place their packs in wagons and then re-form. Following a short advance, the Army fell into line with skirmishers and interpreters forward in case the Prophet wanted to negotiate. The Indians, seeing Harrison rapidly deploying for battle, agreed to a morning meeting between the two sides.

Trusting that the Indians might attack anyway despite their promise to wait, Harrison had his commander of dragoons, Major Joseph H. Daviess, find a suitable place to encamp. Daviess notified Harrison of a dry piece of open oak-covered ground located 10 feet above a marshy prairie that was suitable for a conventional defensive encampment. The Army occupied the area and then formed into a modified conventional hollow square. Due to terrain irregularities, the perimeter resembled a trapezoid with two lines 150 yards apart at the east that gradually tapered to roughly 75 yards on the west. The front line consisted of one company of militia and a battalion of Regulars, with two additional militia companies running east-southwest facing generally southeast toward Prophet's Town. Faced to the northwest, the back line overlooked Burnett Creek, manned by a battalion of Regulars and four militia companies. Closing the trapezoid on the western (narrow) side were the so-called Yellow Jackets (or Spencer's Company) of mounted riflemen. The eastern (wider) side consisted of Well's mounted Kentucky militia. Two companies of light dragoons under Captain Benjamin Parke and one infantry company were situated just behind the front line toward the southeast corner of the perimeter. Sometime during the night, the Prophet decided to attack the American encampment with the expressed objective to kill Harrison.[42]

Under cover of darkness early on the morning of 7 November 1811, the Indians rushed the encamped Americans from several directions. The fighting raged for several hours. The Indians, bolstered by the Prophet's guarantee that the American bullets would magically pass through their bodies, attacked with vigor as individuals or in small, uncoordinated waves. The Americans fought conventionally from their ranks. Harrison was able to maintain the integrity of his perimeter. Gradually, the Indians grew demoralized; they saw many of their "impervious" comrades shot dead or wounded while Harrison rode his horse unscathed despite many attempts to shoot him out of the saddle. Dawn brought a counterattack by bayonet-wielding infantry and dragoons who drove the Indians off. Having lost 20 percent of his force (37 killed and 126 wounded), Harrison did not pursue the Indians into their town. Instead, the Army reestablished a perimeter for the night. The following morning mounted troops entered an abandoned Prophet's Town. The soldiers hauled off what supplies they could carry and burned the rest. In the process, they discovered evidence of British support from Fort Malden in Canada. The Army then withdrew in good order. Although Indian attacks in the frontier did not cease, many tribes scattered at the news of defeat. The Prophet's fraudulent claims of magic ensured his political demise.[43]

What became known as the Battle of Tippecanoe marked the culmination of two centuries of American tactical warfare that had evolved from the informal practice of the early 1600s into the doctrinal *Regulations* of 1779. As the first American Army keystone doctrine, the *Regulations* conveyed a philosophy of duty and leadership for virtually every rank while describing how to fight war at the tactical level. But the publication addressed conventional tactics alone; no guidance was offered concerning the use of Rangers or unconventional warfare. Despite not being recognized within the Army's principal manual, informal tactical approaches to war nonetheless continued in the frontier from 1779 until 1783. Keeping in mind the ongoing *petite guerre* debate in eighteenth-century Europe, one can grasp why the unconventional approach to war held no place in the *Regulations*. In his treatise written soon after the American Revolution, Johann von Ewald summed up the beliefs of many that killing civilians in war was repulsive. By extension, those who practiced unconventional warfare were also repugnant.[44]

The *petite guerre* debate does not, however, fully explain why the *Regulations* took shape as they did. Steuben was certainly deficient in American frontier warfare experience, but he was cognizant of Prussian Jaegers and other European forces designed for *petite guerre* (as demonstrated by his 1784 suggestion to create an American legion suitable for the frontier). Furthermore, as Steuben's commander, Washington influenced the manual's development and might have provided input drawn from his own frontier experience or that of his officers. He elected not to do so, probably due to an understanding that conventional armies represent political legitimacy and Rangers were "more of a plague than an asset." Rather than champion these units in 1777, Washington experimented with European-influenced light infantry companies consisting of men hand-picked from conventional regiments who were trained to move fast and use the lighter *fusee* musket. Steuben wrote of light infantry in the *Regulations* and the manual contributed somewhat to Anthony Wayne's victory at Stony Point, New York, in 1779. But American light infantry fought conventionally. Their skills were in no way comparable to the long-range combat capability afforded by the avant-garde Rangers. Given Washington's envy of the British Army and loathing of Rangers, recognizing unconventional tactics within the Army's keystone doctrine would have acknowledged them as the equal of conventional troops, something Washington was disinclined to do.[45]

Washington's influence and wartime necessity influenced Congress to approve a doctrine that failed to reconcile fully the reality that the American Army fought conventionally *and* unconventionally. Because the *Regulations* did not address how to conduct both forms of tactical warfare, it contributed to the Harmar and St. Clair routs in which American units, applying unmodified conventional doctrine, proved unprepared to face unconventional Indian forces. Inadequate doctrine poorly executed not only shook the Army but also unsettled the confidence of the government and the American people. It also emboldened the British, Canadians, and Indians.

Critics such as Jefferson realized that the *Regulations* were inappropriate for the frontier but no doctrinal revision was undertaken in the 1790s. Instead, Congress changed the Army's organization. Wayne and Harrison subsequently modified the manual on their own to great success, as demonstrated at Fallen Timbers by Wayne (on the offense) and at Tippecanoe by Harrison (on the defense). The two commanders proved that conventional doctrine, properly modified, was more than capable of defeating unconventional forces. Yet, even as the Army had taken its first steps to regulate the chaos of war through doctrine, events were already playing out that would ensure the official demise of the 1779 *Regulations*.

The Decline of the 1779 *Regulations*

Between 1783 and the early 1810s, the majority of America's national security threats had surfaced on the frontier. British troops occupied forts in the Old Northwest, plotted mischief with Indians, and hindered American westward expansion. Spanish Florida encouraged Indian troubles by supplying arms and causing border tensions with Georgia and the Mississippi Territory. For the Army, the modified 1779 *Regulations for the Order and Discipline of the Troops of the United States, Part 1* had ultimately proven good enough to subdue Indians. By the Battle of Tippecanoe in 1811, the Army's 1779 keystone doctrine had weathered many storms and contributed to critical victories.[46]

But America's quasi-war with France in the late 1790s led citizen-soldiers such as George Washington, Alexander Hamilton, and Charles Cotesworth Pinckney to fret over the ascendancy of the French Army in Europe. While war was not formally declared, ongoing ocean confrontations meant preparing for a potential naval blockade or fending off the landing of hostile troops. The Army's Regulars were few in number and militia units lacked the training to withstand well-drilled European troops. Discussion over what to do led national leaders to consider the creation of a military system capable of producing professional soldiers. As inspector general charged with overseeing Army training, Hamilton championed a French-style military academy. He also supported replacing the 1779 *Regulations* with the French Army's tactical doctrine, the *Réglement* of 1791.

Hamilton was taken by the *Réglement* for good reason. After the disastrous French and Indian War (1754–1763), the Bourbons had enacted a number of army reforms. With the outbreak of revolution in 1789, France's Military Committee directed regiments to adopt numerical designations instead of specific names. The French artillery became a separate service on 29 October 1790. In 1791, the militia system was abolished and replaced with a National Guard of volunteers designed to augment the Regulars. Revolutionary fervor filled the ranks with raw recruits leading the French to produce the *Réglement,* a manual with simplified instructions for a rabble army of patriots.[47]

Although the *ordre mince* (Prussian-influenced linear tactics) tended to dominate French Army doctrine, the 1791 manual represented a compromise between advocates of the line formation and those who believed in an ability to shift from column to line and vice versa (Guibert's *ordre mixte*). In executing either formation, a French regiment contained two battalions of nine companies (with one company designated as grenadier) composed of platoons. The prerevolutionary Old Régime three-rank line for giving fire was retained with three variants. The first allowed for the front rank to kneel and fire with the second and third ranks firing while standing. An alternative option permitted the first and second ranks to fire while standing as the third rank reloaded weapons for the first two ranks. A two-rank system was also acceptable where a battalion gave a "rolling" (sequential) fire from left to right after which each soldier loaded and fired at will. Officers preferred the two-rank system, for undisciplined soldiers could master it with relative ease.[48]

To assist commanders in controlling the movement of troops up to battalion level, the *Réglement* contained detailed written instructions and diagrams for virtually every conceivable contingency. In battle, however, commanders retained the latitude to employ line or column without reverting to fixed rules. When marching, a unit might maintain an unregulated step. When maneuvering, a commander could employ either 76 steps per minute (normal cadence) or 100 steps per minute to move more rapidly while retaining order.[49]

Despite possessing a doctrine designed for rabble, the French Army was not prepared for a European war in which opponents formed multinational coalitions. To counter this threat, on 23 August 1793 France initiated the *levée en masse,* a national conscription, which remained in force until 22 August 1795. Underpinning the *levée* was the belief that every French man and woman was obligated to serve the nation in some capacity during crisis. Conscripts swelled the ranks and, as the Army expanded, the *Réglement* enabled draftee battalions to attain sufficient proficiency to fight alongside Regulars.[50]

While French doctrine allowed for both line and column formations in battle, most battles during 1793–1794 were generally fought using a combination of fire and movement. Whole brigades often deployed as skirmishers. Entire units gave fire before rushing the enemy as a fevered mob, despite what doctrine directed. By 1795, experienced survivors tempered these horde tactics. The use of columns, lines, skirmishers, and snipers became the norm. Line troops also learned to employ light infantry methods and vice versa, an important goal of the *Réglement*. During the 1796 Italian campaigns, the French fought well under their doctrine as applied by the emergent military leader and future emperor, Napoleon Bonaparte, whose campaigns soon earned international acclaim. Between 1805 and 1807, Napoleon employed line and column formations to virtually overwhelm opponents on the battlefield.[51]

The rise of the French Army and growing fear of invasion contributed to the birth of a U.S. military establishment. To this end, Congress, through the Peace Establishment Act, established West Point on 16 March 1802. Based upon a French educational model, the school's primary purpose was to educate artillerists and engineers for a young nation with a limited scientific base. Given the French Army's ascendancy, many American officers deemed the 1779 *Regulations* obsolete due to its reliance upon Steuben's Prussian views and Revolutionary War tactics. Still, while the 1791 *Réglement* earned the respect of certain American advocates, many defended Steuben's work as still relevant. Others looked toward Britain, not France, as the greater threat due to ongoing impressments of American sailors, an attack on the USS *Chesapeake* (1807), and continued intrigue with Indians.[52]

Although arguments over doctrine reform were driven by the need to counter the most likely and dangerous threat, French ideas dominated the discussion. France was dangerous. It had developed a tactical system that curbed British Army dominance through a balance between line and column formations and maneuvers that increased speed, flexibility, and firepower. Given the power of France's army relative to Britain's, the American doctrine, dating back to 1779, seemed obsolete.

But it was the raising of an effective army quickly during times of crisis that concerned many Americans. While American Army Regulars in the Old Northwest had more often than not withstood enemy fire both in defeat and victory, the soldiers were limited in number and, given the vast expanse of territory to be patrolled, overextended. Augmentees were another matter. Without sufficient regular troops to fight a war in one location while maintaining security in another, army leaders were forced to employ partially trained militia and raw volunteers. For new arrivals in the ranks, mastering the nuances of the 1779 *Regulations* was a tedious affair. Instruction in the particulars, even under Steuben's watchful eye during Valley Forge, required weeks of continuous drill. On the frontier, where the Army frequently had little time to respond to a crisis, the service often fought as it had trained, not the opposite. Typically given little time to prepare for battle, militia and volunteers found the Army's tactical doctrine unfathomable or too restrictive compared to their more familiar informal practices.

The issue of how to rapidly mold Regulars and augmentees into a cohesive force during chaotic times bedeviled army leaders. The French system, produced in part to turn mobs into effective soldiers, seemed to be a reasonable solution for an American army composed of Regulars, militias, and volunteers. Eventually, French tactical manuals became available in the United States. In 1807, Lieutenant Colonel John MacDonald translated the *Réglement* of 1791, as modified by Napoleon, into English. Later, in 1809, U.S. Army Lieutenant Colonel of Rifles William Duane published an improved translation for his American Military Library series. Duane apparently intended for the U.S. government to adopt his version as doctrine. However, the manual failed to receive official approval.[53]

In addition to undertaking translation work for the American Military Library, Duane combined his journalistic passion with his limited firsthand exposure to military readiness to produce a written appeal for a true military system. This would mean discarding the 1779 *Regulations* and forcing the regular Army and militia to train to one doctrinal standard. Stating that government must assume blame for ill-prepared military institutions lacking "the best principles" in the art of war, the journalist-officer rejected the "useless tract of baron Steuben." In its stead, he called for a national military system designed to accommodate both Regulars and militia and, by implication, volunteers. Duane's scathing criticism warrants partial reprinting:

> [First, U.S. Army tactics] should be such as to be equally applicable in its operation to the militia and to the Army of the U. States, whenever the former are called forth. The second, that every act and duty appertaining to the military establishment should be transacted by none other than men subject to military order, control, and responsibility; and liable to be put in motion or brought to account for delay or neglect in a military manner. These two principles lead to the consideration of

William Duane (The Granger Collection, New York)

what would be an efficient military organization; and here we have a host of formidable enemies, ignorance, a disorderly mass; indolence and idleness, hanging on the flanks, the steady habits of old prejudice ever alarmed for its patronage or its place; all immediate exclaim, would there not be great confusion produced by abrogating some duties and introducing others.[54]

In addressing what many knew to be true, Duane called for an efficient military bureaucracy and army doctrine suitable for both professional soldiers and others who defended the nation from frontier and overseas threats. Since the 1790s, federal military intervention under Harmar, St. Clair, and Wayne had been the solution for aiding westerners with their security issues. But army doctrine had not addressed the informal approach to war characteristic of unconventional forces. Frontiersmen were difficult to muster, train, and equip along conventional norms. Moreover, when territorial forces were formed, it was typically to counter an immediate, unconventional threat. The 1779 *Regulations*, being far too complicated to master in a few days, often bewildered raw troops. To rectify the situation, the nation required an easily understood tactical doctrine suitable for Regulars, militias, and volunteers.[55]

The 1812 *Regulations*

From 1810 into 1811, the search for a suitable army doctrine caused many individuals to produce manuals with the hope of gaining War Department approval. Irenée Amelot de Lacroix, a colonel and former Chief of Brigade in the French Army, translated and republished the *Réglement* of 1791, including changes undertaken by Napoleon. Printed in Boston, Massachusetts, in 1810, the manual attracted some interest. Robert Smirke's *Review of a Battalion of Infantry* also enjoyed some popularity in the United States but more so in England. The following year, Isaac Maltby, a brigadier general from Massachusetts, published a theoretical and practical manual titled *The Elements of War*, which attained popularity in his own state. Although each contributed to the available military literature, these manuals failed to secure federal approval.[56]

After 1811 and the Battle of Tippecanoe, tensions with Britain increased rapidly on the frontier. As the threat of impending war meant expanding the Army once again, the U.S. government now seriously pursued a new army tactical doctrine capable of turning raw material into disciplined troops. But without a mature military establishment capable of producing the intellectual grist necessary to create a doctrine from scratch, French tactics, designed for citizen horde armies, offered an instant solution.

In 1812, with the outbreak of war with Great Britain evident, Colonel Alexander Smyth, the army inspector general under Secretary of War William Eustis, published the *Regulations for the Field Exercise, Maneuvers, and Conduct of the Infantry of the United States Drawn Up and Adapted to the Organization of the Militia and Regular Troops.* As inspector general, Smyth was responsible for the training of soldiers and, by default, army tactical doctrine. Smyth drew from and improved upon MacDonald's translation of French tactics. The president and the War Department approved the manual on 30 March 1812, thereby superseding the 1779 *Regulations*.[57]

Given that the Army was an infantry-dominated force, the French-based 1812 *Regulations* were suitable for conventional Regulars, militia, and volunteers. About 1,000 copies were distributed as the war began. A second edition emerged later that same year. Working rapidly from previous publications, Smyth did not reject the 1779 *Regulations* completely but instead produced a hybrid that drew upon that manual and the French 1791 *Réglement*. The 1812 *Regulations* contained directions for training and maneuvering individual soldiers and groups of men through brigade level. Instruction included guidelines for standing and marching properly, procedures for loading and firing weapons, and descriptions for how and when units were to shift from line to column and vice versa.

Marching and maneuvering men consisted of the common step and quick step. The common step was measured at 2 feet from heel-to-heel, with 75 such paces made per minute. Quick step was 100 steps per minute, with each step also being 24 inches long. To facilitate hearing commands over the noise of battle, the men were arranged in ranks, with each soldier touching lightly the elbow of the individual standing next to him. The head remained upright and the eyes were fixated on the ground fifteen paces in front when moving. As Smyth noted, preserving the sanctity of the line was paramount. This could be achieved only by marching at a regular step, touching elbows, and keeping shoulders square to the front. When the troops were in battle, the manual advocated the French three-ranks method and rolling-fire technique:

> The fire commences by the right file of each company, and proceeds in succession to the left, but only during the first round. Each file will fire when the file next to the right have primed. The first rank loads, and keeps up the fire; the second hand their empty pieces with the right hand to the rear rank, receive theirs loaded, fire them, load and discharge them a second time, then exchange them for their own pieces loaded by the third rank, and so keep up the fire. The third rank is not to fire.

Within the brigades, each regiment contained two battalions. Eight numerically designated companies existed within each battalion, two of which were labeled "grenadiers or light infantry." In battle array, a typical army regiment might have four companies on line, two companies behind as a reserve, and two companies of light infantry in front acting as skirmishers.[58]

Smyth was credited as the sole author of the new *Regulations*. As inspector general, he wrote the doctrine for bureaucratic approval, in this case the president and the War Department. Neither the 1779 *Regulations* nor contemporary French Army doctrine had addressed unconventional warfare; therefore Smyth's work focused on conventional tactics while ignoring instructions for American Rangers. This situation stood in contrast to Congress' decision on 2 January 1812 to raise six companies of Rangers. A seventh company was soon added and, in February 1813, ten more companies were also authorized. Without doctrine to guide them, Ranger companies patrolled the Indiana, Illinois, and Louisiana territories, as well as Ohio, Kentucky, and Tennessee. The fact that President James Madison and the War Department approved a new keystone doctrine without considering Congress' raising of Ranger companies indicates how little the executive branch thought about the legislative branch and Rangers. Ranger tactics remained informal practice.[59]

Beginning on 18 June 1812, when the U.S. government declared war with Britain, army leaders were ordered to use the newly approved *Regulations* and discard the previous 1779 version. Many officers adopted the 1812 manual immediately. Others, however, found it too complex and clung to the previous doctrine. This was probably the case at Fort Detroit in 1812, where Governor William Hull of the Michigan Territory had deployed an American Army of Regulars, militia, and volunteers several weeks before war began. A veteran of the American Revolution, the aging Hull was no doubt familiar with the 1779 *Regulations* but it is unclear which manual he consulted, if any. When war commenced, Hull cautiously advanced his force into Canada only to lose his nerve and withdraw to Fort Detroit. The British and Canadians under General Isaac Brock subsequently laid siege. Hull surrendered the fort, 2,000 men, and numerous civilians on 16 August 1812.[60]

Later that same year, the American Army used both the 1779 and 1812 versions of the *Regulations* to prepare for an assault at Queenston Heights in Canada. Neither manual discussed how to conduct a contested river crossing, which characterized the attack. No doctrine was capable of soothing the competing egos of Major General Stephen Van Rensselaer of the New York militia and Smyth, now a regular army brigadier general with a field command, who despised each other. Although Smyth had written the Army's tactical manual, he proved to be ineffective and stubbornly refused to coordinate his efforts with Van Rensselaer. His disastrous Niagara campaign sullied both his reputation and the *Regulations* he had written. Colonel (later Brigadier General) Winfield Scott attributed the Army's failure to weak leadership, poorly disciplined militia, and badly executed tactics. In light of Smyth's miserable performance, army officers came to reject the 1812 doctrine as tainted.[61]

The *Regulations* of 1813

Wounded during the battle for Queenston Heights and subsequently captured, Winfield Scott was released in March 1813 to discover that the 1812 *Regulations* had been replaced by a new manual written by William Duane. Duane's version, the *Regulations to be Received and Observed for the Discipline of Infantry in the Army of the United States,* was also based upon the French 1791 *Réglement* with some original ideas about training soldiers more efficiently. Partially written in 1808, the completed work did not see publication until 1812, having passed through the hands of Congress and ultimately winning approval. Still, Duane's work had been overshadowed by Smyth, who had the backing of the Secretary of War. With Smyth's reputation sullied, Duane's work now rose to the forefront of army doctrine. It was approved on 19 March 1813.[62]

Duane was an exceptional writer but lacked command experience. His manual proved more philosophical than practical. Duane saw drill as the manual's centerpiece, for it was the essential disciplinary tool for victory. He believed drill had allowed Napoleon Bonaparte to achieve critical successes in Italy and Austria a decade earlier, for "moving men in military order and in a military manner, form the most important of all duties of the military art; they are the immediate objects of good discipline." Although continuing in the established pattern of focusing on conventional infantry tactics only, the 1813 *Regulations* omitted the individual drilling of new recruits in favor of group work. Duane set the common step at 76 steps per minute and quick time at 90 steps per minute, although he retained much of what the 1812 manual had already institutionalized to include the use of line and column. Duane concluded that "50 or one hundred men may be brought to the required habit with nearly the same labor and in nearly the same time as one may by another method," thus rejecting individual training, the bedrock of the 1779 and 1812 *Regulations,* in favor of efficiency. Officers were now to train their soldiers as a group in order to save time when working with raw recruits, an unwieldy concept for instructing thickheaded troops. Given how difficult it was to drill soldiers in groups, most officers came to despise the doctrine outright as "too fancy" and more fit for the parade field. Scott called it a "miserable handbook." He preferred the 1812 *Regulations* or his personal copy of the *Réglement*. By the end of the war in 1815, the 1813 doctrine was all but ignored.[63]

Given the War Department's doctrinal troubles, it is little wonder that the Regulars, Rangers, militia, and volunteers were tactically mediocre from 1812 through 1813. Unable to impose any system of order upon the Army, the War Department's doctrinal dysfunction had combined with weak leadership and poor training to produce disasters at Detroit and Queenston Heights in 1812. The following year proved only slightly better. Victory at York (Toronto) in April 1813 and the death of Tecumseh at the Battle of the Thames in October were offset by defeat. At Crysler's Farm in November 1813, 2,500 half-trained American Regulars under Brigadier General John P. Boyd (a Tippecanoe veteran who had recently assumed command from Major General James Wilkinson) faced 800 British and Canadian troops employing contemporary European tactics. The Americans suffered 350 killed and 1,000 captured. At Prairie du Chien (in present-day Wisconsin), an American garrison reinforced by forty Regulars and sixty-four Rangers surrendered to a British force composed of Regulars, Canadians, and Indians.[64]

During 1813, even as Duane's manual became army doctrine, Congress passed a resolution requesting President James Madison to prepare and bring before the assembly a tactical doctrine suitable for Regulars and militia. No action was taken; the army leadership was far too consumed by the war effort. Instead, every officer was at liberty to train his force as he saw fit using whatever manual he happened upon, a situation that led to tactical chaos.

There was some success, however. In 1814, at the Battle of Chippewa, a victorious American Army demonstrated a modicum of tactical effectiveness because Brigadier General Winfield Scott, under the command of Major General Jacob Brown, had disregarded the 1813 *Regulations*. Using an edition of MacDonald's translation, possibly Smyth's version, and a personal copy of the original French 1791 *Réglement* to train his brigade and one other, Scott employed "schools of the soldier" and the company to drill men under his scrutiny. He personally supervised his officers as they compulsorily drilled their troops three times a day for several weeks. Although some commanders preferred the 1779 *Regulations* or even British doctrine, Scott instructed his brigade in battalion, regiment, and brigade tactics according to a French model, ensuring that the troops (a combination of Regulars, militia, and volunteers) understood how to shift from line to column and back again, as well as how to give fire from either two or three ranks.[65]

Scott also practiced "open order" tactics based upon the French *tirailleurs,* skirmishers or light infantry. Open order tactics employed light infantrymen to fight in dispersed formations as individuals. They deployed in front of the "closed order" concentrated formations of the rank and file. Skirmishers were not mobs, although some tended to fight that way. They often required specialized training to enable them to seize woods, fight in towns, and use firepower to devastate tightly compacted heavy infantry formations. In open ground, however, skirmishers were susceptible to artillery bombardment, cavalry charges, and the massed fire of infantry. As such, deciding how and when to employ them effectively required a keen tactical mind. Between 1792 and 1793, French commanders had made use of *tirailleurs;* sixty-four men per battalion received training in such methods. Scott, mindful of French Army tactical developments, undoubtedly followed suit in 1814.[66]

On 3 July 1814, Brown led an amalgamated force of 3,500 Regulars, militia, and Iroquois Indians into Canada to recapture Fort Erie, which had fallen in December 1813. After retaking the fort, the Army advanced to meet elements of British Major General Phineas Riall's army, which totaled 2,000 Regulars, Canadian militia, and Mohawk Indians. During a furious and bloody affair that included a stationary engagement at a range of sixty paces, the Army compelled a determined British force to retreat in good order. Scott's brigade was particularly impressive. Although the Americans had suffered heavier casualties, the victory solidified Scott's reputation as master tactician and trainer. On 25 July 1814, about 3 miles north of the Chippewa site, a reinforced Riall defeated Scott at Lundy's Lane. After the bloody mêlée, the American Army withdrew to Fort Erie. The American Niagara campaign would ultimately fail. Still, the 1814 battles provided examples of what French Army doctrine employed by competent American leadership might accomplish on a conventional battlefield.[67]

The War of 1812 was far from a purely conventional war, as numerous Indian tribes sided with the Americans and the British. Tecumseh and his followers joined the British to drive the Americans out of the Northwest. Using what Americans considered unconventional tactics, Indians raided Peoria in present-day Illinois. British forces and their Indian allies struck along the Rock River to impede American boats. War parties ransacked and burned American settlements seemingly at will. Fears were heightened after Spain allied with Britain, for frontier settlers realized that no American regular troops existed north of the Mississippi and Illinois River junction. With the Illinois and Missouri militia discharged by late 1813 due to a lack of supplies, the frontier war was left to unconventional Rangers and some Indians who were persuaded to fight against hostile tribes.[68]

Despite the Army's use of conventional doctrine in Canada and other northern locations, the Creek Civil War of September 1813–March 1814 represented both the use of conventional doctrine and an American informal approach to war. Located primarily within the Mississippi Territory (now western Georgia and most of Alabama), the Creeks were a diverse and divided nation split between Upper Town and Lower Town factions. The Lower Creeks were more receptive to cooperation with the U.S. government than the Upper Creeks who, by the early 1800s, were influenced by the Indian revival message of Tecumseh and his brother, the Prophet. In 1812, an Upper Town faction known as the Red Sticks killed several Lower Creeks and raided Tennessee in the belief that a "great spirit" was forging all Indians into a force that would drive white settlers out of the Southwest. In 1813, the Creek nation was torn asunder politically and involved in a civil war. By the summer of that year, the war had broadened in scope. Frightened frontier white settlers mobilized territorial militias in response.[69]

In need of resupply, Red Stick leader Peter McQueen and a party of warriors traveled to Spanish Florida for arms and powder eagerly provided by British merchants anchored off shore. On 27 July, upon return to U.S. territory, McQueen met and defeated a Mississippi militia force led by Colonel James Caller at the Battle of Burnt Corn Creek. McQueen's victory emboldened William Weatherford (Red Eagle) to join the cause. On 30 August 1813, Weatherford and 250 Red Sticks attacked 120 Mississippi militiamen and a large number of settlers at Fort Mims, where they killed and scalped over 250 men, women, and children. Weatherford announced that any Creek not joining the Red Sticks' cause would suffer a similar fate.[70]

Several American armies responded to the Fort Mims catastrophe. They included partially trained Regulars, militiamen, and volunteers from Georgia, Mississippi, and Tennessee and friendly Creeks primarily under Chief Pathkiller. Organized conventionally into brigades along with Indian allies, the forces employed both conventional and informal tactics. On 3 November 1813, a force commanded by militia Major General Andrew Jackson surrounded the hostile Creek village of Tallushatchee. The Army advanced in conventional columns for security and control purposes, then broke ranks to burn the town while many inhabitants were still inside their homes. Most of the resident men were killed but eighty-four women and children were spared.[71]

Jackson's army, a force of 1,200 infantry and 800 cavalry and mounted rifles, later advanced in three supporting columns over 30 miles to engage a Red Sticks force laying siege to friendly Indians at Talladega. At 4 A.M. on 9 November 1813, Jackson concocted an attack plan employing militia infantry and some mounted troops in column on the left and volunteer infantry and mounted soldiers advancing in column on the right. In the center and slightly rearward, 250 cavalry followed as a reserve. About 400 yards to the front, several companies of infantry and scouts advanced with orders to engage and fall back in order to draw the Red Sticks into the center. The columns on the left and right were then to envelop and annihilate the trapped Indians. About 8 A.M., the advanced force succeeded in drawing the Red Sticks to pursue them through the brush. Indian war whoops resonating through the trees caused three militia companies to bolt. Jackson plugged the gap with his reserve. The left and right columns closed as planned and the Red Sticks were shot to pieces. In about fifteen minutes, Jackson's force killed 300 of an estimated 1,000 Indians. American losses totaled 15 killed and 85 wounded. Although some Indians escaped through a small gap in the American lines, the Talladega siege was lifted.[72]

Jackson later attacked Tohopeka, a Red Sticks fort of about 1,000 souls and the center for Indian followers of Tecumseh's spiritual awakening. Using the Tallapoosa River that bordered the village on three sides to their advantage, the Red Sticks constructed shoulder-high breastworks on the open end of the peninsula to allow for crossfire on an attacking force. On 27 March 1814, Jackson's army of 3,000 troops fired on the defenses using artillery to little effect. Cavalry and Indian allies managed to successfully attack the enemy village from the rear by crossing the river. The fort was eventually stormed. For five hours, the American informal approach to war took over. American troops broke ranks and formed into small groups, then burned the village, killing or capturing the men, women, and children. Those who sought cover along the riverbank were ferreted out and eradicated. The savage battle effectively ended the Creek civil war. Soon thereafter, the August 1814 Treaty of Fort Jackson ceded 23 million acres of land to the United States.[73]

For the United States and Great Britain, the War of 1812 ended in a strategic draw. Many Americans nonetheless believed that they had won the conflict, as reflected in the subsequent "Era of Good Feelings." The War Department, however, was in disarray doctrinally. The American Army's method for regulating the chaos of war through published doctrine had resulted in three manuals from 1779 to 1813. Although only one doctrinal manual was officially approved for use at any given time, all three, as well as personal copies of the French *Réglement*, had been consulted by Regulars, militia, and volunteer units. Doctrinal dysfunction had not only caused tactical chaos among the Army, but nearly caused a national setback in war.

Creating the *Infantry Tactics* of 1815

American infatuation with French Army doctrine during the War of 1812 stemmed more from an attempt to find a tactical system capable of turning civilians into soldiers rapidly than obsession with the successes of Napoleon Bonaparte. But Napoleon's emphasis upon climactic battle, his use of mobile artillery to devastate opposing formations, and his organizational ability to concentrate forces rapidly in time and space had earned the French Army respect on both sides of the Atlantic.

Fixation upon French doctrine as suitable for both seasoned and raw recruits had blinded Americans to the fact that European officers had not been idle in seeking ways to counter France's military power. In 1802, the Prussian general G. von Scharnhorst experimented with more flexible tactics and the increased use of skirmishers, the creation of a militia, a *levée en masse*, and, in 1807, a national reserve under the *Krumper* system. Later, in 1813, Prussia adopted the *Landwehr* and *Landstrum* systems. In Portugal and Spain, Arthur Wellesley, the duke of Wellington, successfully combined conventional British infantry tactics and informal practice used by Portuguese guerrillas to defeat the French. As war raged in North America in 1814, Napoleon Bonaparte became mired in Russia and soon suffered near annihilation.[74]

Drafting an acceptable keystone army doctrine for an infantry-dominated Army became a U.S. government priority following the Treaty of Ghent (1815) that ended the War of 1812. Given that war with Britain or other powers was still possible, the service required a manual capable of fielding a mixed force of Regulars, militia, and raw volunteers that rivaled European land power.

Previously, in December 1814, Congress had removed army doctrine from the inspector general's purview alone by appointing a board of officers to create a new conventional army tactical manual. The service's master tactician, Brigadier General Winfield Scott, became the board's president. Although French military power had waned with Napoleon's defeat at Waterloo in 1815, Scott's charge was to prepare infantry drill regulations "conformable to the House of Representatives" and "after the pattern of the Rules and Regulations for the Field Service and Maneuvers of the French Infantry." Congressional backing, general officer rank, a solid field reputation, and stalwart confidence in French tactics allowed Scott to regulate its contents. Under Scott's watchful eye, the board wrote a manual that included precise movements for the American soldier (incorporating Scott's so-called school of the soldier) together with drill instructions for company- through division-sized maneuvers. In following French methods, Scott's instructions were both simple and detailed enough to enable a raw recruit to become proficient in short order.[75]

Scott recognized that the infantry-dominated American Army depended upon muskets as the primary individual weapon. But muskets were notoriously inaccurate, subject to malfunction, and of limited range. Given the shortcomings inherent in the weapon's technology, it was possible to advance to within 50 yards of a musket-equipped enemy defensive line without suffering undue casualties. Scott's offensive tactics therefore focused upon concentrating troops for control and firepower in tightly packed regiments located 22 paces apart. As with the French 1791 *Réglement*, American battalions arrayed in three ranks, the first two firing with the third rank loading weapons. The manual also allowed for the occasional use of the two-rank line. Scott ordered the ranks to be 13 inches apart; men within the ranks stood elbow-to-elbow with shoulders square to maintain order and control. Shifting from line to column and back again was also emphasized; units were to use both the column and line and be proficient in shifting back and forth with ease. In an attack, the regiments advanced in line or column slowly and methodically with bayonets fixed using a 28-inch step at the rate of 90 steps per minute. A direct step was ordered for quick time, or 110 steps per minute. Scott believed the paces to be tolerable for avoiding loss of control and fatigue, the exception being the last few paces before crashing into the enemy line when 140 steps per minute or a run was allowable. Skirmishers or light

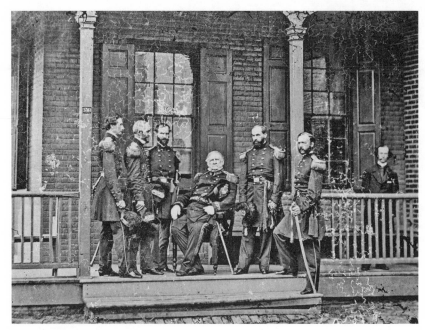

Winfield Scott, seated center (Courtesy of the National Archives)

infantry preceded the advance in loose formations, with no more than two companies per regiment arrayed in that fashion. Defensive formations included standing in ranks against infantry or forming square to fend off cavalry attacks. Scott insisted that the manual of arms, marching, the loading and firing of weapons, the alignment of soldiers into ranks and files, and maneuvers in line and column be explained in detail.[76]

Although critics called the new manual plagiarism in that it was a virtual verbatim copy of the French infantry regulations, Scott had deliberately chosen expediency over originality. His aim was to ensure standardization among a force prone to diverse levels of proficiency. The War Department acknowledged that point when it approved *Infantry Tactics* (also known as *Scott's Tactics*) on 28 February 1815, noting, "With a view toward uniformity throughout the Army, all infantry exercises and maneuvers embraced in this system will . . . be strictly observed." As the archetype master tactician, Scott's ability to replicate French tactics designed for mass conscript armies was so thorough that he later garnered the respect of the Swiss military theorist, Antoine Henri de Jomini. Several states passed laws mandating that *Scott's Tactics* be adopted by their militias; an abridged version for such a purpose was published in 1817, although its use was not federally mandated.[77]

With 1,000 copies in print by 1817, the manual gradually filtered throughout the Army. Of note was the title change from previous army doctrine. Whereas the 1779, 1812, and 1813 manuals had been called *Regulations,* the new title, *Infantry Tactics,* indicated an emphasis upon training and combat procedures. Although administrative regulations existed in various forms prior to 1812, Scott later produced the *General Regulations for the Army* (approved by Congress in 1821) to cover management. Together, these publications represent two distinct army intellectual paths: *Infantry Tactics* instructed how to *fight* an army while the *General Regulations* established how to *manage* one. Implementation depended upon the officers and noncommissioned officers who were widely dispersed throughout the various states and territories. Ongoing missions such as coastal defense, Indians and interlopers, as well as desertions, fatigue detail, guard duty, maintenance, and other mundane tasks often made unit training a hit-or-miss proposition.[78]

In fairness to Scott, copying and improving upon foreign ideas was common practice for American society; the United States was the product of cultural mimesis. As progeny of a culture prone to borrowing and improving upon foreign ideas, the Army's leadership had followed suit in drawing heavily upon foreign concepts when framing doctrine. American Army doctrine was transnational, for its authors imported, translated, and modified foreign ideas of warfare. Without a professional military tradition to produce high-quality officers and enlisted men, mimicking foreign doctrine, be it British, French, or Prussian, was a logical course of action. While army doctrine had embraced British and Prussian ideas in 1779, it was the French system of 1791, with its focus upon training hastily formed citizen armies, that best fit the American concept of employing Regulars, militia, and volunteers in war, albeit conventionally.

From 1779 to 1815, American doctrine writers had focused on a conventional foreign threat. They made no attempt to regulate unconventional warfare by publishing guidelines for how to fight Indians. Two tactical systems therefore came to exist within the Army, a formal one expressed through doctrine that regulated conventional tactical drill and a second consisting of informal practice expressed primarily on the frontier through unconventional warfare. The regular Army, however, occasionally departed from its conventional doctrine to engage in informal practice (particularly when facing Indians) although many civilians considered ambushes, burning, raids, and scalping too repugnant for advanced societies.

In weighing threats to the Republic, Indian tribes and criminals paled in comparison to heavily armed and well-drilled British, French, and Spanish forces. With foreign powers established in Canada and Florida, as well as west of the Mississippi River and in the Caribbean, army doctrine focused upon what Congress perceived to be the greater hazard: a potential European invasion. To face European powers and triumph, army doctrine had to be conventional. As far as unconventional foes were concerned, conventional doctrine, when properly adhered to and implemented, was good enough to defeat them, as Fallen Timbers had demonstrated in 1794. Although the Army continued to fight unconventionally periodically, Ranger units all but disappeared until World War II, a testimony to the dominance of conventional army doctrine.[79]

During its lifespan of four decades, *Infantry Tactics* proved to be beneficial though of limited utility for an army confronted by both frontier constabulary duties and war. Although the War Department was unable to enforce its use among all units, officers and soldiers became comfortable enough with a tactical system that seldom changed. On 12 May 1820, Congress passed an act that required militia units to conform to the regular Army's drill regulations, but the War Department lacked the oversight authority and means to enforce the law.

Army policymakers followed European military affairs closely. When the French modified their doctrine, Americans followed suit. In 1824, the so-called system of 1815 required modernization. A board of officers convened to make the necessary changes. Board members Winfield Scott, Hugh Brady, John R. Fenwick, William J. Worth, and Silvanus Thayer changed the original manual by modifying commands, repositioning officers and noncommissioned officers in formations, placing a light infantry company on the left flank of a battalion as a rifle company, and converting the right-flank company into grenadiers. Although the modifications were minor, the board members provided so many comments that the basic manual grew into two large volumes. Following scrutiny by the War Department, the revised *Infantry Tactics* became doctrine on 5 January 1825. Militia officers, however, soon found the new manual overly detailed for their part-time soldiers. The board then reconvened in October 1826 to produce an abridged version. That manual, based upon regular army doctrine but written specifically for militia use, was published on 5 December 1826.[80]

Although altered periodically, French tactics had changed little since the American's adopted the 1791 *Réglement* in 1815. In 1827, the French

Army leadership conducted a training camp at St. Omar to seek more flexibility to its tactical methods. The result was the publication of a new French tactical manual in 1831. Three years later, on 8 April 1834, the Congress of the United States directed the War Department to once again revise American Army doctrine. And, once again, Winfield Scott was called upon to make the alterations. After several months of work, Scott presented the army leadership with the results on 3 February 1835. This edition placed more emphasis upon a battle line composed of two ranks over three ranks in deference to extending the formation and thus achieving more tactical flexibility. However, the simplicity of the 1815 version had vanished. The 1835 manual was excessively procedural and lacked the flexibility the French had sought.

A master of details, Scott had changed the formation of a rifle company from the tallest men being in the back rank sized from both the right and left (to allow for firing over the heads of front-rank soldiers) to sizing from the right only. In this way, when marching in column for purposes of control and speed, the tallest soldiers were located at the front to act as guides for those in the back. It also prevented shorter men from firing into taller ones if a commander ordered his lines to turn about and give fire toward what had been the rear. Moreover, when turning from column into ranks by executing a flanking maneuver, the tallest soldiers ended up on the right or left and out of the line of fire of shorter men. Scott also preferred countermarching, the use of columns in mass to turn in various directions, as a way to increase speed. However, such a maneuver is cumbersome; soldiers in the back tend to lose gap discipline with the men in front of them, especially in rolling terrain. Only the best-drilled units avoided what contemporary terminology calls the accordion effect, the repeated bunching up and stretching of formations of men when marching. Tactically, countermarching caused officers to frequently lose control of their formations as they strung out over a given distance. This could prove disastrous when under fire. Scott revised the manual in 1839 to correct the situation, but the basic tactical drill movements remained rigid and machine-like. An updated version was published in 1846, but the manual was still complex and driven by drill.[81]

Infantry Tactics in Peace and War

From 1815 until the 1850s, army officers engaged in a number of missions ranging from exploration to manning frontier forts to defending coastlines. Doctrine written to enable the Army to fight conventionally and win tactically proved of limited value in these endeavors. In 1817, Secretary of War John C. Calhoun decided to use the Army to garrison a series of forts beyond the Mississippi River in an effort to facilitate westward expansion. With congressional approval, small posts emerged from Fort Leavenworth, Kansas, to the Yellowstone River and beyond, joining those already established in the Wisconsin territory.

To better prepare soldiers for their duties, Calhoun sought to standardize training, equipment, and weaponry. In 1824, he ordered the establishment of the Artillery School of Practice at Fort Monroe, Virginia, and, two years later, an Infantry School of Practice at Jefferson Barracks, St. Louis, Missouri. Although congressional funding was at times dismal, the French-based schools played an essential role in imposing doctrine upon the service.[82]

For a garrison army, official doctrine succeeded as a means to drill individual soldiers but was less effective in training large unit formations. Frontier regiments were too dispersed in isolated forts and their small garrisons inhibited large-unit training. Competing duties and personnel turmoil also meant that training was a hit-or-miss affair. Even trained infantry units found it difficult to confront mounted Indians operating on the prairies and plains. Beginning in 1816, Congress not only sought to garrison the west, but also determined that the nation possessed a vulnerable coastline. That conclusion spurred, albeit slowly, the construction of coastal fortifications that required army troops.[83]

Where doctrine was deficient in addressing the numerous army missions of garrison life and coastal defense, influential yet unofficial publications emerged to fill the void. These included works written by West Point Professor of Civil and Military Engineering Dennis Hart Mahan. Mahan, an 1824 West Point graduate, spent four years in France where he gained an enthusiastic appreciation for Napoleon's art of war, as well as French artillery and engineering. His West Point courses drew on his international experience. Translated French publications such as S.F. Gay de Vernon's *Treatise on the Science of War and Fortification* and Jomini's *The Art of War* blended with Mahan's views of American warfare into a series of class lectures covering strategy, fortifications, logistics, and engineering. Approaching the military profession as both a science and an art, Mahan's lecture notes appeared in the 1847 *Out-Post*, a publication designed for military aficionados who were interested in developing an American art of war. Given that the Army consisted of a small professional core augmented by raw recruits and militia, Mahan advocated defensive tactics. To Mahan, mob armies incapable of fully mastering the nuances of tactics meant that offensive operations were a callous waste of life. When an attack was warranted, Mahan preferred advancing with a close-order line formation followed by infantry. He also advocated using skirmishers ahead of the main attack who made use of available cover to close to within 200 yards of the enemy and then rushed forward.[84]

In 1846, one of Mahan's former students, Lieutenant Henry Wagner Halleck, West Point class of 1839, produced *Elements of Military Art and Science*. Although not doctrine, the publication influenced army officers because it applied both Mahan's thoughts and Jominian ideas about the conduct of war to the American strategic situation. Halleck, however, was far more defensive minded than Jomini. This is no surprise, for Halleck, along with Mahan, had concluded that the defense was the stronger form of war. Although Halleck acknowledged the importance of offensive tactics, he considered a defensive approach to war more suitable for the mass, nonprofessional armies America raised during crisis.[85]

More than just a manual to prepare troops to fight, *Infantry Tactics* saw use in war, as well. The Black Hawk War of 1832, more a pursuit than a campaign, was of too short a duration to reach any conclusion over the doctrine's use or effectiveness. The same can be said of the Army's doctrinal performance in enforcing the government's Indian Removal Act of 1830. Army doctrine, however, played some role during the three Seminole conflicts fought to resolve decades of border intrigue between Georgia, the Mississippi territory, and Spanish Florida. In those wars, the Army maneuvered at times but also suffered from the lack of written guidance concerning how to fight an unconventional foe. Many past lessons were relearned in battle. When the Seminoles stood and fought, the Army often resorted to firepower and bayonets rather than tactical maneuver.

During the first Seminole War (1817–1818), a 3,500-man force of Regulars, militia mostly from Tennessee, and Creek Indians invaded Florida to subdue runaway slaves, hostile Red Sticks Indians, and British agents. Fought conventionally and unconventionally, the Army's success led to U.S. annexation of Florida in 1821. The Second Seminole War (1835–1842), a conflict undertaken to enforce federal law to remove Indians to the western side of the Mississippi River, also witnessed a mixture of conventional and unconventional warfare. On 28 December 1835, a force of 8 officers and 100 enlisted artillerymen trained to fight as infantry led by Major Francis L. Dade was nearly annihilated near the Wahoo Swamp after advancing through open pine woods in two conventional parallel files with an advanced guard, a main body, a rear guard, but without flankers. Commanders eventually rediscovered that while conventional tactics allowed for moving forces rapidly and in good order, informal practice was often necessary to convince Indians to capitulate. In shifting from large bodies of troops with lengthy supply trains to small self-sufficient groups, the Army's leadership targeted the Indians' dependency upon nature for sustenance. Returning to informal practice established in the early 1600s in Virginia, the Army destroyed food, shelter, and animals. These actions brought about Seminole submission through starvation while reducing army casualties. The Army recruited local guides to reconnoiter entangled areas where doctrinal formations often proved of little use. The Third Seminole War (1855–1858), fought to remove Seminole tribes that had remained after the previous conflict, again proved that success at times meant diverging from or modifying *Infantry Tactics*.[86]

The Seminole Wars ensured that *Infantry Tactics* became embedded within the American Army wartime experience. As the Army formed and reformed in response to various national security crises during three conflicts in Florida, doctrine provided the foundational principles for regulating the Army's tactics during the chaos of war. Although the use of *Infantry Tactics* was often haphazard, the government had seen to it that the manual was updated periodically and contained the most current French military thought. Three decades of army personnel grew comfortable with its contents.

Still, because the army leadership borrowed French ideas with little modification, *Infantry Tactics* continued to mimic European conventional tactics. When these precepts failed to match the wartime realities of fighting Indians, army leaders both modified doctrine and also reverted to informal practice techniques dating to the colonial era. During the Seminole Wars, army leaders used doctrinal guidelines to move and maneuver troops, but also digressed to informal practice through ambushes, burning, and raids. To be sure, doctrine and informal practice complemented each other in a series of successful Indian campaigns.

Of course, doctrine by itself is valueless without individuals sufficiently skilled to turn words into action. For the American Army, that task fell primarily to West Point officers schooled in French military theory and tactical methods, as well as militia and volunteer leaders. However, it was the military academy that had produced a small but competent crop of officers who ultimately imprinted their training methods upon regular army regiments. As the officers advanced in rank and influence, their imposition of tactical doctrine upon the service became a part of army professionalization. But professionalization came at a price, for the Army was tactically more French than American. Nonetheless, in 1846, the French-based *Infantry Tactics* would be unleashed upon Mexico.

Mexico: The War *Infantry Tactics* Imagined

Unlike the conflicts fought against Indians, the War with Mexico (1846–1848) was a purely conventional conflict of the sort that Congress and the *Infantry Tactics* board of officers had envisioned between 1814 and 1815. American expansion and Mexico's dispute over Texas independence in 1836 and statehood in 1845 had not been resolved by 1846. The American government offered to buy Mexican territory, not only to settle the Texas question but also to acquire California. When negotiations made little progress, President James K. Polk ordered Brevet Brigadier General Zachary Taylor's army to the Nueces River in Texas in July 1845. In February 1846, Taylor, having advanced to the Rio Grande, awaited his orders. Mexico ended diplomatic discussions soon thereafter and sent troops to the Rio Grande River. On 23 April, a brief but bloody skirmish was fought between American and Mexican troops along the river. Congress declared war on 11 May 1846.

At the time, the American regular Army included 734 officers and 7,885 enlisted men, an insufficient force to fight a war and also defend the nation from additional threats. Most of the experienced troops had fought Indians or were familiar with frontier garrison duties; Taylor had about 1,500 troops under his command but not all were Regulars. The Texas militia contributed about 1,390 men. To augment the regular Army, Congress authorized 50,000 volunteers to serve at the president's discretion for twelve months or for the war's duration. Of these, 15,000 men joined the regular Army in Mexico.[87]

Because line regiments were understrength in both officers and men, the ranks soon filled with raw volunteer recruits. *Infantry Tactics* proved more than adequate to assist officers in turning rabble quickly into soldiers. Although not a simple system to master, the manual was effective. Officers were able to hone large-formation maneuver skills, which had long eroded since 1815. The 5th Infantry Regiment, which had not drilled together as a whole for nine years, was typical. Unable to properly execute large unit tactical drill, the officers and men of the 5th spent hours learning and relearning individual and unit movements on the parade ground.

One soldier remarked, "Life became nothing but drill and parades, and your ears filled all day with drumming and fifing."[88]

Given the chaos of expanding an army during crisis, the service was fortunate to have a single standard, *Infantry Tactics*. Yet, most senior regular army officers had long ago forgotten its contents. To avoid embarrassment, countless commanders and staff officers turned to self-study. Recent West Point graduates best understood current doctrine, and many juniors educated their superiors in private when not drilling the enlisted troops. Regular army officers also trained their volunteer counterparts, many of whom were more familiar with Brevet Captain Samuel Cooper's 1836 manual published at the direction of Major General Alexander Macomb, then commanding general of the U.S. Army. Written for use by militia and volunteer units, Cooper's manual consisted of four parts—infantry, cavalry, artillery, and regulations—the infantry portion being a near-copy of *Infantry Tactics*.[89]

Although less of a problem than during the War of 1812, doctrinal compliance was not absolute, as certain commanders trained their regiments and battalions to whatever manual they happened across. Despite these variances, *Infantry Tactics* dominated during the war. Soldiers used it to learn the manual of arms for loading and firing weapons and basic skills such as how to stand at attention and march to three rates of speed. "Common time" meant marching at 90 steps per minute with each step being 28 inches in stride. The faster "quick time" caused the troops to march at 110 steps per minute, while "double time" was a run at about 165 steps per minute. Troops were taught to move in columns and shift into a two-rank line, the easiest formation for raw troops to grasp quickly, and to give fire three times per minute. A three-rank line was also authorized. For an army composed mostly of Regulars and volunteers, *Infantry Tactics* proved its worth in forging a hodge-podge force into a cohesive army.[90]

In battle, the American Army attacked by column and rank into a hail of deadly Mexican musket fire despite often-heavy losses. Defensively, they gave fire by ranks and formed into a hollow square formation to counter cavalry charges. A typical encounter took place at Palo Alto on 8 May 1846. There, a Mexican force of about 6,000 men with twelve cannon under Major General Mariano Arista stretched out over a mile in line formation. Taylor commanded a 2,228-man American Army and advanced in columns to within a half-mile of the enemy. As the Mexican artillery opened fire, the Americans shifted into line formation behind their own artillery. With an artillery duel soon underway, a small Mexican force attempted to outflank the Americans. The U.S. 5th Infantry Regiment reacted by forming into a defensive square formation and held off several attacks by Mexican infantry and cavalry. The fighting was desperate until dusk brought the affair to a draw. When the Mexican Army withdrew a few miles south, Taylor pursued to Resaca de la Palma. Heavy underbrush made a mockery of tactical formations, often forcing units to break informally into small groups of a dozen men or so who then advanced in open order. Still, the Americans pushed through the undergrowth with sufficient strength to cause the Mexican line to crumble and withdraw. The Army continued to demonstrate its tactical prowess throughout the war. In the mountain passes and hills of Buena Vista in February 1847, 4,600 Americans under Taylor, mostly volunteers, held off the determined onslaughts of General Antonio Lopez de Santa Anna and a force of 20,000.[91]

Winfield Scott's assault upon Veracruz on 9 March 1847 and subsequent advance to Mexico City marked a doctrinal high point. Transported by elements of the U.S. Navy to the key Mexican port of Veracruz, the American Army conducted an amphibious landing 3 miles south of the city. *Infantry Tactics* did not discuss this type of operation but Scott, the Army's master tactician, had given the matter considerable thought. After landing 8,600 men in specially fabricated boats, Scott extended his line northwestward to envelop the city's 3,600 Mexican defenders and 15,000 civilians. Cutoff and under siege by Scott's artillery and naval guns, Vera Cruz surrendered on 27 March. Scott's army had secured a port for a logistical base and soon advanced westward to face Santa Anna in numerous battles. From Cerro Gordo until Mexico City, the Army periodically marched in columns and ranks, faced down the enemy in line, and circumvented enemy defenders through flanking maneuvers, before using scaling ladders to assault the Mexican military academy at Chapultepec just outside of the capital. When Mexico City fell on 14 September 1847, the U.S. Army was arguably the most doctrinally proficient fighting force in the world. Officers and soldiers alike, writing after the war, stated with confidence that their training had prepared them well for the hazards of combat. In 1848, the Treaty of Guadalupe Hidalgo ended the war with Mexico.[92]

Infantry Tactics had contributed significantly to American Army success in the Mexican War. Doctrine helped to regulate the chaos of battle by providing a common intellectual foundation for army officers to train their soldiers conventionally and rapidly to face a competent Mexican Army. In waging an offensive war of conquest, the Army defeated its opponents by moving rapidly in column, shifting into line as needed, and returning to column with ease. But many operations were improvised. Keystone doctrine offered no guidance for amphibious landings, sieges, or assaulting fortifications. To plan and execute a successful campaign, an individual's experience, innovation, and skill were required to augment written principles. With the war's end, the Army downsized and scattered once again into small units to occupy the nation's forts, coastal defenses, arsenals, and quartermaster depots. Missions included exploration of the newly acquired territory, mapping large areas of land, and building infrastructure while protecting the frontier population from brigands and Indians. When practical, army troops consulted doctrine to prepare for the next war.[93]

Through military action between 1779 and 1848, the U.S. Army had used doctrine and informal practice to assist the government in acquiring over 500,000 square miles of land. This area included what later became California, Nevada, and Utah, most of Arizona and New Mexico, and parts of Colorado and Wyoming. Now a transcontinental republic, the nation had fulfilled what many called "Manifest Destiny." *Infantry Tactics* had made a significant contribution to the endeavor.

By 1861, what many veterans recalled of wartime doctrine had been what they had experienced at subordinate levels of command in Mexico. Scores of Mexican War leaders, primarily lieutenants, captains, and majors, later served as generals at the brigade, division, and corps levels of command and staff in the American Civil War. Their memories of combat consisted of a farrago of *Infantry Tactics*, Napoleonic warfare methods, and fading personal experiences. However, the doctrine of 1861 was not what they had executed in a foreign war fought but fifteen years previously. In 1855, a new doctrine had emerged, one that reflected changes in military technology sufficient enough to end the forty-year reign of *Infantry Tactics*.

FROM FRENCH DRILL TO TEUTONIC INITIATIVE: *INFANTRY TACTICS* AND *REGULATIONS*, 1855–1905

Every war has its surprises.—Arthur L. Wagner quoted in T. R. Brereton, *Educating the Army*

By 1848, when the War with Mexico terminated in an American victory, *Infantry Tactics* had been the Army's tactical doctrine for thirty-three years. Based upon French ideas, the manual had proven its worth assisting the Army attain national policy objectives in service to an embryonic continental power. The bedrock of warfighting practice during an era of national expansion, the manual's precepts allowed army officers to regulate the actions of soldiers in peace and war. For their part, soldiers learned to precisely execute specific tactical movements when directed.

With a life span of just over three decades, the Army's tactical drill system was well-known but not enforced among all regular, militia, and volunteer units. Commanders could use the manual or not, for the War Department was incapable of enforcing any doctrinal standards upon dispersed federal forces and state governments. Army leaders also found that *Infantry Tactics*, a conventional warfare doctrine by design, was awkward when confronting Indians. Without an overarching tactical doctrine to regulate every plausible mission, army leaders habitually combined convention with informal practice, two very different approaches to war that existed simultaneously.[1]

Doctrine was not the concern of newly appointed Secretary of War Jefferson Davis in March 1853, at least initially. As a West Point graduate, infantryman, and Mexican War veteran, he understood that more urgent matters outweighed rewriting army doctrine to cover every conceivable mission. In addressing service organization, strength deficiencies, and foreign threats, his policies included convincing Congress to fund two additional infantry regiments and two more cavalry regiments. To rectify the difficulty of controlling a vast geographic area with overextended forces, Davis sought larger but fewer frontier garrisons. Politics prevented Davis from achieving his goal. Still, he created five military districts where commanders redistributed their forces, as they deemed fit. In regards to threats of foreign invasion, Davis convinced Congress to pay for additional coastal fortifications.

Secretary Davis soon found himself facing additional national-security problems that emerged from European rifle technology. For decades, musket-dominated armies had periodically used rifled weapons. Extensive use of rifles was unheard of, however, for they were expensive and prone to fouling. By 1849, industrialization had helped to reduce manufacturing costs. French Captain Claude Etienne Minié produced a viable rifle-and-bullet combination that others had experimented with in the 1830s. Eventually, the French MLE 1853/54 rifle allowed a small conical lead bullet to be rammed down the muzzle. When fired, the projectile expanded to catch helical grooves inside a rifled barrel. Upon exit, the bullet spun, increasing the projectile's accuracy, lethality, range, and stability. Originally, Minié designed a two-piece projectile, but later refinements by others led to the development of a one-piece, hollow-core expandable lead bullet. The term "Minié ball" or "minnie ball," however, stuck. Used in abundance, this weapon/bullet combination made distinctions between heavy and light infantry meaningless. It also signaled the decline of mounted forces such as cavalry and dragoons.[2]

European rifle technology convinced Davis that warfare was about to change drastically. In anticipation of a dramatic shift in tactics, Davis formed a board of officers in November 1853 to write new doctrinal procedures to replace the musket-oriented *Infantry Tactics*. Davis chose Brevet Lieutenant Colonel William J. Hardee, West Point Class of 1838, to lead the effort. Identified early on as a *wunderkind,* Hardee had attended the French Cavalry School at Saumur, France, in the early 1840s. There he mastered both French and army drill. Hardee's subsequent courageousness in Mexico, coupled with a knack for the nuances of tactical maneuver, earned him a faculty position at West Point. In 1853, he became an academy tactics instructor. As a master tactician possessing knowledge of French military procedures and language, Hardee was more than qualified for his assigned task. Yet, he was a cavalry officer; quite possibly chosen not only due to his experience and intellect but also to devise procedures to lessen the rifle's effect upon American mounted troops. Regardless,

William Hardee (U.S. Army Military History Institute, Carlisle Barracks, Pennsylvania)

the board followed established precedent and drew upon French tactical doctrine, in this case the 1845 version.

In truth, the French had anticipated the development of the eventual "system of Minié" as early as the 1830s. By 1838, French Army tactics included a rifle countermeasure called the *chasseurs-à-pied*, battalions of meticulously drilled, athletic soldiers who fenced with the bayonet and attacked at a jog or "gymnastic pace." On offense, the *chasseurs* were to advance faster than the average soldier by running through the defender's firestorm of lead projectiles. Upon entering the opposing line, the survivors finished off the enemy with the bayonet. When the envisioned rifle-and-bullet combination finally emerged in 1849, French training experiments with their doctrine concluded that assaulting a defensive position at any pace risked obliteration by massed rifle-musket fire. Although France fielded twenty *chasseur* battalions in 1853, French Army leaders discounted the affiliated tactics as folly. Instead, soldiers were directed to attack in close order behind skirmishers, a return to the tactics of Napoleon Bonaparte.[3]

Although the French Army rejected the *chasseurs-à-pied* approach as an impractical means to regulate tactics, the American 1853 board of officers copied the 1845 French doctrine anyway. Although not an exact duplicate, board methods included detailed procedures for deploying from column to line formation, changing cadence for advancing infantry, and the use of skirmishers in front of the main force. The board recognized that the rifle offered a significant advantage to the defense. Any assault had to close the distance between opposing forces as rapidly and efficiently as possible while remaining flexible enough for changing battlefield conditions. Still, the draft manual mimicked the French version to such an extent that any French officer would have felt at home on an American drill field.

As the president of the board, Hardee retained the common step of *Infantry Tactics*, 28 inches at 90 steps per minute and quick time at 110 steps per minute, while adding the double quick step of 33 inches executed at 168 steps per minute and, in times of emergency, 180 steps per minute. In the attack, infantrymen were to use the various steps, as the situation dictated, to maneuver toward the opposing line and run through enemy fire. Hardee calculated that the average unit could cover 4,000 yards under fire in about twenty-five minutes.

Both the double-quick and emergency steps required extremely fit soldiers capable of running in wool uniforms and heavy combat equipment regardless of season, terrain, or weather conditions. While some board members considered this to be wishful thinking at best, Hardee believed it possible. Soldiers who eventually executed the double-quick step nicknamed the pace the "shanghai drill," an analogy to certain situations where "organized chaos" occurred. Board members also included instructions for how to deploy a force from an attacking column into line at the double quick without halting, an alteration to *Infantry Tactics* that required a unit to stop before changing formation. To instill drill and maneuver methods at the increased paces the board envisioned, the manual retained progressive schools of the soldier, the company, and the battalion. Some of Winfield Scott's ideas as reflected in the previous doctrine lived on. Thus, French military thought further imbedded itself into American keystone doctrine.[4]

Despite having the French tactical manual in hand for translation and transcription, the board found writing army doctrine to be laborious and time consuming. Instead of their regular duties, board members spent most of 1854 and into early 1855 writing the manual.

In January 1855, while the new doctrine was under development, Davis formed another board of officers to evaluate the new Springfield rifle-musket. Debates arose over the efficacy of muskets vice rifle-muskets, which ultimately shaped the draft tactical manual. Comments from numerous sources went back and forth to a War Department filled with officers who were also eager to offer advice. Suggestions (often conflicting) numbered in the hundreds. Comments included proposals on the fixing and unfixing of bayonets, the proper stacking of arms, and other details both mundane and weighty. Changing but one word or a single sentence within one section prompted numerous amendments elsewhere. The Army's traditional use of a civilian press, this time Lippincott of Philadelphia, to print the manual led to further frustrations. Frequent government changes failed to mesh with the printer's proofs. Civilian proofreaders did not understand army jargon and periodically made erroneous contextual changes. On 22 March 1855, a frustrated Hardee was in Philadelphia speaking directly with the printers at Lippincott. He wrote the adjutant general to warn that the stereotyped manuscript contained errors "more numerous and important than I anticipated" thanks to the printer's ignorance of convoluted military terminology and conflicting changes in the text.[5]

On 29 March 1855, after weeks of frustration and revisions, Davis approved the new doctrine as *Rifle and Light Infantry Tactics for the Exercise and Maneuvers of troops when acting as light infantry or riflemen* or, as some called it, "Hardee's Tactics." As a new manual written specifically to both exploit and lessen the effects of rifle-musket and conical projectile, the manual was produced with technology in mind. With Hardee appointed as commandant of cadets at West Point from 1856–1860, future army officers became more than familiar with the manual's intricacies. In late 1856, the Springfield Armory began production of the Model 1855 rifle-musket: 9 pounds in weight, 56 inches in length, with a sword-style bayonet, and a caliber .58 barrel to incorporate the Minié ball. These weapons eventually entered service with the Mounted Rifles and the 9th and 10th Infantry Regiments.[6]

The Army and the 1855 Doctrine

For the army bureaucracy, changing the primary infantry weapon from the traditional musket to rifle-musket was no small matter. About one month after the release of *Rifle and Light Infantry Tactics,* Davis ordered the Delafield Commission to Europe to observe and report upon changes in the military systems of the Continent's principal nations, with particular attention given to organization, transportation, medical arrangements, clothing and equipment, arms and their adaptation for use, and fortifications, among other areas of concern. The Crimean War, which had broken out in 1854, was of keen interest. As a secretary who had trumpeted technological change, Davis might have created the commission to bolster his case. However, dispatching army officers to observe foreign military matters in peace and war was nothing new. More than 150 such visits had occurred since Winfield Scott's departure to Europe in July 1815. Davis personally handpicked the commission members. Corps of Engineers Major Richard Delafield led the group. He was accompanied by Ordnance Corps Major Alfred Mordecai and Cavalry Captain George B. McClellan. Curiously, no infantry officer was selected, most likely due to Davis's desire to select officers with scientific backgrounds who could assess technology and a cavalry officer capable of measuring its impact upon mounted forces.[7]

Arriving in the Crimea too late to observe the mass effects of rifle technology at the Battle of Alma and the beginnings of the siege of Sevastopol, the commission was nevertheless able to fulfill its mission and published a detailed commentary soon after members returned in 1856. While each officer wrote a separate report, Delafield and McClellan warned that Britain and France potentially posed a threat to American security and advised to continue focusing upon coastal defense. Ordnance expert

Mordecai commented that most European armies still depended upon smoothbore muskets, but that he expected infantry units to adopt rifles in the near future. He nonetheless viewed rifles as more an auxiliary arm to muskets and, as he observed at Sevastopol, were used with great effect out to 600 yards in what the British termed "rifle pits." While impressed, he cautioned against adopting the rifle wholesale until the Army gained more knowledge and experience with the technology.[8]

Of note was the commission's overall finding that the Russian military system was superior to the French. McClellan went so far as to learn Russian, translate Russian cavalry manuals into English, and offer them as a model for U.S. Army cavalry doctrine. However, the members' enthusiasm for Russian methods ignored the fact that the U.S. Army had just fielded a new tactical doctrine based upon French precepts. In the end, American obsession with France proved too strong to overcome. It would require a catastrophic French military defeat to break decades of American Army dependency upon French military thought.[9]

Where the Delafield Commission was lukewarm in its acceptance of the rifle as revolutionary technology, the 1855 army doctrine was far from revolutionary. Despite being written in reaction to more lethal and accurate rifle-musket technology rather than a smoothbore weapon, column-and-line formations remained virtually identical as previous doctrine. Davis and many army officers understood that the new weapon was lethal at increased ranges, but regulating the behavior of troops in battle remained of primary concern. In war, the noise and smoke of combat necessitated that a commander be able to see his left and right flanks in order to direct his men. Bugle calls, flags, hand signals, messengers, and shouts conveyed orders, which meant that commanders must remain close to their soldiers. These methods changed little from earlier wars, and the 1855 manual continued such practice. Without additional new technology to direct the soldiers in battle (such as tactical radio), commanders had to rely upon the same methods to control troops that had been used since the musket came to dominate warfare. For now, the pace of advance in weaponry had changed but little else.[10]

From 1855 until 1861, the new doctrine and its complementary weaponry saw limited action, primarily in the West. On 1 September 1858, at the Battle of Four Lakes on the Spokane Plain, approximately 600 soldiers under the command of Colonel George Wright deployed into line against 400–500 Spokane, Coeur d'Alene, Palouse, and Pend d'Oreille Indian warriors accused of treaty violations and harboring murderers.

With skirmishers and dismounted dragoons pressing forward, American infantry armed with the rifle-musket lined up in ranks and gave fire. The Minié balls dispatched Indians at a range of 600 yards, well beyond the range of the opponents' arrows, lances, and trade muskets. A subsequent charge by army forces resulted in an estimated sixty Indian warriors killed and dozens more wounded. Without suffering a single friendly casualty, the Army's short-lived and small-scale battle demonstrated the lethality and range of rifled weaponry on an unconventional force, reinforcing the effectiveness of conventionally based doctrine when employed against such forces.[11]

Despite actions such as the Spokane Plain, the Army spent far more of its time during much of the era of the 1855 doctrine as a frontier constabulary force thinly spread among numerous undermanned posts conducting peace operations. Tasks involved border patrol and the intricacies of diplomacy, escort duty, exploration, and law enforcement. These missions continued to be left up to the experience and judgment of men, not doctrine. In 1856, federal troops under Colonel Edwin V. Sumner at Fort Leavenworth, Kansas, and Lieutenant Colonel Philip St. George Cooke at Fort Riley, Kansas, received orders from the Franklin Pierce Administration to deal with recalcitrant settlers in "Bleeding Kansas." The latter being a law-enforcement mission, the Army operated under informal practice in that the 1855 manual did not apply.[12]

The Civil War and a Revised Doctrine

By the eve of Civil War in 1860, American regular troops were prepared for conventional warfare. However, few officers had experience maneuvering any military formations larger than a few hundred soldiers. A handful of senior officers, mostly West Point graduates, and some enlisted men had gained tactical experience fighting in Mexico during the 1840s. Some had fought Indians on the frontier but had not led formations numbering in the thousands of troops. Brigadier General Irvin McDowell, a West Point graduate and the federal commander at Bull Run (also known as First Manassas), remarked that there was not one man in the Army who had ever maneuvered 30,000 troops, although he had seen a demonstration of this scale *once* during a troop review in Europe.[13]

With the outbreak of war in April 1861, the federal and Confederate armies swelled with thousands of unskilled volunteers in need of training. Officers demanded doctrinal manuals at a rate that overstretched a Government Printing Office (GPO) neither organized nor staffed to meet requests. Private publishing firms including Lippincott, Harper's, and Van Nostrand filled the void by reprinting a variety of military literature. Manuals included French-based diagrams showing the manual of arms, drill procedures for parades and ceremonies, and maneuvers from company through corps levels. Also included were obsolete versions of *Infantry Tactics,* Henry W. Halleck's French-derived strategic and operational treatises, and the current 1855 doctrine. West Point texts written by Professor Dennis Hart Mahan proved popular. Swiss officer Antoine Henri de Jomini's *Summary of the Art of War,* an 1838 best seller based upon Napoleon's methods, was also in demand.[14]

In 1861, the opposing armies shared the same doctrine. The 1855 manual was in demand by both federal and Confederate officers because so many West Point-educated graduates were familiar with it. About twelve versions of the manual appeared. Yet, commanders struggled to translate the printed words into tactical actions. Many officers were more familiar with muskets, but Civil War armies were industrial-age entities equipped with rifles, rifle-muskets, repeaters, breechloaders, and improved munitions. Samuel Colt had played a significant role in providing the technology. As early as February 1858, a board of officers led by Brigadier General W. S. Harney formed to "examine Colt's arms with breech attachment and pistol-carbine." In a May 1860 letter to Secretary of War John B. Floyd, Colt requested "the opportunity to make a practical exhibition in your presence so improvements might be made" to pistols and carbines.[15]

Colt and other manufacturers mass-produced rapid-fire weapons of increased lethality that drastically changed tactical warfare and affected army doctrine. Where a trained soldier could discharge and reload the 1855 or 1861 Springfield rifle-musket three times in one minute, breach loaders and repeaters enabled them to fire seven shots or more in the same amount of time, although often with diminished range due to the size of the cartridge. Still, in the attack, commanders and soldiers advanced across the "deadly ground," the area between opposing lines, like sheep being led to slaughter. At Fredericksburg, Maryland, federal Major General Ambrose Burnside lost 12,000 troops in a single day.[16]

The 1855 doctrine had called for soldiers to move faster as an effective measure for countering rifle lethality. But, in the chaos of battle, officers found it difficult to regulate the actions of large formations of troops. Line officers depended upon visually acquiring both flanks of their unit in order to issue commands through voice and bugle calls. Given the noise, smoke, and horror of battle, commands designed for formation control became impossible to hear (let alone execute) when soldiers spread out. Commanders therefore concentrated their units within a short space so they would "brush the elbow" of the person standing next to them. Although employed similarly to the infantry formations of Napoleon Bonaparte's time and the *Infantry Tactics* used in the War with Mexico, attacking soldiers became fodder for defensive forces armed with thousands of breechloaders, muskets, repeaters, rifle-muskets, and modern artillery.[17]

To be sure, France's tactical manuals had regulated American Army doctrine so thoroughly that General F. V. A. de Chanal, a French Army observer watching federal forces maneuver, reported, "Our methods have been copied exactly." De Chanal had recognized that his military's *chasseur* tactics so dominated Civil War battlefields that both sides were *de facto* French in battle. Many commanders were so schooled in French doctrine that they found themselves asking "How might Napoleon have acted?" under the situations they encountered. After the Battle of Shiloh in 1862, Confederate Brigadier General Thomas Jordan recalled consulting Napoleon's 1815 order for battle at Waterloo.[18]

In 1861, the War Department issued a change to the 1855 doctrine, partially because William J. Hardee, the author of the federal Army's manual, had resigned his commission to become a Confederate general officer. But practical matters that predated hostilities were also at play. Although the Army's doctrine was based upon the 1855 rifle-musket equipped with a sword bayonet, eighty-one different shoulder weapons were in use during the war. Most common among the regular Army and the state militia units were 42-inch muskets or 40-inch rifle-muskets with socket bayonets. Regiments armed with differing weapons and bayonet configurations found the 1855 doctrine unsuitable for the manual of arms. Some officers instead used the 1861 *Baxter's Instructions For Volunteers,* a manual that blended *Infantry Tactics* with earlier obsolete manuals based upon muskets. The Army's interim fix, issued in 1861, partially resolved the discomfiture over Hardee's defection, but tactics changed very little.[19]

Casualties, however, demanded additional change. Prior to Bull Run (or First Manassas) in July 1861, both the federal and Confederate armies consisted largely of volunteers averaging sixty days of service. Extensive drill using various manuals prepared the troops to use columns and lines for the offense while adopting the square formation on the defense, particularly to repel the atypical mounted cavalry attack. However, in following tactical principles at Bull Run, federal commander Irvin McDowell sustained 2,706 casualties, or 9 percent of his force. Pierre G. T. Beauregard, McDowell's Confederate counterpart, suffered 1,981 casualties, or 6.5 percent losses. Such casualty numbers provided sufficient impetus to produce a new tactical doctrine.[20]

Changing doctrine during war was not without precedent; army leadership had done so several times during the War of 1812, although with mixed results. The mission fell to Silas Casey, a seasoned senior lieutenant colonel with Seminole and Mexican War experience who had worked on the 1861 War Department doctrinal revisions. He was also a discontented second-in-command officer of an infantry regiment at Fort Steilacoom near Tacoma, Washington. In August 1860, Casey wrote the War Department to complain that despite thirty-five years' service he was the only individual of his rank not to have commanded a regiment. He requested assignment to the Army's recruiting service in Washington, D.C., relaying that his wife suffered from paralysis and that they had two children "who I wish to place at school." Another child, a son, was a navy midshipman aboard the *Niagara*, whom "we have not seen for more than three years." As for professional reasons, Casey also desired to present his tactical manual recommendations in person to a board of officers at the War Department. His efforts led to a transfer and eventual appointment to brigadier general within an expanding Army on 10 September 1861.[21]

Casey relied upon both the 1855 manual and the 1861 doctrinal update in producing a new keystone doctrine. Given the pressure to reduce casualties and the reality that the war was being waged in an overwhelmingly conventional fashion between symmetrically organized and equipped forces, the manual ignored unconventional warfare. Instead, Casey produced a near copy of the 1855 manual, although with the acknowledgment that "a revolution in artillery and infantry weapons necessitated a departure from those processional movements and formations in order of battle which characterized the school of Frederick the Great and now considered as obsolete."[22]

Casey addressed two significant issues. The first was to provide tactical directives for brigades and regiments, not just battalions. Second, officers, volunteers in particular, desired that existing tactical methods be made easier to understand and execute. The 1862 manual rectified these problems through guidelines for brigade and regimental drill and more precise directions for how to deploy a force, where to stand when directing troops, how to shift to a column or line effectively, and other details. Although the 1855 manual allowed for a three-rank line in battle, Casey

Silas Casey (Courtesy of the National Archives)

advocated two ranks to extend a unit's formation and thus theoretically expose fewer troops to enemy fire at a given point. As with the previous manual, he also encouraged soldiers in the offense to run through the hail of defensive fire. In sum, Casey expanded and clarified the 1855 regulations while incorporating the 1861 changes, but made relatively few additions. The president and the War Department approved *Infantry*

Tactics (also known as Casey's *Infantry Tactics*) in two volumes on 11 August 1862.

The 1862 *Infantry Tactics* in War

The Army's 1862 doctrine saw widespread use, even among Confederate units. While many units adopted it, others did not, leading to non-standardized tactics on the battlefield. To its credit, the War Department directed many army regiments to undergo precombat training under a provisional brigade that Casey commanded near the national capital. While 1862 *Infantry Tactics* represented a significant, if imperfect, step toward enforced standardization of tactical doctrine, it failed to stem continuing high casualty rates. The doctrine was more easily understood than its predecessor, but it could not overcome the inherent strength of the defense.

Faced with extensive casualties in the offense, some officers circumvented doctrine by modification. By the end of the war, frequent use of skirmishers and open order had increased. Assaulting troops often fired from the prone position or advanced in short rushes as enemy fire waxed and waned. This technique, used at Fort Donelson, Tennessee, in February 1862, predated Casey's changes. In June 1864, Ulysses S. Grant ordered an unsuccessful frontal assault at Cold Harbor that followed precepts within Casey's manual but produced horrific consequences. In September 1864, at Fisher's Hill in the Shenandoah Valley, a federal brigade attacked in ranks up a hill into a murderous Confederate defensive fire and suffered alarming casualties. Yet, surviving members laid down in their ranks, opened fire, then rose up and rushed forward into the attack.[23]

One officer who gained approbation for inventive tactical thinking was volunteer Colonel Emory Upton. Upton studied the problems of tactical linear warfare while in command of federal troops. At Spotsylvania, Virginia, in May 1864, he directed his unit to revert to three ranks instead of the two ranks approved in 1862. Arranging his regiments in columns of consecutive lines with spacing to allow freedom of maneuver and supporting fire, Upton's brigade quickly seized Confederate breastworks, as ordered. Success came too fast, however. As his attack lacked support from other federal brigades, Upton was forced to withdraw.[24]

Innovative officers are not abnormal in war. But to effect change, certain conditions must exist. Circumstances were in Upton's favor, for he was a proven brigade commander who had used standard doctrine for years. He was no maverick, for his success combined cleverness, hard work, luck, and skill, which earned the confidence of two general officers, Ulysses S. Grant and George Meade. They in turn provided him with opportunity. With general officer backing, Upton advanced his ideas by being an inventive traditionalist. His tactical actions appeared to modify approved principles, not reject official methods.

Innovation also took place within the lower ranks. The 1862 *Infantry Tactics* had not provided specific instructions for the use of cover and concealment in the offense or defense. All the same, soldiers gradually made use of available brush, boulders, fence rails, and trees in seeking protection. Skirmishers and main assault forces tramped through brush and woods to get close to enemy defenses. Although many officers and enlisted men viewed entrenching as cowardly, self-preservation nonetheless prompted soldiers to pile up soil in front of them using bayonets, belt buckles, hands, shovels, or sticks.[25]

While conventional doctrine and a variety of unofficial military manuals regulated tactical warfare on the battlefield, other military operations were conducted informally outside the realm of doctrine. Army tactical doctrine focused more upon individual and unit drill in order to impose a common system of discipline and maneuver upon the force. Its tactics, mimicking French sources, were not mission-specific in design nor tailored to the American continent; the 1855 and 1862 doctrinal manuals lacked guidelines for fortress defense or reduction by assault or fire, trench warfare, or raids, and coping with the nuances of weather and terrain.[26]

By war's end in 1865, army doctrine had produced mixed results. When the manuals were used, both sides found that the tactics suited the defense more than the offense. Civil War units in the attack conducted mostly frontal assaults against entrenched foes armed with not only rifle-muskets capable of three shots per minute in a skilled hand, but also breech-loaders, repeaters, and artillery generating tremendous firepower. No American war had seen such high casualties or the types of horrific wounds produced at battles such as Fredericksburg, Chancellorsville, or Cold Harbor. French-based army doctrine, carefully cultivated and mastered over five decades, was a crucial factor in federal forces losing over 130,000 troops killed in combat with 280,000 wounded. Confederate losses stood at 160,000 dead, with at least 250,000 injured. Army policymakers were surprised by the carnage of modern war. Facing a future where technology enabled the defense to reign supreme, something had to be done.

Doctrine for a Rapid-Fire Battlefield

Army officers analyzed French-based 1855 and 1862 army manuals that had successfully produced a tactical system enabling thousands of seasoned and raw troops to comprehend the intricacies of brigade, regiment, company, and individual drill movements. Soldiers learned to march in column or line and to execute tactical offensive and defensive maneuvers. But both manuals proved to be flawed. Neither considered that the means to regulate a soldier's actions in battle had changed little since Napoleon's era fifty years earlier. Using line and the column formations that Bonaparte would have recognized, soldiers attacked into a maelstrom of lead produced by thousands of defenders armed with smoothbore muskets and rapid-fire rifled weapons. Massed defensive firepower shredded attacking troop formations long before they reached the opposing line.

Complicating matters was a growing national industrial capacity to manufacture rapid-fire weapons in quantity. Given the advantages of reduced reloading time with improved accuracy and lethality, breechloaders and other such arms were certain to become standard issue. Attack-oriented army officers were horrified not only by the number of battlefield casualties, but also by the increasing realization that the defense now dominated war. If victory lay in the offense, there was little doubt that the Army required a new tactical doctrine to overcome the defensive power of modern weapons technology.[27]

In 1865, the horrors of the American Civil War battlefield were nearly repeated when the U.S. government ordered Major General Philip Sheridan and 50,000 troops to the Mexican border. Earlier, in January 1862, Britain, France, and Spain had sent naval forces to the country after Mexican President Benito Juarez suspended interest payments on foreign loans in July 1861. While the British and Spanish withdrew in April 1862, Napoleon III of France deployed 38,000 troops armed with the 1857 rifled percussion musket and similar rapid-fire weapons capable of three shots per minute. Although these actions were a violation of the 1823 Monroe Doctrine, President Abraham Lincoln was in no position to resist a European imperial power reestablishing itself in the Western Hemisphere.

In May 1864, the Austrian Prince Maximilian arrived as the proclaimed emperor of Mexico. France provided troops in support, although they had a tenuous hold upon the country. In 1865, when Sheridan arrived on the United States-Mexico border, he faced a potential offensive operation in support of Juarez against a well-drilled French Army, equipped with modern weapons, numbering over 31,000 troops. Given how French-based American Army doctrine had recently fared against the Confederacy, the French force in Mexico was a potent foe. Thankfully, the crisis ended with the French Army withdrawing, leaving Maximilian to be executed by Mexican troops on 19 June 1867. One can but imagine what might have occurred had the American and French armies faced each other in battle.[28]

As Mexico's internal woes nearly drew America into a foreign war, army policymakers, stunned by the horror of Civil War battlefields, mulled over how to reconcile doctrine with technology. In 1865, both Emory Upton and William H. Morris began seeking solutions to the challenges of contemporary tactical warfare and technology. Although Morris claimed that Upton's methods closely duplicated his own work, each had concluded from his wartime experience that the breechloader was the weapon of the future. There was little doubt that army tactics had to compensate for technology. Doing so meant more than simply attacking at a faster rate of advance, as previous manuals had advocated. To organize, move, and employ men effectively on a rapid-fire battlefield meant revolutionizing the fundamentals of tactical warfare.[29]

By November 1865, army officers were questioning their service's obsession with French military precepts. Individuals writing to the service's unofficial *Army and Navy Journal* expressed their desire for an "American School" of tactics, one that rejected all European influence outright. Upton, however, still relied upon French ideas as the basis for a new doctrinal manual. Decades of French influence and the ingrained routine of borrowing and modifying foreign ideas were difficult to discard. Still, where both the 1855 and 1862 army doctrine called for two or three ranks when giving fire, Upton required but a single fighting line. His reasoning was that as modern weapons were capable of unremitting fire, a second or third rank was no longer essential to create and sustain firepower.

For Upton, modern war necessitated dispersal of forces to present fewer targets in a given location. This meant less commander-directed control through verbal orders, bugle calls, and unit drill and more emphasis upon individual training, trust of subordinates, and initiative. Upton believed that officers must instruct their soldiers to act more independently and to fire at individual targets when presented, not just discharging their weapons in massed volleys in a general direction.

Emory Upton (Courtesy of the National Archives)

Independence of action meant that Upton's regiments contained battalions organized with four companies divided into smaller groups of
four men (later called "squads"), each group having its own numerical
designation. Where a battalion commander could give orders to each of
his companies individually, a company commander could direct each
group of four men independently, depending on the tactical situation.
In the offense, the battalion attacked in a two-part formation. The first
part was the "fighting line" containing two companies, the first company
deployed as skirmishers with the second one behind it in support. The

"reserve line" constituted the second part. It contained the additional two companies, each following behind the fighting line in column formation. As the battalion advanced, the fighting-line skirmishers moved forward as individuals, using the terrain for protection to close with the enemy. As the skirmishers advanced, commanders continually fed troops from the second fighting line company into the skirmish line to increase its size and firepower potential. When the skirmish line of the combined two companies reached a point approximately 150 yards from the enemy defenders, they formed a continuous line of rapid-fire weapons that covered more area, made opposing flanking attacks difficult, and exposed fewer troops to enemy fire at any given point. Using the skirmish line for protection, the battalion commander ordered one of the reserve line companies in column to "wheel" left or right into a single rank and to charge the enemy. If that unit failed to gain the enemy line and secure it, the second reserve company then charged. The two charging companies, supported by the fighting line, alternated advancing and firing in order to panic the defenders and to ultimately seize the opposing line. Upton's system simplified tactical movement, increased flexibility, and reduced the commands needed to deploy the force. It allowed commanders to take advantage of available cover and terrain while employing the breech-loader with maximum effectiveness. By January 1866, Upton believed that his system had achieved the revolution in tactics necessary to restore the offense to prominence. His system, *A New System of Infantry Tactics, Double and Single Rank, Adapted to American Topography and Improved Fire-Arms,* was initially published at his own expense and readied for an army board of officers' review.[30]

Upton's *Infantry Tactics* was not entirely ground-breaking. Whether aware of it or not, he had incorporated evolving French thinking, as expressed by Colonel Charles J. Ardant du Picq. In his study of European army tactics, Ardant du Picq had reached conclusions similar to Upton's concerning the individual soldier and the moral (psychological) factor in war. He also had discussed the increased use of independent-minded skirmishers instead of perfunctory unit drill.

Furthermore, elements of the American Army's 1862 *Infantry Tactics* remained in Upton's work; most of the individual drill movements and other procedures came from that manual. Columns and ranks remained the basic formations necessary to move soldiers in a controlled fashion onto a battlefield. The new manual did not reject attacking in ranks for control but instead favored the independent action of skirmishers, open order, and individual initiative. The single line, open order, and skirmishers were not a panacea, however, for Upton acknowledged that a credible defense often required several ranks of men arranged in depth to defeat an attack.

As with previous army doctrine based solely upon conventional warfare, Upton made no mention of ongoing army missions such as reconstruction and frontier duties. Most army officers despised such assignments, anyway. General Nelson A. Miles, upon his transfer to the West after reconstruction duty, recalled, "It was a pleasure to be relieved of the anxieties and responsibilities of civil affairs, to hear nothing of the controversies incident to race prejudice, and to be once more engaged in strictly military duties." With additional endorsements from General Ulysses S. Grant and a board of officers, the War Department wasted little time approving Upton's manual and directing its full compliance in August 1867.[31]

The 1867 *Infantry Tactics*

Right from the beginning, the 1867 *Infantry Tactics* was a controversial manual. Some, such as the Army's commanding general Ulysses S. Grant, believed that the 1867 doctrine contained unique principles independent of foreign "machine" thought. Many officers found the manual reasonable, for the new formations seemed practical as a way to overcome the effects of modern weapon technology. Others, however, were less convinced. One officer observed that Upton had copied 1,428 of 2,147 paragraphs from the 1862 doctrine. Another complained about the lack of consistent language in describing individual drill movements. For those who were less pedestrian and grasped that the manual was an engine of tactical change, the issuing of orders to individual groups of four men in the heat of battle seemed unworkable. Regardless, the resultant hullaballoo demonstrated that army officers had not only received copies of the new doctrine, but had immersed themselves in its contents.[32]

While the new manual was written for conventional operations, it contained precepts that were suitable for frontier warfare. Since the 1790s, army leaders had developed an informal approach to unconventional wars against Indians that combined both conventional and unconventional methods. Army operations employed Regulars, auxiliaries, and Indian allies doctrinally to march or deploy in battle. However, the tenacity of Indians in resisting the Army meant that the informal practice of ambushes, burning, and raids against them, their villages, and their foodstuffs were often necessary to force capitulation. This often meant directing soldiers to operate in small groups as opposed to more conventional line or column formations. *Infantry Tactics*, with its emphasis on small units led by junior leaders, was up to the task.

Still, Indian wars typically necessitated adopting nondoctrinal approaches. Army officers made use of Indian scouts to track down hostile forces, a technique not mentioned in the 1867 doctrine. George Crook was arguably the most innovative frontier commander, for, by 1876, he had turned mobility and supply into an art form. Using mules to haul supplies, Crook focused upon the proper animal breeding stock, skilled handlers (mostly civilians), pack saddles, and packing design to increase speed of transport. By avoiding the flat country used by army supply wagons, Crook was able to track down Indians in areas that they once had considered safe.[33]

Although the Army was an infantry-dominated service, frontier missions led to discussions over the tactical suitability of cavalry vis-à-vis infantry. Cavalry advocates touted the greater mobility of horse-mounted soldiers, but reality proved otherwise. After several days on campaign, horses and riders tended to wear out to such an extent that infantry often outmarched them. Scouting or reconnaissance was often left to pacified Indians or to civilians hired for such purposes. Moreover, as cavalry officers reluctantly admitted, horses were merely a conveyance for transporting soldiers to the battle where they would dismount to give fire. Cavalrymen were not trained to fire effectively from the saddle, possessed lighter weapons than infantry, and tended to become distracted in combat when horses became unruly. Heavy-infantry columns, however, found moving undetected to be problematic. They were most effective when used in combination with cavalry through continuous pursuit and converging, plodding columns, such as in the Red River War (1874–1875).[34]

West Point curriculum provided instruction in conventional warfare methods using standard unit formations but also stressed small-unit operations. Small-unit affairs dominated frontier actions. A West Point education therefore provided future officers with the tactical knowhow, if properly applied, to strike at unconventional forces through ambushes, night marches, raids, and winter campaigns. For its part, the 1867 doctrine advocated the use of aimed fire, cover, and dispersion, principles compatible with the frontier. By the 1870s, to assist cadets in mastering the nuances of frontier missions, army veterans were assigned to West Point to share their experiences with the cadets.[35]

Through education, army officers had access to an intellectual basis necessary to apply or modify army doctrine to suit the informal practice that dominated frontier warfare. In August of 1877, during the Nez Perce campaign, Colonel John Gibbon was in hot pursuit of a hostile force near the Big Hole River in Montana. In what became known as the "Big Hole Battle," Gibbon ordered his soldiers and civilian auxiliaries to conduct a night march beginning at 1:00 A.M. on 9 August. He moved his force doctrinally in column along a creek that bordered the Indian encampment. At first light, the force scrambled out of the creek bed in skirmish line. A few Indians fired several shots before running off. Gibbon's force then charged the village, killed several men, women, and children, and burned the teepees. During a lull in the fighting, as the soldiers plundered the area, the Indians rallied to drive Gibbon's force atop a wooded knoll. Although the Indians killed several men with well-aimed fire and captured both a mountain howitzer and the army supply train, Gibbon's force held out long enough until a relief column appeared. In this brief but bloody battle typical of the Indian wars, Gibbon lost about ten officers and men killed, seventeen civilians killed, and forty men wounded. Indian losses included fourteen men killed, along with some forty to fifty women and children. As the Indians ran off with pilfered army weapons and ammunition, Gibbon continued pursuit the next day, reinforced by the relief column. Gibbon's actions represented a combination of doctrine and informal practice that typified hundreds of actions within the frontier from 1865 until 1890.[36]

Although the Army planned and executed numerous campaigns against Indians, the War Department ignored unconventional operations within the Army's approved doctrine. Without army official guidance in conducting engagements against such foes, officers took matters into their own hands. Several officers wrote publications that drew upon their frontier experience. The most noteworthy book was *The Prairie Traveler*. Published in 1859 with War Department approval, the author, Captain Randolph Marcy, crafted a campaign manual to be used when confronting Indians in the West. His conclusions from analyzing foreign endeavors drew upon Turk and French experiences in North Africa against nomads. He remarked upon the foibles of friendly-force overdispersion in countering guerrillas, superior mobility as essential, and night marches necessary for surprising opposing forces in their camps. Above all, he concluded that continuous offensive operations were crucial for success. Offensive operations countered the inherent advantages of indigenous foes, such as popular support, familiarity with terrain, and climate acclimation. Other semiofficial works such as *Mountain Scouting* (1881) and *The Indian Sign Language* (1885) joined *The Prairie Traveler* and various army professional journal articles in providing practical guidelines for frontier dilemmas.[37]

Such publications supplied army officers with needed assistance in confronting frontier realities. However, Lieutenant John Bigelow and others believed that the Army had already become a constabulary force with duties that moved the service away from war preparation. He wrote: "Other work than waging war may incidentally devolve upon an Army without derogating from its efficiency, but when other work is its only work, and the only one for which it is fitted, the so-called Army is nothing but a police force." His view defined ongoing service debates of the 1870s. Should the Army prepare for conventional warfare or focus upon conducting unconventional operations? Moreover, should doctrine address these operations? Ultimately, the army leadership accepted that regardless of what the service was actually doing, producing doctrine for the next war was more important.[38]

Debating the 1867 Doctrine

Soon after the publication of the 1867 *Infantry Tactics,* global military circumstances led to intense scrutiny of the document during the 1870s and 1880s. A series of European conflicts, the Austro-Prussian, Franco-Prussian, and Russo-Turkish wars, saw further advancements in weaponry, including magazine-fed rifles, machine guns, smokeless powder, and increasingly accurate breech-loading artillery. Moreover, the army leadership became more concerned with producing one doctrinal manual to standardize the drill of artillery, cavalry, and infantry. In light of these developments, army officers questioned whether or not the 1867 doctrine had provided for sufficient force dispersion. If not and more was required, how were junior officers to command and control their forces in the pandemonium of battle? Given the resultant casualties brought about by modern weapons, how were officers to motivate their men to cross the deadly ground and thus break the opposing line? Such questions were highly debated and, for the most part, left unanswered.[39]

American officers soon became acutely aware that the Europeans were engaged in their own tactical debates, from which two schools of thought were materializing. Following a humiliating defeat at the hands of Prussia in 1870, leaders of the venerated French Army determined that élan, the spirit of the offense, required more attention than fire superiority.

Ardant du Picq and Ferdinand Foch were proponents of the idea that no defense could withstand the onslaught of well-motivated, offensive-minded soldiers who were culturally superior to the enemy. From both social Darwinist and psychological perspectives, France's military leadership rejected dispersion in favor of concentrated attacking formations. Motivated troops armed with bayonet-tipped rifles, they believed, would impress the determination, ferocity, and superiority of the French soldier upon the enemy. Germany, however, drew a different conclusion. Dispersed or "extended order" became the norm, with battalions deploying small units to rush forward using cover, alternately firing while other groups advanced. Commanders and noncommissioned officers were given greater responsibility in managing small formations of men. This eventually led to the adoption of a squad system similar to that of the 1867 U.S. Army manual. While the Germans also appreciated the advantages of cohesion and aggressiveness attained by encouraging élan, they considered firepower more important.[40]

In 1869, the War Department appointed a board of officers at Fort Leavenworth, Kansas, to ensure that the infantry-dominant 1867 doctrine would be incorporated within artillery and cavalry tactics. Upton had his doubts as to the efficacy of assimilating three different service branches into one tactical system, but work commenced in January 1873. Joining him in this endeavor was artillery expert Henry A. Du Pont, Colonel John Eaton Tourtellotte (General William T. Sherman's aide), and Captain Alfred E. Bates, a cavalry authority from West Point. In late-1873, a revised version of *Infantry Tactics* emerged with some minor changes to drill uniformity, but the basic ideas remained, such as the "fours." *Cavalry Tactics* appeared in 1874 and *Artillery Tactics* in 1875.[41]

Although *Infantry Tactics* had been updated, many army officers still found the manual unsuitable for future war. In 1874, Upton delivered a paper to West Point's Thayer Club in which he analyzed recent Prussian tactics. He concluded that the Army's doctrine differed little from the Prussians with their extended order. He saw nothing new in what Prussia had accomplished in battle but believed that their solution to future tactical conundrums was far superior to that of the French.[42]

Certainly, Prussia, with its ability to defend unsecured frontiers and having won a decisive military victory over France, impressed Upton. But the collapse of France at the hands of the Germans also demonstrated that a small state with limited resources profited tremendously from a professionalized military, mass-produced war material, and an intricate national mobilization system. In the aftermath of the German Empire's founding in 1871, the country's military system became the model for expansion-minded Western powers.[43]

American military leaders quickly sensed Germany's ascendancy on the Continent. Coinciding with the shocking results of American Civil War battles, France's decline and Germany's rise to dominance sparked an internal American Army reform movement driven by a small group of military intellectuals and reformers. The most influential members included General William T. Sherman, commander of the Army from 1869 to 1883, and his protégé, Upton. Sherman championed the establishment of a military educational system for all ranks, one that developed specialized knowledge within branches such as artillery, cavalry, and infantry.

As a prelude to reform, Sherman directed Upton to visit European and Asian armies in late-1875 through most of 1876. He was particularly interested in gathering information regarding the British colonial experience in light of American campaigns against Indians. Upton, however, had little personal interest in fighting native people. Although he visited British forces and observed their military operations in Asia, his evaluation concluded that the troops were praiseworthy only to the extent that they copied European continental models. He ignored any lessons to be gathered on the subject of unconventional warfare.[44]

After traveling across China, India, parts of the Ottoman Empire, and Russia, Upton arrived in Germany in 1876. In Berlin, he became infatuated with the German war machine, which most European armies were attempting to emulate. In his view, the German system reflected what the American Army needed: a professional military based upon sound doctrinal principles, military schools and institutes, and little government interference. Upton especially admired German military schools' emphasis on providing entry for middle-class citizens, which expanded the officer ranks of the country's mass army by displacing the old aristocracy with a new officer corps based on knowledge, not birth. Upton left Germany determined to reform U.S. Army doctrine along German lines.[45]

Having ignored Sherman's guidance to glean what he could from Asian armies, Upton became an ardent champion of the German system. He believed that its general staff, mass army, and freedom from civilian interference stood in welcome contrast to the problems that beset the military during the American Civil War. He defended his views in *The Military Policy of the United States,* a widely circulated manuscript subsequently published in 1904 by then Secretary of War Elihu Root. Although Upton was no friend of the militia or National Guard, he advocated a professional "expandable Army" built upon conscription, for "every American citizen owes his country military service." As proof of German effectiveness, Upton noted that the Prussians had defeated Austria in six weeks in 1866. In 1870, the Prussians crushed America's doctrinal icon, France, in three and a half months.[46]

Many senior army members shared Upton's regard for German efficiency, but they understood that the American government and people were unlikely to abandon civilian control of the military or support a peacetime expandable army. Upton, however, was zealously determined to fight for his vision of military reform. Consumed by the effort, he aimed to fend off critics by revising the Army's tactical doctrine a second time. The strain proved more than he could bear. He committed suicide in March 1881. Although his work was not completed, Upton's ideas contributed to the creation of an army school system, influenced by German notions regarding the relationship among the state, the military, and the people in war.[47]

In the wake of Upton's suicide, attacks upon the 1867 doctrine seemed to intensify. One prominent voice was Lieutenant Colonel Henry M. Lazelle, who wrote to Sherman that any single- or double-rank formation advocated under current doctrine was doomed to failure in the face of modern weaponry. The *Army and Navy Journal,* once a staunch Upton supporter, printed Lazelle's letter and declared that the Army's tactical manual was dead. In response, Sherman undertook Lazelle's advice to appoint another board of officers to revise army tactics. Sherman thought a board of officers to be unwieldy, however, and advocated that a sole author accomplish the task. This in turn triggered a race among would-be theorists who sought to lay claim to status as Upton's heir apparent. Dozens of new ideas soon flooded the War Department.

Leading the reformist charge were Lieutenant John Bigelow and Captain Arthur L. Wagner. Reprising ideas first advanced by Washington and Steuben in 1779, each believed that foreign ideas must adjust to a Republican army fighting in North America. Bigelow lambasted Prussian ideas within his 1880 article "Mars-la-tour and Gravelotte," where he cited incidents in which German commanders had made serious tactical blunders. In a separate study published in 1889, Wagner also criticized the Prussians. Focusing on the 1866 campaign at Königgrätz, he cautioned the Army against blindly accepting European models when there was so much to learn from America's historical experience.[48]

The debate over a future army doctrine continued well into the late 1880s. The militia or newly titled "National Guard" had a voice. As self-supporting, localized volunteer forces, state militia units had long avoided total compliance with federal army doctrinal guidelines. State units were subject to the vagaries of funding and, as demonstrated during the Civil War, were weakened by the national conscription of potential members into the regular Army. The National Guard movement of the 1870s did not produce truly "national" forces but a hodgepodge of diverse units that varied by location. Many units elected their officers and formed units for athletic or social purposes, as much as military ones. Federal Army attempts to impose doctrinal principles upon National Guard units were anemic at best.[49]

By 1885, regular Army-National Guard cooperation had improved to some degree. Guard officers earned membership in the Military Service Institution of the United States, an organization founded in 1878 to promote professional discussion. Guard officers soon contributed articles, joining Bigelow, Wagner, and a growing list of intellectuals now publishing regularly about army tactical nuances. In 1884, regular Army officer N. B. Sweitzer separated "drill" from "tactics" by distinguishing "maneuver tactics" and "battle tactics." Maneuver tactics, he argued, referred to the administrative movements necessary to bring a unit onto the battlefield where battle tactics then took over. In 1888, citing improvements in modern firearm technology, regular Army officer Lymon W. V. Kennon commented, "The distinction between maneuver tactics and fighting tactics is more marked now than ever before."[50]

Individuals such as Bigelow, Sweitzer, Kennon, and Wagner made significant contributions to the tactics debate. Bigelow's ideas emphasized strict adherence to principles of war, a trait traceable to the works of Jomini. Sweitzer changed definitions and ideas about drill, while Kennon merely endorsed those remarks. Wagner, however, was both practical and philosophical, stressing the importance of flexibility and initiative in applying military axioms, an idea more in line with the Prussian military theorist, Carl von Clausewitz. Similar to Clausewitz, he observed that no two battles were the same and, as such, what proved useful in one case might be disastrous in another. War, he concluded, manifested a diverse nature that resisted bending itself to enduring principles.[51]

Toward a Teutonic Army Doctrine

To date, the ongoing doctrinal debate consisted of arguments over fire-power and maneuver, speed and shock action, formations and tactical control, and the advantages or disadvantages of massed fire versus individual marksmanship. In 1887, the debate changed direction once General Philip H. Sheridan succeeded Sherman as the Army's commanding general. Sheridan, who had visited Europe in August 1870, had been impressed by the German system that he observed. His conclusion was remarkable given that he "saw no new military principles developed, whether of strategy or grand tactics, the movements of the different armies and corps being dictated and governed by the same general laws that have so long obtained, simplicity of combination and manoeuver, and the concentration of a numerically superior force at the vital point." Regardless, Sheridan sought to end the ongoing debate over tactical reform by appointing a board of officers to explore the issue. The board was directed to review numerous reform proposals languishing in the Office of the Adjutant General and to scrutinize foreign military developments. Sheridan desired that the process terminate with the development of tactics that were both appropriate for a modern battlefield and in conformity with American tradition.[52]

The officers constituting the Board of Revision, as it was known, met in Washington, D.C., in February 1888. In addition to reviewing copious manuscript submissions from army officers, the board members acquired drill manuals from Austria, France, Germany, Great Britain, and Italy, as well as twenty-three additional works on tactics. Many of these documents were translated into English for the first time. Board members pored over the materials and, in April 1889, moved to Fort Leavenworth, Kansas, to access the library of the newly formed Infantry and Cavalry School.[53]

Relocating to Fort Leavenworth allowed Arthur Wagner, an instructor at the school, to enter the discussion in an unofficial capacity. Wagner's reputation as a military intellectual was based upon his analysis of *The Campaign of Königgrätz* and his having visited the German *Kriegsakademie* in Berlin. In March 1889, the *Journal of the United States Cavalry Association* published his work, "The New German Drill Book and Some Deductions Therefrom," a detailed analysis of the methods used by the German Army. Wagner stressed that drill was a necessary preparatory step before

Arthur Wagner (U.S. Army Military History Institute, Carlisle Barracks, Pennsylvania)

fighting. However, he believed that once hostilities commenced, fluidity should take precedence on the battlefield. In recognition of this fact, German military leaders had adopted flexible methods in creating a training system that sought to replicate the chaos of battle. Conversely, where the Germans had eliminated the battle line, allowing their soldiers to advance in small squads using cover and firepower, American Army doctrine had relied more upon mastering repetitive drill movements deemed suitable for all conflicts. Formation control was more important to Americans. To Wagner, the Germans had solved the riddle of crossing the deadly ground by advancing small units, and the American Army must do so, as well.[54]

In May 1890, the board produced a new manual containing many of Wagner's ideas. But unlike previous manuals that were often affiliated with primary authors such as Alexander Smyth, William Duane, Winfield Scott, and others, Wagner's name was not attached to this one. Instead, the new manual was considered the product of the Army's corporate mind. As with the 1812, 1862, and 1867 army doctrine, the manual retained some of the past while looking to the future. The double rank of fours for drill purposes was preserved. But the extended order was adopted for combat. Clearly, the board had divided drill and tactics into two categories, drill for training and tactics for war. Underlying the manual's concepts, however, were German ideas such as individualism and initiative in battle while dispersing forces over a wide area.

The 1891 *Regulations*

Less radical than some had thought, the board maintained much of what Upton had incorporated into the 1867 *Infantry Tactics*. Soldiers' marching step was lengthened from 28 to 30 inches to improve speed. Marching cadence went from 110 steps per minute to 120, and the old common time of 90 steps per minute was abolished. Numerous changes were made for the manual of arms and various commands. Tactics were explained in clearer language than the previous doctrine, skirmish movements simplified, dispersion advocated, and a section of two squads became a unit. Divisional and brigade movements were deleted. Prior to the manual's publication, the board had sent the new regulations to senior personnel within the Army for comment. After making additional alterations, the Army's commanding general, John M. Schofield, directed immediate implementation. He also invited criticism in an effort to attain doctrinal "perfection."

The results were published as the 1891 *Infantry Drill Regulations*. The new doctrine not only dropped "tactics" from its title but also provided more guidance on how to fight than just prepare. Still, the manual caused an immediate uproar. Francophiles voiced their displeasure in adopting Teutonic methods. In pointing out that the French had rejected the German idea of dispersion in favor of élan, they called into question the validity of extended order formations. Many officers, more French-thinking than not, raised doubts about German dispersion as the answer to the defense, arguing that troop diffusion meant less firepower at a given point on the battlefield. Most critics, however, were more than familiar with the American soldier and ridiculed the idea that a corporal was capable of leading troops in battle. In effect, the 1891 doctrine had altered service culture by diminishing officer responsibility and placing more accountability on noncommissioned officers in combat. Major James Chester, an officer with little faith in the leadership ability of enlisted men, was among the loudest voices of protest.[55]

Structurally, the 1891 *Infantry Drill Regulations* contained several snags. Congress failed to approve a change in force structure allowing army regiments to match the German organization of three battalions of four companies each. The board members had assumed that Congress would act immediately; however, the change did not occur until after 1898. Another issue concerned the deletion of the single rank in the offense. Board members believed that the extended order made the linear rank, single or otherwise, obsolete. Skeptical officers, however, argued that the single rank must be retained to maintain control. From their perspective, the extended order was best employed by skirmishers who advanced in front of the main body, a well-regulated formation composed of successive single ranks. Although it seemed reasonable for the extended-order formation to be capable of suppressing enemy fire, many argued that it still took soldiers arrayed in single ranks to ultimately assault and break a defending line. By arguing for both extended-order and single-rank arrangements, critics sought to retain ideas from the 1867 doctrine while easing their personal anxiety of losing tactical control due to the increased interest in soldier individual initiative.[56]

Criticism poured in from officers seeking clarification of questions both mundane and profound, from parade-ground issues to arguments concerning French methods and the German system. Part of the blame for the seemingly endless controversy rested squarely upon the War Department, whose senior officers had failed to sell the new doctrine to the force as a whole in advance of publication. The officer corps of the 1880s was far more educated and professionally aware than in previous generations. Typical of the group was Arthur Wagner, an individual less critical of the manual *per se* but more concerned about the lack of American examples to justify change. Although Wagner had contributed to the 1891 manual, he complained of "Prusso-maniacs" who insisted upon drawing lessons from recent European wars without considering the American

Civil War experience. The War Department got the message and offered Wagner the opportunity to provide such justification, which he gladly accepted.[57]

Wagner eventually produced a two-volume set of works, *The Service of Security and Information* (1893) and *Organization and Tactics* (1895). With their publication, Wagner became the legitimate intellectual successor to Upton. Both publications had doctrinal implications, the first because it sought to justify the 1891 *Infantry Drill Regulations* from a historical and theoretical perspective; the second for its delineation of strategy ("the art of moving the Army in a theater of operations") from tactics (the process of "mov[ing] troops on the battlefield"). But there were other reasons, as well. Leavenworth students were too dependent upon outdated warfare and tactical texts, most of them British editions. Wagner emphasized American military history as a laboratory for encouraging tactical originality and developing textbooks within a national cultural framework.[58]

A stickler for detail, Wagner insisted on defining terms, a trait that would become an intrinsic part of army doctrine. Although Wagner discussed drill for training purposes, his views represent a shift in service thinking. Since 1778, manual writers had advocated drill as a critical tool for disciplining soldiers. Wagner, however, rejected that idea. "Many individuals believe that drill promotes discipline," he wrote, but "many militia units drill well, but lack discipline." "Drill," he added, "means instruction, but does not necessarily promote well disciplined soldiers." He suggested that the best indicator of discipline could be seen in the "endurance of hardships by the soldiers, and in the willing, energetic, and intelligent efforts to perform their whole duties in the presence of the enemy." By emphasizing performance under fire rather than maintaining perfect parade ground formations, Wagner helped to overturn the Army's longstanding cultural belief that discipline came from drill.[59]

Wagner's writings about human behavior and war suggest that he was familiar with the work of Carl von Clausewitz. Clausewitz had explained that to understand war, one must first understand human beings. Echoing this notion, Wagner wrote: "You must understand human nature to appreciate what men can and cannot do, understand national characteristics [culture]," to employ soldiers within their capabilities, an observation noted not only by the Prussian master but also by George Washington and von Steuben one hundred years previously.[60]

The fruits of Wagner's labor led to the 1890s' development of semiofficial manuals at Fort Leavenworth, Kansas. The Leavenworth Board manuals were the first American Army publications to delve into the distinction between the offense and the defense in war, particularly at various echelons of command. Battalion and company commanders were advised on "how to fight," as separate units or as part of a larger formation. The rifle company was reorganized to include the platoon, section, and squad. Institutionally, the Army placed more emphasis upon small-unit leadership by segmenting the company into smaller units led by lieutenants, sergeants, and corporals.[61]

The 1895 *Infantry Drill Regulations*

Although the 1891 *Infantry Drill Regulations* raised an intellectual ruckus among officers, the doctrine caused few operational difficulties. The Army was at peace; the frontier had officially closed with the 1890 national census. In 1895, however, the Army replaced the caliber 45.70 single-shot black-powder Springfield rifle with the Krag-Jorgensen, an internal magazine-fed bolt-action rifle using a smokeless powder cartridge in caliber 30–40. The design and operational differences between the two rifles were substantial enough to warrant a change in the manual of arms. As an update to the 1891 manual, the 1895 *Infantry Drill Regulations* not only modified individual drill movements to account for a bolt-action weapon, but also allowed for the single-rank formation in deference to previous officer criticism. This configuration was used through the 1898 War with Spain but was later dropped.

By 1897, although many thought the Germans had theoretically solved how to cross the deadly ground and restore the offense to primacy, questions remained. In an effort to study the situation further, the Army conducted training exercises in which small-unit leaders were to take the initiative while stressing bayonet drill and target practice. While valuable lessons were learned, no exercise provided hard evidence suggesting which tactics were optimal in assaulting a well-prepared defensive position while preventing substantial casualties. Indeed, the Army's 1895 doctrine offered a choice between closed or open formations in the offense. Officers could employ a closely packed attack formation of successive single ranks that maintained control but invited annihilation or use an extended-order arrangement with less command restraint that promised to spare lives. For micromanaging officers prone to believe that what can be controlled will be controlled, choosing between more tactical checks or fewer was a difficult choice to make. Regardless, despite the Army's adoption of the German system, debates continued over how much tactical regulation to enforce in attempting to negate the chaos of battle. This issue was still not resolved when a foreign war brought the debate to a temporary end.[62]

One Doctrine for Two Different Wars

The 1898 War with Spain and its spin-off conflict in the Philippines proved significant challenges to army doctrine. For several decades in the late nineteenth-century, the U.S. government and public in general had worried about Cuba, a country under Spanish imperial domination 90 miles off the coast of the United States. On 25 April 1898, political misgivings gave way to war and the Army prepared to invade the Caribbean island. At the time, the Army consisted of 28,000 regular troops deployed across eighty forts augmented by 100,000 National Guardsmen. President William McKinley asked for an increase to 65,000 Regulars plus 125,000 volunteers, the latter subsequently growing by an additional 75,000 men. Owing to fear of tropical disease, 10,000 so-called "Immunes" (men supposedly resistant to such contagions) were also added to the ranks. By August 1898, the regular Army stood at 55,000 troops with volunteer strength about 215,000.

Modern rifle technology was used by both Americans and Spaniards. American Regulars were issued the five-shot Krag-Jorgensen rifle, though the National Guard and most volunteers carried obsolete Springfield trapdoors. Spanish soldiers were armed with clip-fed, bolt-action 8 mm German Mauser rifles. Faster to fire and reload than American rifles, the smokeless powder, heavy Mauser bullets often penetrated even the thickest cover. In Cuba, Lieutenant Colonel Theodore Roosevelt of the 1st U.S. Volunteer Cavalry or "Rough Riders" reported, "[I] was standing behind a large palm with my head stuck out to one side, very fortunately; for a bullet passed through the palm, filling my left eye and ear with the dust and splinters."[63]

Gathered from across the country, the American Army formed at numerous federal and state camps. At Port Tampa, Florida, about 25,000 men, mostly Regulars, drilled on the beach. An additional 30,000 men were located at Camp Cuba Libre, near Jacksonville. At Chickamauga Park, Georgia, 60,000 men arrived at Camp George H. Thomas. Camp Alger, Virginia, held another 23,500 by mid-August 1898. Other troops, left behind when the Cuban expedition departed, were scattered among

Mobile, Alabama, Miami, Florida, and Tampa. San Francisco, California, became the point of departure for troops headed for Manila in the Philippines. Many volunteer units remained at their state camps.[64]

Soldiers headed for Cuba trained to both 1891 and the updated 1895 doctrine, depending upon the weapons they were issued. Tactical drill to instill discipline was practiced at Port Tampa by the Fifth Infantry, at Chickamauga Park with the Twelfth and Twenty-Fifth Infantry Regiments, and the 2d Illinois Volunteers at Jacksonville, Florida. Units also practiced extended order as skirmishers, with soldiers advancing in small units and simulating the firing of their rifles from the prone position. Other soldiers moved from column into firing order while standing within a single rank. Rather than emphasize individual marksmanship, commanders, cognizant of wasting ammunition, trained their units to fire in volley, the dominant technique used in combat in Cuba, Puerto Rico, and the Philippines. Under cover of the main body, troops trained as companies and battalions to mass-fire while several smaller units were ordered to break ranks and advance by running. Soldiers ran while carrying their weapons either over their shoulder or in one hand. Battalions within the regiments rotated from being the main attacking force to the reserve. Calisthenics and rifle drill also filled the training day, a scene repeated by thousands of army troops at over two dozen camp locations.[65]

In San Francisco, beginning in May 1898, a mixed assembly of Regulars and volunteers consisting of Major General Wesley Merritt's Eighth Corps assembled at Camp Merritt. Raw recruits formed into "awkward squads" to receive instruction in individual movements, marching, and physical fitness. Commanders drilled their units on various maneuvers, including extended order and the use of skirmishers at the squad, company, battalion, and regimental levels. Although mandated at a minimum of three hours per day, training priorities were left up to the discretion of commanders who knew their units best. Division and brigade commanders trained their subordinates to include force-on-force mock battles with some units imitating Spaniards. While commanders had latitude in directing training exercises, the 1891 and 1895 *Infantry Drill Regulations* underpinned their efforts.[66]

On 22 and 23 June 1898, 17,000 members of Major General William Shafter's Fifth Corps invaded Daiquirí, Cuba. Doctrine did not address the nuances of conducting an opposed landing from the sea. Fortunately, the Spanish offered no resistance. As units moved inland, junior leadership and small-unit actions soon dominated tactics in accordance with reigning doctrine. Army officers centralized planning and decentralized the execution of orders; both 1st U.S. Volunteer Cavalry Colonel Leonard Wood and his second in command, Lieutenant Colonel Roosevelt, formed their plans and gave general guidance to their subordinates who were expected to operate within the established framework. Company-grade officers (captains and lieutenants) supervised units, but noncommissioned officers often led patrols, headed work parties, and handled the digging of entrenchments, tasks that a generation before had been directed by commissioned officers.[67]

With its thick undergrowth, Cuba's rough terrain made it difficult for even small groups to maneuver effectively. Rather than disperse, soldiers under fire tended to cluster behind any available cover, obscured from their commanders and other units. Advancing in multiple columns with flankers for security was difficult, as the terrain and density of foliage forced entire regiments of several hundred soldiers to follow narrow trails in single file. Consigned to a regimental attack often one-soldier wide, commanders used scouts or an advanced guard to precede the strung-out files. These men were to detect the enemy and provide early warning for the main body to deploy in the jungle. The work was strenuous because of the tropical heat and prevailing fear of eminent battle. Temporary halts to provide relief often found exhausted troops scattered along the trail. War correspondent Richard Harding Davis reported that the enemy "saw the advance and began firing with pitiless accuracy into the jammed and crowded trail and along the whole border of woods. There was not a single yard of ground for a while to the rear which was not within the zone of fire." Frequently unable to spot the enemy because of the smokeless powder of the Mauser rifle, soldiers often shot back in a general direction. It was difficult to sight the enemy and engage individual targets, as doctrine had directed.[68]

In the early afternoon of 1 July 1898, Army Regulars and one volunteer unit, the 71st New York, formed into a firing line facing the well-defended San Juan and Kettle Hills near Santiago, Cuba. While awaiting artillery support, the troops witnessed the devastating firepower unleashed by Captain John H. Parker's battery of four Gatling guns. Acting as a suppressive fire upon the Spanish defenders, the guns swept the crest of the hill. Infantry and cavalry (fighting dismounted) advanced up the hill in loosely organized lines. Private Charles Johnson Post, 71st New York Volunteers, recalled that the American attack at Kettle Hill went up the rise in "no nice order, no neatly formed companies crossing that plain or mounting the slope." In truth, the Army advanced in extended order in small groups. Post reported that the assault seemed like the crowd at a football stadium, "when the game is over and a mess of people are straggling across [the field.]" One observer, Richard Harding Davis, described the scene: "[They] had no glittering bayonets; they were not massed in regular array. There were a few men in advance, bunched together, and creeping up a steep, sunny hill, the tops of which roared and flashed with flame."[69]

Army tactical principles demanded that assaulting troops break the enemy line. But the Spanish were already withdrawing as the Americans closed in upon the crest. Private Post and his unit reached the top of the hill in the sweltering heat only to watch the Spanish run down the reverse slope into Santiago. The American soldiers then lay "on their backs . . . or with their elbows on their knees and panting for breath." Too drained to pursue the enemy, the men watched the Spanish regroup and prepare fortified positions. After eight hours of fighting in which they fended off several counterattacks, the American Army secured San Juan Hill. The action cost over 1,300 American casualties. On 10 and 11 July, both forces fought the last battle for Cuba, an artillery and rifle fire duel directed from opposing trenches.[70]

The American Army placed Santiago under siege while other forces assaulted Puerto Rico on 25 July 1898. Some 17,000 officers and men composed of elements of the First, Second, and Fourth Corps supported by rapid-fire guns and artillery landed virtually unopposed at Guánica near the southwest corner of the island. There, General Nelson Miles formed the Army into four assault columns designed to move rapidly inland to outflank approximately 8,000 Spanish defenders blocking the mountain passes leading to San Juan. American forces executed the plan efficiently. On 12 August 1898, the island was secured at the cost of four dead and forty wounded.[71]

As American soldiers fought in Cuba and Puerto Rico, the Army was also engaged in the Philippines. An occupation force since 30 June 1898, American troop strength had grown from an initial small presence to 800 officers and 20,000 men (with 77 officers and 2,338 men in Cavite or in transport vessels off the coast). Other army forces garrisoned Manila, while another 16,000 troops were shaped pentagon-fashion in a defensive line that extended for about 16 miles around the city. Outside this perimeter was the Filipino Army of Liberation, a force of about 15,000 to 40,000 men armed with a variety of weapons, including the Spanish Mauser. As the terrain varied within the American perimeter, the regimental commands of the U.S. 8th Corps, separated by swamps and other terrain features, were arrayed in semiautonomous areas.[72]

Negotiations with Spain culminated with the Treaty of Paris on December 1898. Cuba gained its independence. The United States gained Guam and Puerto Rico. In exchange for $20 million, Spain transferred control of the Philippines to the United States. Emilio Aguinaldo had previously declared the Republic of the Philippines, and his forces clashed with American troops on 5 February 1899.

Doctrine had not addressed combat in cities. Instead, as the fighting escalated, army forces assaulted Filipino positions in extended order, advancing platoons of men by rushes and alternating covering fire, as doctrine prescribed for a general assault on a defending enemy. After securing Manila by March, the Army went on the offensive to destroy Aguinaldo's Army of Liberation. Through most of 1899, the two armies fought a conventional war across difficult terrain composed of bamboo thickets, paddies, and jungle. They were treated to incessant heat, insects, disease, and exhaustion. The summer saw the Americans conduct minor or limited operations because of monsoon and the withdrawal of volunteer units. By fall, the Army was once again engaged in conventional operations, methodically moving northward and southward along the island of Luzon. Gradually, Aguinaldo came to accept that fighting the Americans conventionally was an unwinnable proposition. On 13 November 1899, he shifted to guerrilla warfare.

Guerrilla warfare is often confused with insurgencies. Guerrillas are conventional troops fighting behind enemy lines for military objectives. Insurgents fight to attain political, not military, ends. In conducting what was really insurgency and not guerrilla warfare, Aguinaldo's *insurrectos* formed into regional commands that included full-time insurrectionaries and part-time militia members. Some wore uniforms; others did not. However, both waged a war of ambushes, assassination, fund raising for supplies and weapons, intelligence gathering, intimidation, and raids. Unable to defeat the American Army in the field conventionally, the insurgent target became the will of the U.S. government. Aguinaldo hoped to prolong the war and fuel the public outrage expressed by the many Americans who opposed President McKinley and his foreign policy. Filipinos were keenly aware that 1900 was a presidential election year and that the celebrated William Jennings Bryan was running as an anti-imperialist candidate. McKinley's subsequent reelection on 14 November 1900 severely undermined Aguinaldo's strategy.[73]

Both the 1891 and 1895 *Infantry Drill Regulations* were conventional warfare manuals, continuations of the orthodox thinking that had dominated army doctrine since 1779. While the pages lacked information concerning how to counter guerrillas or insurgents, the publications stressed the extended order formation and Arthur Wagner's views of the "wave attack" in which small groups of soldiers acted independently. Such guidelines proved effective for an army that had come to rely heavily upon railroad and telegraph technology since the Civil War. Although the telegraph was available in the Philippines, insurgents often cut the lines and ambushed repair crews. Without reliable communications technology, senior officers drew upon messenger methods used in remote frontier outposts. Officers, following the doctrinal precepts of initiative, went about their business, trained their soldiers, and learned from mistakes. Brigadier General Theodore Schwan commented that the situation in the Philippines resembled the era of the Indian Wars. In both instances, the Army had to vanquish the foe and then demonstrate fair and just treatment through pacification.[74]

During 1899, army officers followed what became known as a policy of attraction characterized by a new government, roads, schools, and education programs to win over the Filipinos psychologically. In doing so, the Army turned to a mixture of doctrine and informal practice. Shunning large operations for the most part, army units engaged in "hikes" in which small columns of fewer than 100 soldiers searched for *insurrectos* and their camps. Relying upon leaders of small units to execute senior headquarters directives, American soldiers conducted night operations, ambushes, and raids (called "roundups") designed to keep the insurgents off balance while denying them sustenance. Superior mobility was important to army success. Mule trains and local carts with water buffalo teams moved equipment and supplies.[75]

As with the Apache wars of the late-nineteenth century, the use of locals to augment army forces proved essential. Fifteen thousand Filipinos became the Philippine Scouts (light infantry) and the Philippine Constabulary (paramilitary police), as well as indigenous policemen and several volunteer militia units. The Army also recruited spies and informants, while a special translation agency was established to interpret captured documents. The Army issued identity cards and passes, took various censuses, and built files on insurgent leaders to control the local population.[76]

Despite tremendous effort as late as 1900, American government benevolence had failed to end Aguinaldo's resistance. Most Filipino peasants found American culture to be alien and often offensive. Elites appreciated American societal norms but believed that they could form a modern nation on their own. As the war dragged on and the benevolence policy failed to produce results, the Army shifted informal practice to chastisement, punishing the population for supporting the *insurrectos*. Many officers found the situation surprising, for soon after military governor Major General Elwell S. Otis departed the Philippines in May 1900, he had declared the war to be over.[77]

Shifting tactics from attraction to chastisement was initially an incremental affair. As early as 1899, some army officers had ordered the burning of homes and villages in retaliation for ambushes. But by mid-1900, many officers had concluded that benevolence was not working, despite what Otis had claimed. Invoking the more castigatory portions of the War Department's General Order 100, army officers reverted to informal practices reminiscent of American Indian Wars. Local leaders were held accountable for insurgent actions, and communities suffered fines if they did not report anti-American activities. Army leaders punished entire villages for crimes and property damage. If American soldiers or their supporters were harmed, homes and settlements were burned in retaliation. Nighttime curfews were imposed to curtail insurgent movement.[78]

While the degree of public punishment varied by commander and area, military governor Major General Arthur MacArthur further defined General Order 100 on 20 December 1900, after McKinley's reelection. Individuals taking action against the American occupiers were labeled as "war rebels" or "war traitors" and subject to military tribunal. Those who acted apart from an organized army were denied the rights of prisoners of war. Belligerency in any form was not tolerated.[79]

Army officers now also made the cost of the war intolerable to all Filipinos, not just *insurrectos*. Social elites were arrested and incarcerated. In areas where resistance was strong, army troops swept through the countryside burning homes, villages, facilities, and crops. They also slaughtered livestock and destroyed boats and wagons. These tactics served two purposes: the first was to cause insurgency supporters to reconsider the personal cost; the second was to deny the enemy food and shelter. In the wake of the destruction of entire areas of countryside, displaced and hungry refugees rushed to the American-controlled towns. Most went into stockades, where the population was separated from the insurgents. Placing civilians into a confined area smacked of a "concentration" policy, a political nightmare given that the Spanish had followed a similar *reconcentrado* practice in Cuba. Instead, the Army referred to such places as "zones of protection" or "colonies."[80]

American soldiers, their Filipino auxiliaries, and insurgents also used torture to extract information. Although the number of incidents was small, many captives underwent the "water cure" in which large amounts of the liquid were forced down throats or the "rope cure," which involved suspending people in the air. Others were denied food or sleep and placed in overcrowded conditions. Though such practices were officially condemned, some army officers justified them because they often attained results. The *insurrectos* acted in a similar fashion. Some American prisoners were buried up to their neck with their heads covered in molasses to attract ants. Other soldiers were mutilated upon capture; one captive had his eyes gouged out while another had his feet amputated. In isolated incidents primarily committed by individuals, Filipino civilians were tortured or shot for supporting one side or the other.[81]

Gradually, the policy of chastisement succeeded. In February 1902, American soldiers captured Filipino General Vicente Lukban. Soon thereafter resistance on the island of Samar ended. Brigadier General Miguel Malvar, the head of the Batangas militia, surrendered in April. On 4 July 1902, the war was officially declared over.

In the War with Spain, army doctrine had provided a system of discipline suitable for Regulars, National Guard, and volunteer units. The precepts contained within the 1891 and 1895 versions were sufficient to allow the Army to defeat two diverse foes separated by thousands of miles. Although limited by Cuba's terrain, Teutonic principles of small-unit leadership and initiative enabled the Army to operate conventionally in land more suited for decentralized operations than centralized control. The same held true for the American Army in the Philippines during the conventional war period from February–November 1899. When the war turned into an insurgency, doctrine still provided the means for small units to conduct tactical actions that led to the capitulation of Aguinaldo's forces. Still, doctrine had not addressed the specific nuances associated with unconventional insurgent activity. Army officers thus fell back upon informal practice, procedures that had existed in American warfare since the colonial era.

The Boxer Uprising

While Army troops were pacifying the Philippines, an international crisis provoked American military action in China. During the nineteenth century, European powers and Japan had extorted trade rights from a weak Manchu government leading to the creation of various areas of influence, colonies, and protectorates. Foreign presence led to growing Chinese outrage and the beginnings of the I Ho Tuan or "Boxer" movement, an uprising consisting of poorly armed peasants. With their antiforeigner stance, the Boxers gained enough momentum to gain support from the Manchu court. In 1900, a large contingent of rebels invaded Peking, where they burned churches, killing both foreigners and Chinese Christian converts. With the Manchus encouraging the attacks, the foreign diplomatic legations in Peking were soon under siege. Boxer leaders declared a state of war upon all legations, including the United States.[82]

The American government reacted by alerting nearly 15,000 troops of potential duty in China. About 5,000 eventually deployed, although only 2,500 appeared in time to participate. Major General Adna Chaffee commanded the 9th Infantry, 14th Infantry, and 6th Cavalry, all arriving from the Philippines. Chaffee's contingent became part of a second international relief force of 19,000, the first or "Seymour Expedition" having been repulsed in early June.[83]

While the Army operated no differently tactically than if fighting a conventional war under its current doctrine, the conflict differed significantly from the War with Spain. In China, the Army operated as part of a multiservice and multinational force. Chaffee not only commanded army infantry, cavalry, and artillery forces, but a battalion of marines, most of whom had little experience. International troops included members of the Austrian, British, French, German, Italian, Russian, and Japanese armed forces.

Neither the 1891 nor 1895 doctrine had addressed collaboration in war with other services or nations. While this mattered little in terms of cooperating with American marines, the situation was much different from the standpoint of coalition warfare. On 13 July, at Tientsin, a fortress manned by Chinese soldiers equipped with modern rifles, confusing orders and language problems led to the 9th Infantry Regiment attacking in the wrong location. Instead of a straightforward assault on the fortress walls, soldiers advanced in small groups across open ground scored by ditches, mud, and ponds. Fighting in water up to their waists, the soldiers endured a withering fire. The 9th eventually was forced to retreat after losing 25 percent of its members, including its commander, Colonel Emerson H. Liscum.[84]

Eventually, the international force took Tientsin and advanced under various levels of resistance from Chinese troops and Boxers to Peking. Arriving at the city on 14 August 1900, the American Army experienced another facet of international warfare: prestige. Each army wanted to be the first to relieve the besieged legations, thus military operations became more of a race to the finish rather than an organized assault. The Americans were forced to scale the 30-foot-high Tartar Wall without ladders, by working their way up using hand and foot holes in the brickwork. In the end, a British Sikh unit reached the objective first and ended the siege.[85]

The fight for Peking was not yet over, however. American soldiers joined their foreign counterparts to blast holes in gates and walls using artillery before rushing into the void. Here, tactical formations gave way to massed assaults and heavy fighting between walls and within various buildings and rooms. Once the city was cleared of resistance, the Chinese capital was divided into zones, each occupied by one of the powers participating in the relief expeditions.

In the absence of doctrine that detailed the nuances of military government, Chaffee commanded the American zone, an area of several square miles and 50,000 inhabitants, according to his experience. Having served as an Indian agent, Chaffee understood civil administration. Moreover, his second in command, Brigadier General James H. Wilson, had experience from Reconstruction duties in the American South after the Civil War. The two officers established law and order using Chinese police and courts, and enforced sanitation and inoculations against disease. In addition, they institutionalized public works, built hospitals and schools, and closed opium dens and gambling halls.[86]

New Century, New Doctrine

From 1898 through the early 1900s, the Army learned valuable lessons from its military governance and multinational experiences in the Caribbean and Asia. Still, no effort was made to significantly alter army doctrine to accommodate that knowledge. The 1891 and 1895 manuals, the direct descendants in a line traceable to the 1779 *Regulations,* had been written to guide an infantry-dominated force. When change came once more, it was again technology that drove it.

Following minor changes to the 1895 manual in 1901, a new doctrine took form as the Army rejected the Krag-Jorgensen rifle in favor of the Springfield model 1903, a caliber 30.06 bolt-action, magazine-fed rifle with an improved Mauser-type action. Chief of Staff Nelson A. Miles ordered new infantry doctrine to be written in anticipation of fielding the 1903 Springfield. In truth, the manual was an update, not a new doctrine, for it incorporated tactics derived from both the 1891 and 1895 manuals. Also included were extended order-drill revisions as published under War Department direction by Thomas H. Ruger in 1898. Adopted on 23 June 1904, the doctrine included how to form a line from small groups of individuals. Bayonet exercises were discarded as many officers had concluded that the weapon was outdated on a modern battlefield. Instead, each soldier was issued a rod, a device with no remarkable value. In April 1907, after significant fuss, army leaders changed their minds and provided for a new bayonet.[87]

As Secretary of War Jefferson Davis had suspected in the early 1850s and army theorists further explored in the post–Civil War era, the catastrophic lethality of rifled and rapid-fire weaponry upon the offense eventually terminated centuries of precision battlefield drill and controlled formations. In envisioning future war, army doctrine writers of the 1890s had shunned outmoded French-based ideas. Gone were the numerous commands necessary for soldiers to give fire, as well the complex directives needed to cause brigades, regiments, and battalions to maneuver as a collective whole. Instead, the Army's manuals embraced Teutonic ideas advocating the decentralized execution of orders, initiative, and small-unit actions, particularly in the offense.

To be sure, the Army's Teutonic-based methods had discounted senior officer control in favor of more authority by leaders at lower echelons. The burden of combat success, previously resting on the more senior, higher-ranking, and experienced, now sat squarely upon the shoulders of junior and less experienced army officers, both commissioned and noncommissioned. For the Army to succeed tactically in war, doctrine required its lowest-ranking officers to comprehend general guidance from those above them and then execute it. That they did so successfully given the horrific conditions characteristic of contemporary Caribbean and Asian warfare was remarkable. More noteworthy, however, was that their efforts enabled America to take center stage as an emerging global power. The Army's doctrine made a significant contribution to that end.

While effective in combat, army doctrine was not perfectly suited for the American way of waging modern tactical war. Intellectual residue remained from a previous era in the form of column and rank formations. Although soldiers were now to attack in small groups, columns were still necessary to move rapidly in a general direction. Ranks served to concentrate soldiers in physical space in order to allow for mass covering fire as small squads of soldiers maneuvered. And as it had been since 1779, informal practice augmented doctrine primarily when facing unconventional forces, for the Army's manuals ignored them. Although debates continued over such matters, doctrine would soon be altered more by international affairs than tactical considerations.[88]

CHAPTER THREE

DOCTRINE FOR ARMY OPERATIONS:
FROM *FIELD SERVICE REGULATIONS*
TO FIELD MANUAL 100-5, 1905–1945

The American soldier today is a part of a great machine
which we call military organization; a machine in which,
as by electrical converters, the policy of government
is transformed into the strategy of the general, into
the tactics of the field and to the action of the man
behind the gun.—Elihu Root, Marquette Club address,
Chicago, Illinois, 7 October 1899

Even as the 1904 doctrine was attaining government
approval, pressing national affairs were conspiring to ensure the content,
direction, and purpose of future editions would be altered. By January
1899, the War Department oversaw the administration and security of
overseas possessions acquired from the recent conflict with Spain. Having
just witnessed the Army's bureaucracy struggle to mobilize the service for
an overseas war, President William McKinley realized that War Department reorganization was necessary to support an army of occupation.

On 20 July 1899, the reform-minded McKinley offered Elihu Root the
position of secretary of war. A lawyer by discipline and without military
experience, Root assumed office on 1 August 1899. He had much to contemplate, notwithstanding the War Department's performance during
the recent War with Spain. Army generals now governed Cuba, Puerto
Rico, and the subjugated areas of the Philippines, an area measuring just
over 10,100 miles from San Juan to Manila. The regular Army consisted
of 100,000 men with 66 percent of them overseas; several thousand volunteers augmented them. In total, nearly 30,000 army troops were either
already engaged in combat or deploying to join in the fray.[1]

Secretary Root quickly determined that conflict with Spain had created new roles and missions for the Army. Decades of frontier duty had

Elihu Root (Credit: U.S. Senate Historical Office)

resulted in an outpost system with small army garrisons scattered about the country. The commanding general of the Army held a titular rather than authoritative position. In actuality, bureau chiefs administered and supplied the Army, bypassing the commanding general to report directly to the secretary of war. Bureau staff officers were accustomed to holding their permanent, highly coveted positions. Many had lost touch with the field force long ago. Bureaucratically, the War Department was caught within a vortex created by its own operational success.

Doctrine and the Root Reforms

Transformation was not only necessary but urgent, given the new realities of overseas national responsibilities. In addressing Chicago's Marquette

Club in 1899, the secretary stated his intention in these terms: "The machine [Army] today is defective; it needs improvement; it ought to be improved." By describing the Army in mechanistic terms, Root conveyed his intent to create an army system. This inclination meant replacing the War Department's emphasis on individual talent and independent departments with a new philosophy promoting group work.[2]

War Department reforms soon broadened into a general restructuring of the Army. Doctrine played a role in realizing Root's vision. Collectively called the "Root Reforms," two initiatives directly shaped the future course of army doctrine. With congressional support, Root changed the War Department's organization by supplanting the independent-minded commanding general and bureau chiefs with a general officer "chief of staff" who directed an "army general staff." Designed to maximize group input in producing war plans and providing the service with the wherewithal to fight, doctrinal development also fell under the purview of the army staff who would function as service caretakers.

No less important, in 1904 Root established a progressive hierarchy of army schools encompassing pre-commissioning to senior service: West Point and post schools, the School of Application for Infantry and Cavalry (1881), later known as the Infantry and Cavalry School, at Fort Leavenworth, Kansas, became the Army School of the Line, focusing upon tactics and command and staff functions up to division level. The second-year General Staff School, attended by about 40–60 percent of Leavenworth's first-year class, was added and stressed large-unit general staffs, operations, and logistics. The two courses consolidated (1923) to become the Command and General Staff School and later the Command and General Staff College (1946). The Army War College (1904) capped off the school hierarchy.

A planning division reportable to the general staff was located within the War College. Although the army school system with its fixation upon German tactical principles was not perfect and required time to mature, the War Department now controlled two essential elements of doctrinal production: an educational system to furnish officers with the requisite intellectual grist and an army staff "brain trust" to write the manuals.[3]

Since 1779, the Army's keystone doctrine had conveyed a conventional and tactical approach to warfare that emphasized individual and unit combat procedures. While the War Department continued to publish various doctrinal manuals addressing those matters, the service's dominant publication changed in form and scope. No longer just a tactical manual, keystone doctrine now served at least four purposes: to convey government legislation that affected the Army, to disseminate the secretary of war's approved directives as crafted by army staff officers, to educate regular, militia (National Guard), and volunteer officers in their responsibilities, and to prescribe certain actions necessary to control field army operations in armed conflict.

Doctrine also became the means by which the army leadership would further regulate more of what was regarded as informal practice. Although doctrine cannot account for every possible mission, the War Department sought to anticipate as many future requirements as possible in order to provide guidance to army leaders. The augmented scope of the 1905 doctrine marked the beginning of the army leadership's progressive effort to eliminate or reduce informal practice by increased regulation.

The 1905 *Field Service Regulations*

On 1 February 1905, the new keystone doctrine emerged, as the *Field Service Regulations* or *FSR*. Approved by President Theodore Roosevelt under the signature of Secretary of War William H. Taft, the manual was the product of the Army's general staff and the considerable influence of Captain Joseph T. Dickman. A member of the Fort Leavenworth crowd, Dickman drew upon voguish German military thinking. As the *FSR* title indicated, service emphasis shifted from pure tactical matters to regulating service behavior in the field, albeit within a conventional warfare framework. The content, form, and style of this manual became the basis for subsequent army keystone doctrine editions in the twentieth century.[4]

The new doctrine contained twelve articles (chapters). These articles included definitions and instructions for the land forces of the United States and the militia, a straightforward reference to the 1903 Dick Act, as well as stipulations for the size and organization of army forces, their purpose, and the duties and responsibilities of unit commanders and staffs. The War Department general staff did not envision fending off a foreign invader in the near future. Still, any conflict that involved defending the continental United States meant moving troops over difficult terrain and poor roads. To overcome these difficulties, the army corps formation, a unit that had existed since the American Civil War, was replaced by the division. Where a corps occupied up to 35 miles of space, the smaller infantry division required but 11 miles. More compact and easier to maneuver, divisions became the basic combat organization. The *FSR* prescribed that a division include three infantry brigades composed of at least two infantry regiments each, one cavalry regiment, an engineer battalion, a signal company, and four field hospitals. Each division contained one provisional artillery regiment consisting of nine field artillery batteries to deliver fire support. An ammunition column, a supply column, and a pack train, all manned by civilians, supplied the force. Although the regiment was the largest permanent unit in time of peace, provisional divisions and brigades were established during field exercises so that regiments and battalions could train for war.[5]

Given Dickman's oversight of the *FSR*'s contents, it was no surprise that theoretical elements of campaign design were included, such as instructions for establishing bases and lines of communications. In doing so, the army leadership demonstrated that it accepted abstract constructs previously advanced by the nineteenth-century theorist Antoine Henri de Jomini within his book, *The Art of War*. Rather than simply occupying a small part of West Point or Leavenworth school curriculum, Jominian ideas concerning operational design now became an intrinsic part of the service's intellectual basis for how to wage war.

Additional instructions included staff organization, a concept that had not been previously discussed in past doctrine. Administrative instructions delineated the purpose, format, and method of delivery of field orders and the gathering and dissemination of information. Ammunition, transportation, and security issues also warranted separate articles. In deference to the realities of the Army occupying overseas possessions, a section on military government was also included. In addressing this subject, army doctrine now included specific instructions for military missions that had long been left up to the ingenuity of commanders and, thus, informal practice.[6]

Tactical action in war was the focus of Article VI, *Combat*. Here, the *FSR* distinguished between the offense and defense but noted that decisive results typically rested in offensive operations. Although the defense was viewed as the stronger form of war, defenses were considered temporary measures until the offense could be reestablished. When attacking, an army was to make contact with as little force as possible by using skirmishers, a holdover from previous doctrinal manuals. Skirmishers advanced in front of the main body, made contact with the enemy, and then provided fire support for the main body when it was unleashed in an attack.

In combat, army forces were to operate through three phases: the *preparatory stage, decisive action,* and *completion.* In the *preparatory stage,* army forces were to prevent the enemy from withdrawing prematurely, forcing opposing reserves to commit, and sowing doubt as to where the main attack might fall. Next came *decisive action,* which meant striking a blow at the decisive point in a defending line (another reference to Jomini's conception of Napoleonic methods). Lastly, *completion* consisted of pursuing fleeing enemy troops. The *FSR* was careful to explain that while the three phases might be brief, particularly the *preparatory stage,* combat nonetheless tended to follow the set sequence.[7]

Firing weapons was necessary to attack or defend successfully. For army doctrine writers, however, there was no formula for achieving fire superiority over an enemy force other than achieving the highest rate possible. Thus, the manual categorized weapons fire efficacy into four categories—*distant, serious, effective,* and *decisive*—articulated in yards and dependent upon rifle, light-artillery, and heavy-artillery range capabilities. For example, a rifle was considered to provide *distant* fire value at more than 1,800 yards, *serious* at 1,800 to 1,200 yards, *effective* at 1,200 to 600 yards, and *decisive* at less than 600 yards.[8]

When conducting offensive operations, commanders were advised that they had the advantage of initiative, for the defender must react. In attacking, commanders were to have a specific objective in mind; the defender had only a general objective, to repel the attacker. Attacks were envisioned to occur at multiple points as feigned strikes or the main blow, but the defender was required to respond to each one regardless of purpose. Attackers were to use superiority of numbers and fire to instill confidence and to look forward, not backward. Defenders were considered shaken by a confident attack, their line broken by retreat long before it was turned or penetrated.

As powerful as an attack may be, the manual also discussed advantages of the defense. Defenders were to choose a sheltered position that caused attackers to cross open ground. Field fortifications reinforced terrain and protected defenders from enemy fire. Trench construction reinforced other units and enfiladed an attacker with fire. Ranges were to be determined and marked for improved weapons effectiveness and ammunition positioned for ease of access. When defending, soldiers were less likely to become casualties if they had protection and were able to mass fire more easily with better accuracy. The 1905 manual indicated that morale was generally higher when defending, for soldiers used their weapons for a longer period of time and caused considerable damage to an attacking force.[9]

In recognition of the recent experiences in Cuba and the Philippines, the *FSR* acknowledged several aspects of the modern battlefield. Fire superiority was essential. Frontal assaults against defenders were to be avoided due to the high cost in lives. The manual also pointed out the difficulties of enemy detection due to smokeless powder and the increasing use of field uniforms in neutral tones. A small section discussed the nature of guerrilla conflicts, but offered no specific guidance on how to respond to them, again leaving matters up to the informal practice expertise of field commanders.

To negate the inherent power of the defense, superior rifle fire and artillery were viewed as vital. Overcoming the defense also required tactical surprise and skilled skirmishers possessing strong intellect and courage, excellent marksmanship, and cognizance of how to use cover. When the main body conducted an assault by passing through the skirmishers and their covering fire, the attacking force relied upon "moral stamina" to break the enemy line of defense.[10]

In a change from previous doctrine, the 1905 *FSR* stressed the interaction of infantry, cavalry, artillery, and engineers. This "combined arms" approach again reflected Fort Leavenworth thinking, for the school had stressed the importance of employing all army forces together to enhance capability since 1881. However, doctrine held that the infantry remained of primary importance. In the attack, infantry forces were intended to form an advance guard that would make initial contact with the enemy. Horse cavalry protected the flanks of the main body, which was made up primarily of infantry. While the infantry and cavalry advanced, artillery would fire upon enemy positions, its primary purpose being the destruction of the opposing artillery. Commanders would then move forward to determine where the enemy defense was the weakest, either the defending line or a flank. After a preparatory fire from advanced guard rifle fire and artillery at the point of attack, the main body would assault the enemy force, rushing forward with bayonets fixed upon their rifles. Upon seizing the enemy position, infantry forces could then prepare to repulse any counterattack.[11]

In the defense, commanders were to choose their ground carefully and use engineer officers to determine how best to strengthen the location through trenches and obstacles. Labor, however, was to be provided by line troops. When attacked, the defender was advised to reveal as little of the defense as possible and to make effective use of artillery and reserves. Attacking an enemy flank was especially important to stop offensive momentum. Defending troops were advised to make local counterattacks that might turn into a general offensive. Regardless, the object was to avoid a retreat and to gain the initiative.[12]

In addition to describing attacks and defense in daylight, instructions were provided for night operations. During darkness, it was especially important to coordinate the activities of infantry, cavalry, artillery, and engineers. Instruction included ensuring that extensive reconnaissance was conducted in daylight, using distinctive markings to recognize personnel at night, and to incorporate "watchwords" to challenge individuals approaching defensive lines under nighttime.[13]

General Chaffee ordered the Army's garrison schools to use the 1905 *FSR* as a textbook. Moreover, he demanded that both the regular Army and the militia use the manual when operating in the field. This directive soon proved its worth, when, in late 1906, President Roosevelt dispatched more than 5,000 troops to Havana to assist the Cuban government in controlling an internal rebellion. Designated as the Army of Cuban Pacification, U.S. forces remained in Cuba until early 1909. While there, the units had the opportunity to put the *FSR*'s guidelines concerning military government and jurisdiction (Chapter XII) into practice. A 1908 update placed the subsection explaining the "details of staff organizations" under the more general subsection of "details of organization," but the intellectual thrust remained unchanged.[14]

The 1910 *FSR*

In 1910, the Army produced a new keystone doctrine for a service whose primary weapons continued to be the rifle and pistol. Although many of the manual's observations concerning views of future war remained the same as the previous edition, a number of changes in modern warfare practice were introduced. These changes were inspired by the reports of officers who had visited Manchuria in 1905 to observe foreign armies during the Russo-Japanese War. The observers became enamored with both the Japanese and Russian armies, especially the aggressiveness of their infantry units, the role of field artillery and the machine gun, and their demonstration of tactical adaptability under changing combat conditions. "The Japanese," wrote one officer, "[were] expected to largely limit the use of the machine gun to the defensive, but experience soon taught them to widen its field, and later it was used to great advantage in the offensive." Observers with the Russian Army reported that they were tenacious fighters, for "the Russian infantry is composed of as obstinate troops as there are in the world. It is hard to drive them back if they are ordered to stay."[15]

For American observers, the ultimate symbol of a motivated and determined soldier was the bayonet, for "the war demonstrated that even if bayonets are not crossed the moral effect produced by them must be frequently brought into play." A soldier motivated by *esprit* ran across the deadly ground between trenches and drove the enemy away with a bayonet. That such a soldier was deemed worthy of emulation reflected late-nineteenth-century European military thinking (the French in particular), which regarded a soldier inspired by national élan to be worth ten of the enemy in battle.[16]

Many observer findings from the Russo-Japanese War were controversial. The bayonet and the spirited soldier behind it was often celebrated and taken as fact. But the modernization of service doctrine by increasing the emphasis upon artillery firepower became a thorny issue. The recent Russo-Japanese conflict demonstrated that field artillery fire, not the infantryman with rifle and bayonet, dominated the tactical battlefield. Yet, the American Army leadership refused to subordinate the infantry to the field artillery. Since the establishment of the first American militias, the infantryman had served as the tactical backbone of the Army. In contemplating which was now more important, the human element (represented by the infantry) or technology (through artillery), observers were loath to admit what experience revealed. Despite the catastrophic results of artillery upon infantry formations, observers reported, "Nothing in the Russo-Japanese War demonstrated that, in field battles, the infantry has lost its supremacy." To counter the power of artillery fire, attacking infantrymen were simply encouraged to accelerate their speed by reducing the weight of their clothing and assorted kit so they could run faster through the maelstrom. In sum, the observers revisited the same trite solution to crossing the deadly ground in the offense that had been repeatedly advocated by army leaders since the American Civil War.[17]

Despite controversy, the Russo-Japanese War reports had some effect upon doctrine, at least as far as élan and machine guns were concerned. The 1910 *FSR* encouraged commanders to attack with fervor, for "the assailants have the confidence of their numerical and moral superiority. When the die is cast and the attack is on, they no longer mediate upon the consequence; they look forward not backward." Doctrine writers had assumed enemy defenders would break when facing determined American soldiers, owing to "the effect of impending contact and of the resulting enfilade of reverse fire." Élan was considered essential for producing confident soldiers capable of imposing a shock effect upon the enemy. Yet, attacker success was possible only if commanders were "willing to pay the price of victory" in casualties. As far as the 1910 doctrine was concerned, such a price had to be paid to achieve victory.[18]

In regards to machine guns, a detailed organization for machine gun company personnel authorized one company per infantry or cavalry regiment. Appendix E explained the road spacing required when moving a machine gun company with infantry or a machine gun troop with cavalry. In the offense, machine guns joined with the field artillery to provide fire support. Machine guns were considered most useful as defensive weapons in convoy protection, covering river crossings, and defending bridges, defiles, and other manmade or natural features until reinforcements arrived. Still, while machine-gun advocates such as Captain John H. Parker sought to give the weapon its due, maintenance and reliability issues kept the weapon from becoming a dominant tool by an *FSR* fixated upon the offense.[19]

The 1913 *FSR*

The 1910 version of the *FSR* lasted only three years before its replacement appeared in 1913. Administratively, the new manual's chapters were partially reorganized. The overall thrust remained the same, however, which was to promote the army division in wartime as the basic unit of organization. Two divisions constituted a field army. In peacetime, the service continued to depend upon the regiment for its basic formation in three varieties—infantry, cavalry, and artillery. A coastal artillery corps was also mentioned, as well as various other branches of the service to include corps of engineers, chaplains, medical department, ordnance department, quartermasters, signal corps, and other entities provided by law. Minor changes were introduced to the articles and new emphasis was placed on such areas as the role of the provost martial general.[20]

The term "mobile Army" was introduced for the first time. While the expression appeared only once and was not precisely defined, it implied the use of a conveyance to increase the gathering of information. Within the article entitled "the service of information" and specifically the section applying to aeronautical reconnaissance, the *FSR* discussed the advantages of using "aeroplanes" over dirigible balloons as a reconnaissance asset within a mobile army (owing to the latter's need for storage to protect them from the elements).[21]

Although previous *FSR* versions had addressed the role of balloons, the 1913 edition was the first to discuss aeroplanes despite the fact the Army had created an aeronautical division within the signal corps in August 1907. While several air-minded officers had subsequently argued that the future of warfare was through the air, the army leadership had ignored the use of machine-powered flight in its 1910 doctrine.[22]

According to the latest *FSR,* aeroplanes, along with balloons, were to be employed in undertaking two forms of aeronautical reconnaissance, strategic and tactical. Strategic reconnaissance consisted of using single-pilot scout aircraft capable of flying about 200 miles. Tactical reconnaissance scout aeroplanes carried a pilot and an observer, as well as a radio for communicating with the ground and an automatic machine rifle for protection. Their primary purpose was to observe and control artillery fire. Pilots were advised to fly at least 2,000 feet above the enemy unless protected by fog, haze, or approaching darkness. Aircraft personnel were to communicate with the ground by radiotelegraphy, as well as dropping messages and marked maps. Army troops were to clear airfields in the vicinity of camps, with continuous communication established between the airfield and the army headquarters. No mention was made of the aeroplane as an offensive or defensive weapon.[23]

Machine guns continued to be emphasized, as had been the case within the 1910 *FSR.* One addition, however, concerned their use in dirigibles. As a security weapon, machine guns were also to be used in defending ground approaches to friendly positions and to stop enemy advances. In marches and convoys, machine guns were seen as a protective weapon for shallow draft watercraft. Although the weapon was to be used along with artillery for fire support in the offense, the primary purpose was defensive.[24]

Philosophically, the 1913 *FSR* remained fixated upon the offense as the primary means of decision. In that regard, little had changed since the 1905, 1908 update, and 1910 versions.

The 1914 *FSR*

The Army's latest doctrine lasted but one year as pressing organizational matters forced a replacement in 1914. On 24 February 1913, President William Howard Taft directed the newly formed 2d Division to the U.S.-Mexican border. In answering the call, the army leadership found it necessary to change the peacetime organization for training purposes from the regiment to the division. A "triangle" division structure emerged, one based upon the concept of "threes." Each division consisted of three combat regiments composed of three battalions. A new *FSR* was required to articulate the changes associated with the Army's "triangle" division.[25]

The 1914 manual was far more comprehensive in regulating army behavior than previous versions. Internally, the manual was now divided into four parts, each containing various pertinent articles (chapters). Part I *Organization* contained articles on the role and purpose of the Army, both domestic and foreign. The "mobile army" concept was now specifically associated with the division, which *FSR* writer Major James A. Logan defined as "a self-contained unit made up of all necessary arms and services, and complete in itself with every requirement for independent action incident to its ordinary operations." In Part II *Operations*, the War Department general staff grouped together various articles (information, security, orders, marches and convoys, combat, and shelter), with the whole explaining how to employ an army in war. Each article began with general principles introducing the reader to the overall intellectual thrust, followed by more specific subsections. Part III *Administration* discussed how the general staff viewed supporting an army in combat. Part IV contained several appendices and, for the first time, a quick-reference index to assist the reader in finding specific information. The manual's organization indicated that staff officers and those involved in the education system had given considerable thought to finding commonality among various wartime activities and grouped them according to tasks and purpose.[26]

Theoretical concepts infused the 1914 *FSR*, a reflection of the continued influence of the army school system. Doctrine articulated how to employ a field army in a "theater of operations," the newly designated term for where wartime activities occurred. Field army commanders were to deploy their subordinate divisions no more than a day's march from each other for mutual support and to ensure that an enemy force could not interpose itself between them. When advancing, the column was the preferred formation, as it allowed better control, speed of movement, and improved security. As the advancing force moved closer to the enemy, subordinate divisional units such as brigades and regiments were to deploy into smaller, parallel columns that allowed for rapid deployment into combat formations.[27]

Descriptions of combat actions were more refined than previous *FSR* versions, although there was much carryover from the 1913 edition. Commanders were introduced to the concept of the "rencontre," a surprise encounter with an opposing force. Friendly-force commanders were to contemplate in advance how they would respond to unexpected enemy contact. In anticipating rencontre combat, the advance guard commander was crucial for assessing and then developing the tactical situation. Keeping in mind the preconceived plans and intentions of the supreme commander, the advance guard was either to attack immediately or hold its position so it could maintain the initiative.[28]

Tactically, doctrine offered numerous forms of maneuver, although all were dependent upon achieving fire superiority through a heavy volume of weapons fire drawing the enemy's attention to his immediate front. Depending upon the tactical circumstances, commanders were expected to understand the advantages and disadvantages of each form of the attack. *Frontal attacks* were assumed to result from any form of attack as a matter of course. An *enveloping attack* struck an enemy flank, forcing the defender to overextend defensive lines and thus weaken the center of the defense. A *turning movement* was limited to forces larger than a division. Here, a field army separated into two parts, enabling one to conduct a frontal attack while the other might execute a frontal or enveloping attack. Conceptually, this meant that the turning force drew enemy troops from their positions or "turned" them toward a direction that they were not prepared to defend. A *holding attack* attempted to gain contact with the enemy to prevent its movement elsewhere while another friendly force either assaulted the opposing line or conducted envelopment or turning movement. Regardless, any attack culminated with an *assault* followed by *pursuit* of the fleeing enemy force.[29]

If forced to defend, commanders were given several options, although the *FSR* stressed that any defense was temporary until the offense could commence. One option was the *passive defense* in which a force held its position with or without combat. Holding a position influenced a battle because the enemy had to remain cognizant of the units' location and potential threat. Another option was *the defense seeking a favorable position,* the "only form of the defense that can secure positive results." In this case, an offensive-minded commander might be forced to defend to gain time or an advantage over the enemy, perhaps by holding a strong position where enemy forces would squander resources if attacking. Regardless, a defense often led to a *counterattack* to regain the initiative and the offense. Counterattacks were dependent upon reserves, artillery, and machine-gun fire, as well as precise timing to avoid enemy detection.[30]

Commanders were further advised on such matters as how to select defensive positions, where to place a general reserve, how to withdraw if needed, how to use covering positions during withdrawals, and how to assemble withdrawing forces. In the case of a retreat under enemy pressure, advice was given on the proper manner for conducting delaying actions. Patrols, night combat, and post-engagement actions were also explained in detail. In conveying a plethora of operational and tactical concepts both grandiose and mundane, the War Department general staff had gone farther in regulating the behavior of a field army at war than any army doctrine published to date.

In concert with the mobile army concept espoused in the manual, increased emphasis was placed upon the "use of the combined arms." The doctrine's authors believed that "success [in modern war] may be achieved only by all branches and arms mutually helping and supporting one another in the common effort to attain the desired end." Understanding the capabilities of the infantry, artillery, cavalry, signal, engineers and the like was vital, as was knowledge of the potential of "balloons and aeroplanes for reconnaissance and observation of fires." Regardless, the manual clearly stipulated that the infantry remained the principal and most important arm, so much so that its role in battle was recognized as the core of the entire force.[31]

Philosophically, the 1914 *Field Service Regulations* placed considerable emphasis upon mobility and firepower, and the offense as the decisive form of war. That said, army leaders had yet to determine how to overcome the power of the defense. Doctrine advocated closing with the enemy quickly to avoid excessive casualties, but this was no guarantee for avoiding a failed assault. The *FSR* could only offer that "troops advancing to the attack must understand that the best protection against losses is afforded by an uninterrupted and vigorous advance toward the enemy's position." Soldiers were to rush through the fire rapidly, although they might "use such natural cover as the ground offers."[32]

As early as the 1910 edition, doctrine writers had been more aware of the catastrophic effects of the machine gun; even so, no American field army had yet experienced the weapon's effects firsthand on a large scale. Still, the 1914 *FSR* advised advancing troops to attack machine guns in small groups to reduce casualties. When using machine guns, troops were instructed to employ them only for a short time because they attracted attention and were targets for immediate destruction by hostile fire. In a change from the 1910 edition, the 1914 manual noted that the machine gun was a means to obtain fire superiority in the offense, while in the defense they were to be used "against large targets visible for a short time only."[33]

The Punitive Expedition and the 1914 *FSR*

While the 1914 *FSR* was in effect, the United States-Mexico border developed from a festering sore for both the William Howard Taft and Woodrow Wilson administrations into a call for military action. In mid-1911, previous internal discord within Mexico had led President Taft to increase the numbers and size of border patrols. He later ordered the so-called Maneuver Division to San Antonio, Texas, for four months. Divisions were also formed at Galveston and Texas City. Eventually, a Mexican civil war erupted, with several rivals vying for power. In 1913, Woodrow Wilson succeeded Taft as president. President Wilson decided not to recognize Mexico's President Victoriano Huerta and imposed an arms embargo upon the country because Huerta had assumed power in a nondemocratic manner. When Huerta's forces began to crush the opposition anyway, Wilson lifted the embargo and supported the opposition led by Venustiano Carranza.[34]

In late April 1914, a U.S. force intervened, occupying Veracruz in response to the arrest of American sailors in the port of Tampico and Germany's attempt to deliver arms to Huerta via commercial ships. Army Major General Frederick Funston soon controlled the city with nearly 3,500 marines and 4,000 army troops. Wilson and Huerta resolved the issue without war when Huerta resigned and Carranza eventually assumed power. American troops withdrew after seven months of occupation duty.[35]

Carranza's ally, José Doroteo Arango Arámbula (a.k.a. Francisco "Pancho" Villa), soon parted ways, the Wilson administration's backing of Carranza instead of Villa being partially responsible. Outraged, Villa responded by ordering the murder of American citizens working in Mexico and, on 9 March 1916, leading 500 to 1,000 men against Columbus, New Mexico. The subsequent burning and looting of the town, as well as an engagement with portions of the 13th Cavalry, produced fourteen dead American soldiers and ten civilians. President Wilson responded by sending Brigadier General John J. Pershing, commander of the 8th Brigade, into Mexico with orders to assist the Mexican government in capturing Villa. Pershing's orders included that he was not to rile the local population (wishful thinking given the nature of the mission and Villa's popularity). Secretary of State Robert Lansing received a preemptive diplomatic telegraph from Carranza warning that the Americans would be considered an invading force by many Mexicans.[36]

On 14 March 1916, Pershing organized a provisional (temporary) division, the "Punitive Expedition, United States Army." The division was structured for the mission, with an emphasis on maximizing mobility. Pershing organized his force around two cavalry brigades (two cavalry regiments and a field artillery battery each), one infantry brigade (two regiments and two engineer companies), an ambulance company with a field hospital, a signal corps detachment with an attached aero squadron, and two wagon companies for supplies. He also had at his disposal motorized trucks, armored vehicles, and more than 9,300 horses. The plan called for chasing the bandits with cavalry across the barren and hostile northern Mexican landscape while infantry protected the line of communication back to El Paso, Texas. Removing several thousand troops from Texas left the border area less secure, so, on 9 May, Wilson mobilized National Guard units from Arizona, New Mexico, and Texas. As this still left insufficient numbers along the border, the remaining National Guard was eventually called up, a financially expensive but necessary decision.[37]

Pershing's operational plan followed the tactical precepts spelled out in the 1914 *FSR*. His forces trained to that manual. Where Villa was located remained a mystery, thus the force followed doctrine by moving in two parallel columns, one from Hachita (2,000 men) and the other (4,000 troops) from Columbus, New Mexico. The Second Cavalry Brigade acted as an advanced guard while the two columns were to march to Ascención and then assess the tactical situation. The advanced guard operated in three parallel columns to prevent Villa from moving either westward or eastward of his suspected location, Casas Grandes. In keeping with the mobility concept loosely advocated by the 1914 *FSR*, Pershing used available Mexican railroad transportation to move elements of his force quickly southward. Supply transport consisted of rail, wagons, mules, and motorized trucks. Some combat troops made use of armored vehicles. Eventually the force headquartered at Parral, a small town straddling a primary road and railroad line 400 miles south of Columbus.[38]

Army aviators flew more than 540 missions in support of military operations. Although doctrinally employed as a reconnaissance asset, none spotted enemy forces from the air. Of the original eight Curtis JN-3 aircraft, only two functioned reliably. Still, they provided valuable messenger service as liaison aircraft.

Villa was never captured, although the American Army fought several scraps with his forces. One typical engagement occurred during 28–29 March 1916 at Guerrero between *Villistas* and American cavalry tactically following the *FSR*. The Americans initially advanced in parallel columns. One contingent blocked the western edge of the town while the main force swept in from the east in an envelopment maneuver. Terrain, however, delayed the main attack. This led to the trap being sprung too quickly when soldiers using a combination of rifles and machine guns prematurely opened fire. Most of Villa's men escaped in the ensuing confusion.[39]

Faced with the internal pressure of an unpopular American incursion on Mexican soil, Carranza rushed his troops northward to contain Pershing's force. For his part, Pershing obeyed orders and directed his troops to avoid confrontation with the Mexican Army. Still, several incidents led to deaths on both sides. On 12 April 1916, Mexican civilians and Carranza's forces attacked a column of the 13th Cavalry near Parral. Later, on 21 June, a column of the 10th Cavalry encountered a Mexican Army force near Carrizal. The two opposing leaders conferred and the American commander attempted to advance through the town. The Mexican commander had been ordered to prevent such a move and a brief but bloody melee erupted. The American force advanced in columns and then deployed on line to sweep the area. Intense machine-gun fire inflicted serious casualties before the Americans assaulted following doctrine and captured the machine gun. It was a pyrrhic victory. In ninety minutes of fighting, the 10th Cavalry suffered upwards of 42 percent casualties and was forced to retreat. Twenty-four Americans surrendered. The Mexicans lost forty-five officers and men killed and fifty-three wounded. Given the near destruction of the 10th Cavalry, a second war with Mexico was barely avoided.[40]

After the Carrizal affair, Carranza ordered his forces to find Villa. In January 1917, Carranza's troops weakened Villa's forces sufficiently enough to give Wilson an excuse to call off the expedition. More pressing issues, foremost among them being Germany's announcing the return of unrestricted submarine warfare in the Atlantic Ocean, prompted Wilson to order Pershing out of Mexico the following month. Pershing reentered U.S. territory at Columbus, New Mexico, on 5 February 1917.

Upon returning to El Paso, Texas, Pershing held a victory parade and declared the unsuccessful mission an excellent training experience. The Punitive Expedition had indeed been worthwhile in that it had tested the

1914 *FSR*'s intellectual framework by moving, fighting, and sustaining a large, mobile formation of troops 400 miles into hostile territory. Mobility as a concept, however, was premature. Horses, mules, and ground-pounding infantry often outperformed underpowered motorized trucks. Still, Villa's raid on Columbus created the conditions for the army leadership to test its doctrine during a national emergency and to learn valuable lessons on the eve of global war.[41]

The 1914 *FSR* and the Great War

On 6 April 1917, the U.S. government declared war on Germany. The Army soon commenced preparations for war using the guidelines established within the 1914 *FSR*. In the course of doing so, the manual underwent seven minor changes before 17 August 1917 to correct typographical errors or to provide elaboration where deemed necessary. By the time the armistice was enacted on 11 November 1918, the 1914 *FSR* had seen four more changes (for a total of eleven since original publication). None of these minor revisions represented a significant change in the War Department's prewar view of army operations or tactical combat.[42]

Army doctrine had not addressed the rapid reorganization and expansion of the Army in time of war, but circumstances mandated both. On the recommendation of British and French advisors, the War Department modified the prewar triangular division structure to the "square" division, an organization composed of two brigades, each one containing four regiments. The division grew from 18,000 to just over 28,000 troops. "Squaring" the division ensured more infantrymen within regiments from the prewar authorization of 2,002 enlisted men to 3,720. Larger units were intended to increase the division's ability to withstand significant combat losses and still function, at least in theory. Additional restructuring included machine-gun battalions at the brigade and division levels. To increase the size of the Army, Congress established a draft on 18 May 1917. Among the 4 million men inducted (and thousands of volunteers), approximately 2 million deployed to France. Of those, 1.3 million saw combat.[43]

Command of and responsibility for preparing the Army Expeditionary Forces (AEF) fell to Commander in Chief General John J. Pershing. Pershing's approach was the product of his life-long experience as a regular Army officer and his most recent experience with the Punitive Expedition against Pancho Villa in Mexico. In preparing for Mexico, Pershing had trained his command in musketry and battle tactics from brigade down to platoon level while applying a heavy dose of both offensive and defensive instruction. Pershing ordered that his troops be instructed in trench warfare, but emphasized a war of movement. He deduced that the American soldier must be taught to shoot, to use terrain effectively, to entrench rapidly if needed, and, most important, to drive the enemy from their trenches and then defeat them in the open; in other words, precisely what the 1914 *FSR* advocated.[44]

In seeking to avoid the war of attrition and trenches that had characterized combat operations in France since 1915, Pershing used the term "open warfare." Some officers were right to claim that nothing in the *FSR* mentioned the term specifically. Lieutenant General Hunter Liggett, commanding general of the U.S. First Army, published his own open-warfare guidelines. Yet, for those who were familiar with the 1914 *FSR* through either formal military education or happenstance, the underpinning concept was clear enough within the text. In truth, open warfare was nothing more than semantics. As both doctrine and Pershing noted, the concept embraced 1914 mobile warfare principles, for it involved driving the enemy out of their positions and into the open by engaging them in a war of movement. Individual and group initiative, resourcefulness, and tactical judgment (without neglecting preparations for trench combat per se) were salient to both open warfare and mobile warfare. Both required an army to train vigorously for the offensive.[45]

As commander in chief, Pershing executed the War Department's concept of a mobile army based upon a multidivision organization that exploited mobility to an advantage. In 1917, mobility was not the same as maneuver, for the 1914 doctrine did not discuss a war of maneuver. Where maneuver seeks a positional advantage over the enemy, mobility has more to do with how a conveyance (e.g., an aeroplane, a horse, a motor vehicle, a soldier on the march, or a naval vessel) interacts with its environment (air, ground, or water). In describing heavy artillery, for example, the 1914 *FSR* noted that "the limited mobility of heavy artillery renders its use inadvisable in any position from which the conditions of combat may require a hasty withdrawal." Army doctrine writers understood that heavy artillery was slow to reposition and thus less mobile than lighter guns. Cavalry, however, was deemed to have a significant advantage as a reserve, due to its mobility. Yet, mobility is relative, for under certain terrain and weather conditions, ground-pounding infantry can be more mobile than

John Pershing (Courtesy of the National Archives)

aeroplanes, horses, or motor vehicles. For Pershing, open warfare meant that his force had to move faster than the enemy on the ground (and in the air) to force them out of position.[46]

Pershing's reliance on mobility and forcefulness, coupled with individual and group initiative in the offense, echoed both the 1914 *FSR* as well as the German-based tactical procedures previously instilled within doctrine by Emory Upton and Arthur L. Wagner by the 1890s. Both Upton and Wagner promoted initiative and an offensive spirit, ideas that had carried forward through several *FSR* editions. In the 1914 version, commanders were advised that aggressiveness and initiative usually won and only the offense achieves decisive results. Indeed, the manual went so far as to decree that the actions of any subordinate were inexcusable if they knew their superior officer's tactical plan but lacked initiative in executing it. *FSR* directives expected subordinate officers to understand higher-command intentions and to drive the enemy out of its positions, possibly through the combination of a frontal and flank or envelopment attack. Regardless of the tactical method used, the *FSR* stipulated that unless the enemy blundered or was inferior, all attacks eventually became frontal assaults at some point. Given that frontal assaults were the culmination of any form of the attack, successful ones required a skilled commander and extensive fire support from artillery and machine guns. The object of a vigorous assault supported by firepower was to drive the enemy into the open. From that point, the Army was not to remain satisfied with merely occupying ground but needed to destroy the opposing force through pursuit, thus ending the campaign. In framing his "own" ideas about open warfare, Pershing followed those doctrinal principles precisely.[47]

Creating a cohesive fighting force capable of executing Pershing's version of tactical doctrine included expanding the Army's officer corps. About 200,000 officers were required; just over 5,700 were on the rolls. Many commissions came through accelerated graduation from West Point, as well as National Guard and reserve sources. Some noncommissioned officers received commissions. Officers also came from universities through the creation of the Reserve Officers Training Corps (ROTC), an outgrowth of the National Defense Act (NDA) passed by Congress on 3 June 1916. The War Department also created Officer Training Camps (OTCs), a three-month course of instruction that many apprentice officers found to be a wasted effort, for the instructors often knew little more than the students.[48]

Under the scrutiny of officers and sergeants, some enlisted recruits received a healthy dose of Pershing's view of open warfare from dawn to well into the night. Others were less fortunate, as training standards varied by location and were universally unenforceable. Still, instilling an aggressive spirit within the ranks followed doctrinal principles. Soldiers underwent individual marksmanship training to produce confidence. Aggressive behavior was enhanced by ideas gleaned from experienced British and French advisers who taught the nuances of using trenches as jumping-off points for assaults. Soldiers also underwent rifle drill and bayonet practice to the point of dashing across open fields screaming wildly as they leapt into trenches to finish off the foe. In practicing larger attacks as part of companies, battalions, and regiments, officers often encouraged soldiers to move tactically in small groups. At times, they also directed their men to stay on line in a concentrated formation. It was not unusual to see battalion and regimental officers trailing behind on horseback with their sabers drawn.[49]

Because stateside instruction varied, AEF training was at best rudimentary. Little time was available to master the nuances of Pershing's approach to tactical doctrine. Once in France, AEF leaders established additional staff and tactical schools to instill open-warfare standards and to familiarize soldiers with both European weather and terrain. Gradually, AEF training reduced the emphasis on trench warfare. Instead, Pershing's open-warfare approach became more the norm, so much so that the French and British advisers discounted it as ludicrous. Indeed, Europe lacked the open terrain of the United States that the 1914 doctrine writers had not only based their ideas upon but also what Pershing had experienced in Mexico. Further, the 1914 manual had been based upon the triangular division, which had fewer troops to move about and sustain than the square division. Still, Pershing held that open warfare was possible in Europe, arguing that the Allies' approach was too defensive to achieve victory.[50]

The 1914 *FSR* in Battle

As with the 1914 doctrine, Pershing sought tactical mobility as a means to dislodge the Germans from their entrenched positions. That idea carried forward into battle. Although American forces received some combat experience in 1917, the AEF's first significant attempt at executing open warfare occurred at Cantigny in May of the following year. The elaborately planned assault consisted of a heavy-artillery barrage in advance of a combined infantry-tank attack. Under cover of a smoke screen, a method not addressed in the *FSR*, American infantry picked their way across the landscape, advanced 1 mile and drove the Germans from their defensive positions. Unable to continue pursuit due to extensive German counterattacks, the 28th Infantry Regiment of the 1st Division lost 38 officers and 908 men killed or wounded in three days. These casualties were met with additional losses in June and July 1918, when the 1st and 2d Divisions fought an offensive action against German machine guns and artillery in the Aisne-Marne sector. In one day, the 3d Infantry Brigade of the 2d Division lost 68 officers and 3,184 men, far more casualties than the 28th Infantry Regiment suffered in three days at Cantigny.[51]

As the war progressed, most AEF units prepared for and executed open warfare principles. Yet, the realities of the machine gun and well-entrenched German defenders often stymied efforts to move rapidly on the battlefield. Combat pounded the terrain into a muddle of barbed wire, corpses, mud, poison gas, ruins, and shell craters. As units attempted to employ enveloping attacks or turning movements, the Germans responded with overwhelming firepower from carefully prepared defensive positions and reserves. American infantry often resorted to élan. Rising up from their jumping-off positions with bayonets fixed, lines of infantry picked their way across open ground without regard for concealment or cover. Losses were horrendous.

Pershing's open-warfare concept followed doctrinal principles. Army leaders led their men in assaults upon well-prepared defensive positions with tremendous human cost. Tactical adjustments proved difficult, for most officers found no assailable flank available and thus struck a defensive line by frontal attack. This situation was not new. The Army had faced a similar quandary between 1861 and 1865 when massed-produced modern weapons were first unleashed upon soldiers operating under obsolete tactics. So it was for the AEF in France, 1917–1918, although the 1914 *FSR* had described what might be expected of modern technology, especially artillery, in future conflict. Doctrine, however, could explain only what might happen; only experience could validate or reject theoretical assertions.[52]

Combat in France also meant that the AEF faced a variety of challenges to doctrinal and informal practice; the role of aeroplanes was one. Doctrine had envisioned aeroplanes to be used as a reconnaissance tool but aircraft capabilities and potential had advanced well beyond 1914 *FSR* precepts. Pershing had recognized what the Europeans had already discovered by 1915; wartime realities had transformed aviation from merely being a reconnaissance asset. Instead, he separated aviation into a separate force, appointing Brigadier General William L. Kenly to be Chief of Aviation in August 1918 and ensuring that Colonel William "Billy" Mitchell held jurisdiction over aviation within the zone of advance. Mitchell, who had undertaken private study with allied aviators before his appointment, argued for both strategic and tactical aviation, as the British and French had done. Efforts to equip and organize accordingly, however, were not very successful. An evolving form of warfare, aviation lacked a doctrinal foundation within the Army and, thus, became subject to the whims of advocates and informal practice. Since the 1914 *FSR* had not considered aeroplanes in aerial combat, American aviators attended flying schools in England, France, and Italy to acquire such skills. Though Mitchell had Pershing's authorization to attack targets beyond German lines, he never received the thirty pursuit and bomber groups he actively lobbied to receive. Even so, Mitchell mulled over aviation theories that included the potential for parachute training and a combat drop using the First Division and a multinational bombing campaign accompanying Allied powers' bombers into Germany at night. Mitchell's creative mind and fanatical zeal only increased after the war and dramatically affected the course of future army keystone doctrine.[53]

Similar questions remained regarding the tactical employment of tanks and poison gas, neither of which had existed before the war. As with combat aviation, no guidelines other than informal practice governed their use as assets in the offense or the defense. Pershing, for example, viewed tanks as an infantry-support vehicle, while advocates, such as Captain (later Lieutenant Colonel) George S. Patton Jr., believed that armor was a force unto itself.

Despite the challenges associated with employing the Army's open-warfare tactics in France, the AEF suffered no significant tactical setbacks. Its purpose and missions had been achieved in about five months of sustained combat. The cost in human life was high, however, as more than 130,000 were killed and 200,000 wounded in a short period of time. The ideas contained within the 1914 *FSR* and its two updates concerning the regulation of the Army's behavior through offensive, mobile warfare contributed significantly to both victory over Germany and the horrendous combat losses endured to attain that end.

The service's follow-on mission involved a brief period as an army of occupation. Unlike earlier *FSRs* (such as the 1905 version) that addressed certain aspects of occupation duties, the 1914 edition did not. Thus, when American soldiers occupied the Rhineland area surrounding Coblenz, with a force that peaked at 262,000 troops by February 1919, the AEF was left to informal practice in organizing and conducting its mission. Staff officers devised a contingency plan for further combat operations in the event the Germans refused to sign a peace treaty. But, as the AEF became the American Forces in Germany (AFG) with subsequent drawdown in manpower, it became apparent that no further hostilities would ensue. The last 1,200 AFG troops departed in January 1923.[54]

The 1914 *FSR* in Russia

As the AEF fired its last shots in anger in 1918, two American Army forces were reluctantly engaged in combat operations in Russia. In seeking to appease British and French demands for American participation against the Bolsheviks (and perhaps to have some influence over future events in Russia), President Woodrow Wilson sent 5,500 men under the command of Colonel George E. Stewart to guard the northern port of Archangel and 10,000 men led by Major General William S. Graves to the Pacific port of Vladivostok. Although ordered to remain neutral, Stewart was outranked by his British commander, Major General Frederick C. Poole, and soon found his force in active combat. In his sector, however, Graves, whose mission was to protect the Trans-Siberian Railroad, was senior and in a better position to hold to his orders until Wilson altered them in early 1919. In protecting the railroad line, a vital logistical link for the White Russian forces, Graves deployed his men in small, vulnerable outposts. Ultimately, Stewart's troops left in August 1919; Graves's force remained until April 1920.[55]

Determining whether or not the 1914 *FSR* had any impact upon combat operations is problematic. Still, in Northern Russia, Red Army forces often fought conventionally, allowing for Stewart's forces to engage in open warfare. At other times, the Reds operated as partisans, as Graves discovered firsthand, a form of war that the *FSR* had not considered. In late June 1919, an American outpost at Romankova was attacked by Red Army partisans, who killed or wounded forty-four of the seventy-two Americans present before disappearing.

Graves lacked doctrinal guidance in tracking down and annihilating partisans. His efforts, thus, were based upon informal practice. Graves directed that each post and detachment have an intelligence officer who gathered local information of a cultural, military, and political nature. These efforts proved only partially effective, for it was difficult to communicate without language skills and to gather information as foreigners. Moreover, for company-grade officers such as Captain Robert Boyd, informal practice often meant creating tactics on the fly. In one instance, near the village of Toulgas, 250 miles southeast of Archangel, Boyd's command of 300 soldiers was forced to engage in urban fighting, a style of war that the 1914 *FSR* had not envisioned. Unlike the trench warfare of the Western front, Boyd's unit found itself dashing between buildings and firing down streets while subject to snipers. Machine guns swept to clear the alleys and roads, as artillery fire shattered roofs and turned structures into rubble. Fighting often was as savage as open warfare in France.[56]

Postwar Doctrine Revision and the 1923 *FSR*

In the early 1920s, General Pershing challenged the service to investigate what it had just experienced in France. The result was a War Department analysis examining the effectiveness of mobilization and open warfare, as well as the battlefield effects of aircraft, barbed wire, indirect artillery fire, machine guns, radios, poison gas, and tanks. The 1914 *FSR* and its changes had but briefly touched upon the roles of aeroplanes and machine guns. Moreover, both the War Department and AEF veterans had produced numerous manuals and documents while the war was underway, to include copied versions of French and British trench warfare manuals. Between government and civilian publishers, the Army had compiled a disparate collection of doctrinal pronouncements. These required synthesis into a single, up-to-date warfighting manual.[57]

Responsibility for assessing these documents fell to the twenty panels convened by the General Headquarters, AEF. From December 1918 until June 1919, the so-called Superior Board on Organization and Tactics received reports on topics such as air service, artillery, cavalry, chemical, engineer, infantry, and machine-gun operations, as well as more specialized activities such as medical, motor transport, and the postal service. Findings and recommendations were passed on to the General Headquarters, AEF, for review.[58]

The War Department considered not only doctrine revision due to combat experience but also Congress's mandated changes in organization under the National Defense Act of 1920. In 1922, Pershing observed that current doctrine needed to be revised to bring it into line with newly adopted types of organizations. In addition, the War Department general staff had matured to the point where it could properly direct training in addition to writing doctrine. Pershing and his senior leaders soon concluded that open warfare, with its emphasis on the infantryman and mobility, was the best tactical way to train for future combat.[59]

Although European armies were either reconsidering or revising their own warfighting methods, the army leadership discounted their efforts, choosing to draw upon American practice alone. Colonel Hugh A. Drum, commandant of the Leavenworth School of the Line, put it this way: "Our experiences in the European War have been sufficient and the results so creditable that we have little or no need to borrow tactical doctrines from a foreign country. The tactical principles and doctrines heretofore recognized and taught at the Leavenworth schools have been tested in the European war and have been found to be as sound today as heretofore."[60]

Pershing's desires to strengthen the War Department's oversight of operations ensured that the new *FSR* regulated the operations of large *and* small units, army groups to divisions, while conducting land operations. More important, in keeping with increasing War Department authority over training, the manual was the first doctrine to become the basis for all War Department training regulations, an intellectual wellspring from which other training guidelines would flow.[61]

The process of writing a new manual began with a letter from Major General William G. "Bunker" Haan, director of the General Staff Plans division, directing the commandant of the General Service Staff Schools to revise the *FSR*. On 7 December 1920, Haan signed the order establishing that the *FSR* should cover the doctrines of large and small units (armies, corps, and divisions), but not doctrines related to the arms of the service such as infantry and cavalry, whose branch chiefs were responsible for ensuring that their manuals complied with the *FSR*. The new doctrine's emphasis was to be on field forces within a theater of operations from groups of armies to divisions and including expeditionary forces. The manual was intended to serve as the authoritative reference book for field officers. To avoid postpublication debate over mundane details among the officer corps, the manual was to contain general principles and directions only. Elaboration and rationale for certain procedures would be provided in textbooks and other instructional material serving to educate officers.[62]

Brigadier General Hugh A. Drum, commandant of the General Service Schools at Fort Leavenworth, Kansas, was responsible for the manual's creation. He immediately assigned his assistant commander, Colonel Ewing E. Booth (cavalry), as president of the newly formed Field Service Regulations Board. Other members included Colonel Robert H. Allen (infantry), Colonel Willey Howell (infantry but served as the intelligence officer for the AEF First Army in France), Major Thomas E. Catron (infantry), and Major Condon C. McCornack (medical corps). All were Fort Leavenworth school instructors. Collectively, the board sought to create an *FSR* governing the preparations and conduct of war, strategic and tactical principles for doing so, and principles for guiding the management of field forces in a theater of operations. The new manual would apply to the regular Army, the National Guard, and the Organized Reserve. By June 1922, eighteen months after receiving the mission, Drum sent fifty copies of the final draft, modified by his own hand, to the War Department as "Training Regulations No. 15."[63]

With the draft *FSR* serving as its fundamental doctrine, the War Department G3 (operations) Training Branch became responsible for overseeing 565 related training regulations in pamphlet form. Eventually, the War Plans Division transferred responsibility for reviewing the draft *FSR* to the G3 Training Division. The manual soon went out for comment to the other divisions within the War Department, the army schools, and the service branch chiefs. Although the draft had been written to avoid debate, the resulting onslaught of comments and conflicting agendas ensured that the final text was less cohesive than intended. By November 1922, a War Department G3 Training Branch committee was formed consisting of Colonel Edgar T. Collins (infantry and chief of Training Branch), Colonel John L. DeWitt (quartermaster) from the War Plans Division, and Major George A. Lynch (infantry), a staff officer assigned to the Military Intelligence Division. To ensure uniformity of language, Lynch alone wrote and edited the final draft. It proved so refined that reviewers unanimously forwarded the document without comment. Acting Chief of Staff Major General John L. Hines approved the document by order of the secretary of war on 2 November 1923, thus providing the service with the distilled wisdom of its experiences in the "Great War for Civilization."[64]

The 1923 *FSR*

On 2 November 1923, the War Department published a new *FSR* that emphasized open warfare with more precision than what the 1914 *FSR* had deemed to be mobility. Given that the Army's experience in France dominated the manual, the contents fixated on conventional and not unconventional warfare. From 1923 until 1941, forces overseas were left to informal practice, not doctrine, in fulfilling their legal obligations. This situation applied to a 1,000-man garrison in Tientsin, China (1912–1938), a similar force ordered to Shanghai in 1932 for five months, and troops in Panama as well as in the Philippines.[65]

The doctrine resulted from AEF board findings, experience, and army schools such as Fort Leavenworth, Kansas. Pershing noted that the schools had made good use of "a great [foundation] of unrelated information acquired during the war to the end that a sound military doctrine has been brought up to date."[66]

Such an acknowledgment by the Army's senior leader demonstrated his progressive desire to justify doctrinal reform. In the early 1900s, members of the American Progressive movement lobbied for societal change through the application of social-scientific studies and statistics. Pershing and his followers mirrored the Progressive's practice of using reams of evidence to substantiate doctrinal change. As with Progressives in general, studies were used to advance and validate an agenda, not simply to shed light upon a subject. Review boards generated paperwork and reports that confirmed Pershing's view of open warfare: that the British and French had faced defeat because of defensive-mindedness and a lack of aggression. Although technological advances had resulted in appalling casualties, Pershing believed the infantry could still dominate the battlefield. Thus, in principle, warfare was no different in 1918 than hundreds of years before. Decisive victory would be the end result of an aggressive and mobile infantry-fixated force assisted by other arms unleashed through the offense. While the devil was in the details of execution, Pershing's progressive-minded officer corps had proven to themselves at least that open warfare was a fixed principle for tactically regulating the Army. Fixed principles in warfare were similar to a general conclusion reached ninety years earlier by the Swiss military theorist, Antoine Henri de Jomini.[67]

The 1923 *FSR* was quick to expose officers to such thinking. On page iii, the War Department instructed, "While the fundamental principles of war are neither numerous nor complex, their application may be difficult and must not be limited by set rules. Departure from prescribed methods is at times necessary. A thorough knowledge of the principles of war and their application enables the leader to decide when such departure should be made and to determine what methods should hurry success." Although the manual writers acknowledged that a flexible mind was necessary in battle, principles were the immutable truths that anchored the intellect. One did not deviate, without peril, from operational concepts based upon the principles. Pershing understood this, for he noted, "The war has developed special features which involve special phases of training but the firm fundamental ideas enunciated in . . . field service regulations, and other service manuals remain the guide for both officers and soldiers and constitute the standard by which their efficiency is to be measured."[68]

Organizationally, the *FSR* contained two parts. Part I *Operations* consisted of twelve chapters focusing on how to regulate an army's behavior in war. Part II *Administration* consisted of four chapters that prescribed instructions for supply and replacements, hospitalization and evacuation, military police, censorship, and the military postal system. Eighteen appendices included forms of field orders, march tables for units, abbreviations, and diagrams.

Much of Chapter II was given to the distinction between command and staff. Command was defined as "the authority which an individual in the military service lawfully exercises over subordinates by virtue of rank and assignment." Staff was more for assisting the commander in the control of the Army. In units larger than a brigade, the staff was either a general staff or a technical, supply, and administrative staff. Within the general staff was the chief of staff, who, as the principal adviser and personal representative of the commander, coordinated four staff divisions, "as prescribed in the Staff manual [Training Regulation 550-5]." The technical, supply, and administrative staff included officers of those specialty branches who might be assigned to a headquarters. Chapter IV identified the "combatant arms" as infantry, artillery, cavalry, signal corps, engineers, and air service. The 1923 *FSR* was clear that combatant arms were not to fight alone. Combining the arms was "essential to success" as it enabled commanders to take advantage of the unique capabilities of each. However, the infantry was supreme. All other arms were auxiliaries meant to assist the general mission of the entire force. Chapter XII described not only nighttime "special operations" (consistent with previous *FSR* versions), but also the nuances of woodland combat. Fighting in wooded terrain provided protection from observation and weapons' fire, but increased the hazard of gas concentration because of the density of the undergrowth. In the offense, it was preferred to move around small wooded areas. If an attack was necessary, the edge of the wood line was to be secured as per any attack then used as a staging area. In sparse woods, the attackers would disperse, but in dense areas they were to move in column. Defending a wooded area meant using the terrain as an obstacle to the enemy advance, although the use of "flat-trajectory weapons is seriously impaired, the fire of high-angle weapons is not equally affected." Still, the defender was to consider "tiers of fire," extending the defense by use of sharpshooters (snipers) and placing machine guns among the trees. Given the intensity of the fighting in Belleau Wood (1918), the doctrine writers clearly saw the need to improve upon army tactics in forested areas.[69]

The main topic of Part I, however, was combat, the object being the destruction of the enemy's armed force in battle by concentrating superior ground and aerial assets "at the decisive place and time." Reference to the decisive place and time again reflected Jominian thinking, for the Swiss theorist had argued in the 1830s that a good general must possess *coup d'oeil* (roughly "battle vision"), an innate ability to see combat unfolding, identify the decisive point where the enemy was most vulnerable, and concentrate forces at that specific place to destroy the foe. Success meant application of certain principles of war, especially surprise, which was essential to demoralizing an enemy and a reference to Clausewitz.[70]

Given the AEF's wartime experience with large formations, the employment of armies and corps received considerable attention. Directions included how to establish a successive front through assigning march objectives to subordinate commands. Commanders and staffs were tutored in the procedures required to conduct both offensive and defensive operations, including the use of aviation and cavalry for reconnaissance. Details spelled out the physical space allowances for subordinate units in the attack as 400–800 yards for an infantry battalion and 2,400–4,000 yards for a division. In the defense, frontages were expected to increase and even more so when conducting a delaying action. Achieving formational depth was important, for it allowed a commander to compensate for unforeseen developments and to position reserves. The 1923 *FSR* represented a significant step in doctrinal thinking. The War Department general staff now perceived the battlefield in three dimensions: depth, width, and height.[71]

FSR authors envisioned combat as a series of orchestrated events that developed into contact with the enemy. Attacks began in an assembly position to allow for overall observation and reconnaissance, with the troops facing the general direction of movement. Artillery fire would strike enemy forces 6 miles out, while units advanced in route columns. Each column was to move within a controlled zone of action toward a march objective. Zones allowed commanders to narrow or widen an attacking formation, as the terrain and situation dictated. When contact became imminent, units changed from columns and deployed into smaller combat formations that made use of terrain. In the defense, units were assigned sectors, a means of dividing the battlefield into areas of responsibility. When a defensive area was more formidable due to the advantage of position, the defending force could extend its lines. Conversely, a narrow defense was required for a weaker position.[72]

Offensive, open warfare was stressed above all. A carryover from the 1914 *FSR*'s vaguely defined concept of mobility, the open-warfare concept now had specificity. The concepts of a "meeting engagement" and "attacking a stabilized front" were added. Meeting engagements were situations where the attacker and defender stumbled unexpectedly upon one another. Often quick to develop into a battle, these events required a commander to rapidly assess friendly and enemy force strengths and vulnerabilities. Time was paramount, for while a force commander was contemplating an attack, the enemy might be doing the same or preparing a hasty defense. Commanders were expected to act aggressively, precisely what Pershing sought in open warfare. Stabilized fronts meant launching an attack aimed at forcing the enemy into open ground with "subsequent defeat by the application of the methods of open warfare." This idea was identical to what the 1914 *FSR* had previously advocated.[73]

The 1923 *FSR* not only advanced ideas published in 1914, it also accounted for new technologies such as poison gas and tanks. Poison gas compelled an opponent to use protective masks, which restricted vision and reduced fighting efficiency. But gas was dependent upon terrain and weather conditions to maximize effect, thus it was deemed a temporary means to disrupt an opponent rather than a useful weapon for inflicting casualties. Tanks were described as a means to reduce the amount of time field artillery required to prepare for an offense, owing to their shock value and lower vulnerability to small-arms fire than exposed infantrymen. The *FSR* advised that the main line of a defense should incorporate obstacles and consider optimum avenues of approach for attacking enemy tanks. Moreover, more depth was required when defending, to organize various weapons so as to provide effective antitank protection.[74]

In incorporating major lessons of the Great War, the War Department demonstrated its capacity in conducting an extensive investigation of recent battlefield results. In doing so, the general staff drew upon not only the experience of field officers, but also the ideas of key educators, many of whom were either graduates of or affiliated with Fort Leavenworth or the Army War College. The 1923 *FSR* thus blended the practical and the theoretical to a greater degree than previously seen in doctrine. It was the first doctrine to realize Elihu Root's vision of 1905, an army that not only fights effectively but thinks critically.

The 1923 *FSR was* the intellectual core of the Army, for all service manuals were required to be in accord with it. In their drive to regulate chaos in the aftermath of the Great War, the War Department leadership now shaped ideas contained within the service's principal manual, while also holding sway over the manuals affiliated with each branch of the service. Through doctrine, the War Department had furthered its authority over the service as a whole. Yet, without a war to test it in battle, the 1923 *FSR* was obsolete by the mid-1930s. Evolving military matters nibbled away at the manual's relevancy.

Toward a New *FSR*

The period from the mid-1920s through the 1930s represented an era of considerable change for the Army and its doctrine. Since 1898, mobilizing the American citizenry for national defense had been a thorny issue. The service succeeded in war but found it difficult to expand in size rapidly to meet national emergencies. When war came, the peacetime-based regular Army drew upon the National Guard, a draft, or volunteers. Filling the ranks was one thing, but having trained officers and enlisted men was another, given that the War Department had to plan for war while overseeing the expansion, training, deployment, and employment of a field army consisting of millions of personnel and vast amounts of equipment. Various congressional acts and army programs had been implemented from the early 1900s until 1916. All of them fell short in some ways, thanks in part to a lack of congressional funding. In 1920, Congress passed another National Defense Act, this time delineating among the regular Army, the National Guard, and the Organized Reserves to create a more robust peace establishment. Despite significant changes, the continuing inadequacy of congressional funding from 1921–1935 meant that the act once again failed to measure up to its legal requirements. The regular Army was maintained at half its authorized strength, the National Guard less so. The Organized Reserves was little more than a list of officers. The Army, however, had to obey federal law, thus the act continued to shape keystone service doctrine.[75]

Through the Army War College's various general staff courses, the War Department began a systemic analysis of foreign doctrine. Within the G2 (intelligence) course, officers studied foreign manuals, theoretical writings, and tables of organization and equipment. The G3 (operations) course included extensive comparative studies of military systems (foreign doctrine, training, technology, mobilization plans, and staff procedures), all of which was intended to improve organization, efficiency, staying power, and flexibility. Such studies demonstrated that the educational system had created a learning environment concerned not only with fighting wars, but with understanding fundamental principles from a global perspective.[76]

War College committees studied France, Germany, Great Britain, Japan, and the Soviet Union, as well as other nations including Italy and Switzerland, all filtered through an American cultural lens. If foreign doctrine did not mesh with American political and societal norms, as well as military values, it was often discounted. Still, many foreign ideas were scrutinized and some took root. Throughout the 1920s, virtually every foreign power (save Japan) was concerned with motorization, not mechanization. Global trends in motorization led army leaders to develop a divisional reconnaissance troop, antitank units, and larger artillery pieces of high trajectory, while retaining horse cavalry. Where European armies organized rifle squads around light machine guns, however, the U.S. Army improved the rifle and developed light machine guns as part of a heavy-weapons platoon. Every major nation had adopted the triangular division formation before 1939. This formation removed tanks from the infantry division to create a separate tank division, allegedly to improve mobility and flexibility. The U.S. Army was the last nation to follow suit, in 1939.[77]

In 1928, British theorists including J. F. C. Fuller and Basil Liddell Hart had announced the demise of the infantry because of advancements in armor. Fuller envisioned Britain's experimental Tank Corps as the means for crossing the "deadly ground" that had bedeviled American military experts since the American Civil War. The British theorists saw armor as an offensive weapon designed to drive deep and quickly into the enemy's rear defenses, thus producing not only havoc but delivering a shattering psychological blow that could destroy the will to fight.[78]

Secretary of War Dwight Davis, lobbied by armor enthusiasts Adna Chaffee and George Patton, directed the creation of a tank force based upon the British model. While the Americans borrowed British ideas freely, the armor experiment languished. Peacetime meant lack of money was a major concern for developing mechanized forces and equipment. The service made do with obsolete stockpiles left over from the Great War. The experiment ended after a few months. It served little purpose other than to demonstrate that the Army lacked mechanized forces.[79]

Along with technological changes, force structure modifications made portions of the 1923 *FSR* obsolete. Aviation advocates lobbied for a separate air force. The 1920 National Defense Act had envisioned a robust Air Service, but its size in 1923 remained little more than had been originally authorized. Still, the Lassiter Board, headed by Major General William Lassiter, proposed a force of 2,500 airplanes augmented by airships and balloons with latitude for wartime expansion. In 1925, President Calvin Coolidge appointed Dwight W. Morrow to examine aviation matters. The Morrow Board recommended establishing an undersecretary of war for air activities, with aviation sections housed in the War Department. The addition of two brigadier generals was also recommended, along with the institution of flight pay and the rebranding of the "Air Service" to "Air Corps," thus presenting a more martial image. Ultimately, the 1926 Air Corps Act established an assistant secretary of war for aviation, providing not only civilian oversight but funding within the War Department budget. It also authorized three brigadier general officer positions, air sections within each of the War Department general staff divisions, and a five-year plan for aviation expansion. Although competing with advancements in mechanization, artillery, antitank concepts, and antiaircraft weaponry, aviation benefited enough from patronage and increased spending that a legitimate aircraft procurement program had emerged by 1937.[80]

Force structure, however, was a significant catalyst for a new doctrine. Experimental training maneuvers from 1937–1939, which incorporated the testing of various equipment, resulted in the adoption of the triangular division. Numbering just over 15,000 men, this force was organized along the concept of "threes" reminiscent of Emory Upton's "fours" in the late nineteenth century: three rifle squads per platoon, three platoons per company through the regimental level, with the brigade headquarters eliminated. Perceived as a means to increase mobility as it consisted of fewer infantrymen than the square division of the Great War, the triangular division was more dependent upon motorization. In war, the division was expected to assault with one regiment, be supported by a second, and hold the third in reserve. Although it possessed fewer soldiers than the previous army division, heavy weapons and antitank units imparted the new formation with increased firepower. The triangular division was smaller and less cumbersome, but arguably more lethal than the square division.

In 1929, General Charles Perlot Summerall selected Major General Frank Parker and Colonel Samuel C. Vestal to bring doctrine in line with changes in legal obligations, organizational modifications, and modern technology. Parker initially advocated the publication of a pamphlet on large operations (field armies and corps) to augment the 1923 *Regulations,* believing that this earlier manual was too fixated upon small units. In his opinion, future war required doctrine that described "in orderly and comprehensive fashion the functions and needs of large units and the specific responsibility of their commanders." To that end, he circulated a draft for the service schools to consider.[81]

Major General William D. Connor, commandant of the Army War College, and Brigadier General Edward L. King at the Command and General Staff School immediately took umbrage with Parker's ideas. Although doctrine had been highly influenced by German tactical thought since the late nineteenth century, both general officers wanted nothing to do with "foreign" ideas. They believed that Parker had borrowed too liberally from the French. And, in fact, French Army influence had dominated Parker's thinking. His draft pamphlet was a near-verbatim translation of the French War Ministry's *Provisional Instruction on the Tactical Employment of Large Units, 1921.* A second draft was written that aimed to address the school commandants' objections by incorporating parts of the 1923 *FSR* but retained French ideas. The draft, *A Manual for Commanders of Large Units,* remained unofficial but was nonetheless used as a textbook to instruct students at the Command and General Staff School.[82]

Because portions of the 1923 *FSR* were included in the pamphlet, the Army now had two competing doctrinal manuals within the schoolhouse, a situation similar to the War of 1812 when four doctrinal manuals had been in use at one time. Problems arose from the fact that the 1923 *FSR* stressed open warfare—the offense, maneuver, and the dominance of infantry—while the French-based *large units* favored attrition, firepower, and the defense. By late 1935, the War Department, Fort Leavenworth, and War College leadership deduced that having incompatible official and unofficial doctrine baffled faculty and students who struggled to grasp the intricacies of operations and tactics. The War Department was now under pressure from within the Army to produce a new doctrine by 1935. Eventually, in 1937, Colonel Edmund L. Gruber headed an effort to produce a new manual. Gathering up all available material from within the service and reports from ongoing experimental programs, Gruber requested that units and organizations offer conceptual input. The responses ranged from detailed analyses to total apathy.[83]

Three draft volumes, which built upon the submitted commentary, appeared in 1939. As tentative *FSR*, the draft doctrine was divided into Volume I *Operations,* which regulated combined arms operations and warfighting, and Volume II *Administration,* which governed the management of the force overall. A third volume, *Large Formations,* considered strategic, tactical, and administrative employment of units in war, such as corps and armies. The drafts eventually made their way into the hands of Brigadier General Lesley J. McNair, now the commandant of the Command and General Staff School at Fort Leavenworth. McNair and his staff painstakingly reviewed everything chapter by chapter. McNair was involved to the point where he authored many changes himself. His personal comments also took shape in a formal reply to the War Department. The general staff adopted most of his ideas.[84]

These draft manuals were subjected to far more scrutiny than any previous *FSR*. Because the War Department general staff and the school system had evolved into mature institutions by the late 1930s, both general staff officers and faculty vied for control of content. As with many previous manuals, army organizations were also involved in the review process, which served to stimulate discussion while furthering their agendas. Thus, doctrine writers both benefited and suffered from intense scrutiny and often-competing concepts that required careful handling. Still, the army-wide buy-in approach, a technique dating to at least the early 1850s, meant that every senior officer and school commandant had the opportunity to shape doctrine. The scope of involvement slowed doctrine publication considerably.

Tentative 1939 *Field Service Regulations* Field Manual 100-5 *Operations*

On 9 September 1939, by order of Secretary of War Harry Hines Woodring, General George C. Marshall, the chief of staff, approved the 1939 *FSR* as tentative doctrine. Marshall's approval was historically significant, not only because it represented the arrival of yet another keystone doctrine, but because it led to a change in format from a stand-alone manual to a series of publications. It was the first significant conceptual change since Winfield Scott intellectually separated tactical manuals from administrative regulations in the early 1800s.

The tentative 1939 *FSR* now contained numerically designated "field manuals" or FMs, including FM 100-5 *Operations*, FM 100-10 *Administration*, and FM 100-15 *Large Units*. The subdivision of this particular *FSR* into separate manuals and just one publication with internal parts was significant. The service leadership had not only grouped knowledge and guidance by specific educational purpose, but also acknowledged that there was sufficient difference among tactics, administration, and the employment of large formations to warrant separate publications.[85]

Because of the army leadership's decision to compartmentalize the *FSR* into three distinct manuals, the core of keystone doctrine for regulating the chaos of war shifted from the *FSR* as a single source to the principal warfighting manual, FM 100-5 *Operations*. Unlike the previous *FSR* from 1905 forward, 1939's FM 100-5 contained a warfighting philosophy that considered national strategy, not just operations and tactics. Acknowledging the influence of Clausewitz, the manual was the first to link the political objects of government in waging war with the Army's role in achieving them. War was defined as "the art of employing the land forces of a nation in combination with measures of economic and political constraint for the purpose of effecting a satisfactory peace." Future war required flexibility of mind to adapt to potential change, thus the manual noted that conflict was "based on the skillful adaptation and application of the techniques of modern science. New means are always giving it a different form. These developments must be anticipated, and their influence must be correctly evaluated and promptly exploited."[86]

There is little doubt that service leaders maintained an offensive mindset in advocating that war's ultimate objective was enemy destruction. But the service also admitted that this endeavor was complicated by large and widely dispersed modern armies. Thus, "there may be intermediate objectives to do so [destroy the enemy] rather than one, tremendous attack." In a change from the 1923 *FSR,* the 1939 FM advocated conducting a series of major operations rather than one attack to shatter the enemy's will.[87]

A bold change from the 1923 *FSR* was the increased emphasis placed upon sociological and psychological considerations in war. "In battle," explained the authors, "man is governed by instinct rather than reason. He prefers to fight in a group. He has a fear of the unknown, especially at night and when alone." National ideals and service values were intended to overcome fear and soldiers were to be imbued with "symbolic ideals emplaced by tradition and national culture. [The soldier] will fight for these ideals when he is aroused. His instinct of self-preservation will entice him to flee from danger but he is deterred from flight by the disgrace he feels in the eyes of his comrades. He wants to earn their respect and esteem as measured by the standards of military conduct accepted by the group." In addition, "in the training of the individual soldier, the essential considerations, therefore, are to integrate the individual into a group and to establish for that group a high standard of military conduct and performance of duty." Doctrinally, the army leadership had recognized Elihu

Roots's concept of the individual soldier as part of a larger machine that functioned in accordance with service culture ideals of duty, honor, and country.[88]

The 1939 FM 100-5 also redefined the concept of offense. Tactical objectives no longer required firepower at the point of penetration of a defending line. Instead, the manual acknowledged that "an objective may sometimes be attained through maneuver alone; ordinarily it must be gained through battle." The statement played to mobility advocates who saw speed as a way to encircle opposing forces quickly, although the manual still encouraged "fire and movement" in the attack. As with the 1923 *FSR*, the service preferred to envelop a flank rather than undertake a frontal assault, arguing that the main attack should occur on a narrow front using assaulting infantry reinforced by artillery. Tank units supported infantry assaults while cavalry, now both in motorized and horse versions, scouted ahead or attacked the enemy's flanks and rear. As with the 1923 version, the attack ended with the pursuit. But instead of an aggressive pursuit to destroy the enemy, the 1939 manual advised that it "must not be premature. You must first decisively defeat the enemy." Increased mobility often entailed bypassing pockets of enemy resistance. This meant that an overly aggressive commander had to weigh such matters carefully. An enemy force was considered defeated by "the capture of critical objectives, reports of front line troops taking prisoners, capturing abandoned equipment, reduced enemy artillery [fire]," and the withdrawal of enemy forces.[89]

While primarily an infantry-dominated doctrine, FM 100-5 acknowledged that all arms and services contributed to success in war. One significant contributor was the tank, viewed primarily as an offensive weapon to be given specific objectives and used in mass. Another was mechanized cavalry (distinguished from horse cavalry), a force equipped with armored vehicles and radios. Mechanized cavalry was seen as an intervening force that contributed toward reaching a decision in battle. In that regard, mechanized cavalry was not considered capable of holding an objective for long by itself. Its purpose, as with horse cavalry, was reconnaissance, counterreconnaissance, and security for large units. However, it could also be moved rapidly to the decisive point where its automatic weapons might produce a powerful effect.[90]

The air corps reflected the maturation of its evolving doctrine through more specificity of purpose and missions. Its general mission, as part of the field force operational plan, was to take advantage of its inherent mobility and extreme range of firepower by operating "in three dimensions." Its basic tactical functions were labeled air attack (striking ground targets), air fighting (destroying opposing aircraft in flight), and air reconnaissance and observation (gathering information from the air visually and through photography). Military aviation was further subdivided into principal roles: combat, reconnaissance, observation, liaison, transport, and training and special purpose (not affiliated with combat). Combat aviation missions included pursuit to intercept hostile aircraft or to escort bombers, and bombardment (light or heavy) depending on the target.

Chapter 7 introduced troop leading procedures, which described how a commander analyzed the mission and the situation to determine how to conduct an operation. Once a decision was reached, however, it was not to change without a compelling reason because so many smaller actions depended upon it. At the same time, commanders were advised to avoid stubbornly adhering to a previous decision because to do so might result in failure. As the manual noted, "the art of troop leading also lies in knowing when to make a new decision."[91]

The Army's view of the conduct of both offensive and defensive conventional combat operations changed little from the 1923 edition, though instructions for delivering the main attack were clarified. Main attacks were to occur in narrow zones supported by artillery and heavy infantry weapons. Fire was to be concentrated on a series of objectives, areas to be taken and secured. Timing and the use of tanks and reserves was considered important. If the commander was uncertain as to when and where to attack, commanders of large units were permitted to disperse their forces in depth while holding out reserves and tanks. Artillery was to be used in a more centralized rather than decentralized manner to ensure more control over the ordnance. This was designed to enable artillery to mass fire on the enemy as the situation became less muddled and, thus, allow the main attack to commence.[92]

Guerrilla warfare, as informal practice, had previously received little to no consideration in keystone manuals. In Chapter 13, however, guerrillas were described as recognized belligerents whose strengths lay in using mobility and smaller numbers to strike a superior force or suppress an uprising or rebellion. Guerrillas required enterprise and reliability to be effective and were considered to operate best in small detachments. They attacked after careful planning, constant enemy observation, and through ambushes. After an attack, guerrillas were expected to withdraw rapidly along various routes to a prearranged rendezvous point, taking prisoners for interrogation. Combating guerrillas involved constructing roads and trails to provide access to remote areas where guerrillas operated. Establishing an efficient intelligence service, one that "studied the national characteristics of the people, their leaders, their political attitude, and religious practices," was essential.[93]

Because the secretary of war had endorsed the new doctrine as a tentative manual, Chief of Staff General George C. Marshall directed major field commanders to comment on its contents. Seventeen thousand copies were distributed together with a letter requesting feedback by 1 March 1940. Unlike the well-received 1923 *FSR*, field commanders and their subordinates lambasted the new doctrine's contents. Comments ranged from criticism that the manual had been written without a detailed revision to complaints that its contents were too academic. Air Corps advocates blanched at being relegated to a supporting arm of the field commander and not an independent air force. Marshall and his staff were sufficiently shocked by the reaction to call for revision under McNair and the Leavenworth faculty. Marshall established 1 January 1941 as the suspense date for a new draft.[94]

Compounding the issue of a lack of enthusiasm for the 1939 initial draft was the War Department general staff's inability to blend the overarching service doctrine with those of the individual branch schools. Service schools developed their own tactics related to their specific branches, which, in theory, dovetailed with the keystone doctrine. However, the War Department G-3 rarely held the various branch schools to task, preferring to publish training directives instead of requesting school doctrine for a thorough compliance review. Given differing priorities within the War Department and the schools, the various doctrinal manuals were not "nested" with FM 100-5. The result was a keystone doctrine that failed to integrate the needs and missions of branch schools.[95]

After war officially commenced in Europe on 1 September 1939, service comments and the success of the German Army moved doctrine in a different direction. In initiating what the press called *Blitzkrieg* (lightning war), the Germans had confirmed what an American attaché had concluded after observing their maneuvers in 1924: "The [German] Army, according to our standards, [is expected] to be above average in training, discipline and fighting efficiency; superior in orders, staff work and high command."[96]

As the epitome of combined arms warfare, *Blitzkrieg* tactics led to stunning victory in defeating Poland. American Army officers, however, initially discounted German success as a fluke; surely a "first-rate" opponent would have done better. Once Germany unleashed its war machine against France in spring 1940, service leaders came to understand what the future held. The War Department ordered that maneuvers be held in Louisiana to test the 1939 tentative doctrine, large-scale operations, and the incorporation of new technology. On the day the so-called "Louisiana Maneuvers" ended, the French Army collapsed. Stunned, Marshall called for the creation of an armored force and appointed Adna Chaffee commander of an experimental armored division.[97]

Field Service Regulations FM 100-5 Operations 1941

In light of the Roosevelt administration's efforts to prepare for a potential war in Europe, the War Department and Fort Leavenworth produced the 1941 version of FM 100-5, along with FM 100-10 and FM 100-15. The new field manual, approved on 22 May 1941, was the combined product of the 1939 version, service criticism, German *Blitzkrieg*, and lessons taken from the Louisiana Maneuvers.

In the introduction, Marshall informed the service that "knowledge of these doctrines and experience in their application provide all commanders a firm basis for action in a particular situation." Flexibility, however, was needed in applying doctrine, for "set rules and methods must be avoided" because "they limit imagination and initiative." In asserting that principles of war exist but they must be applied in context, Marshall combined the thinking of both Clausewitz and Jomini.[98]

The manual consisted of sixteen chapters. The most significant alterations from the 1939 manual were Chapter 3 (Leadership), Chapter 4 (The Exercise of Command), which specified command in battle or "battle command," Chapter 7 (Halts and Security During Halts), Chapter 8 (Troop Movements), Chapter 11 (Retrograde Movements) as a stand-alone chapter, and Chapter 12 (Special Operations), which dropped guerrillas in favor of partisans, and added jungle and desert operations. Chapter 14 (Air Task Forces) catered to air advocacy, while Chapter 15 (The Division) reflected new formations: infantry, motorized, cavalry, and armored. Chapter 16 (GHQ Tank Units) was also added.

Chapter 1 began with an explanation of how to organize the nation for war by establishing function-based geographic areas. The Army was deemed part of a larger national defense establishment in which the United States became the "zone of the interior," home to the mobilization and industrial bases for war. Extending from this home base was the "theater of war," the land, sea, and air space in which wars occurred. The theater of war was then subdivided into various "theaters of operation," areas the War Department designated for conducting military operations. Within each theater, "combat zones" corresponded to areas where a field army, corps, and divisions conducted operations. Linking the national base through the various theaters to the combat zone was the "communications zone," an element of the theater of operations contiguous to the combat zone. Its indirect purpose was to provide a command relationship from the battlefield through the various levels of command to the president and secretary of war. Directly, however, it established supply and evacuation channels, as well as a way to support and maintain an army in the field through various agencies. The 1941 manual was the first to articulate theoretical elements of operational design for war in a practical way.[99]

The new doctrine emphasized the concept of combined arms warfare, a restatement of the 1939 tentative doctrine. Offensive in nature, combined arms allowed the commander to distribute forces into two or more tactical groupings, one for main or decisive attacks in which the greatest offensive power was concentrated to bring about a decision, and a secondary attack (or holding attack) whose sole purpose was to support the main effort. The main attack was to seize a designated objective and destroy hostile forces. The secondary attack was to hold the enemy in position and force commitment of reserves, thus denying the opposing force commander the freedom to reinforce the area in front of the main attack. Main attacks reflected current German thinking, for the attack was to be situated on a narrow front using infantry, artillery, tanks, and other supporting weapons to include combat aviation and deep echelons of reserves. Conversely, secondary attacks were limited in depth, contained few reserves, and maximized firepower to fix the enemy and deny repositioning of forces.[100]

The 1941 FM 100-5 classified attack maneuvers into two categories, envelopments and penetrations. Single or double envelopments as well as turning movements directed the main attack to strike the enemy's flank or rear area while the secondary attack struck the enemy's front. Surprise and superior mobility were essential, the object being to surround forces and destroy them. A penetration passed through a portion of the enemy's defense with the purpose of opening a gap in the enemy line so as to rush through additional forces such as cavalry, armored, and motorized units to strike into rearward enemy positions. Surprise was also a factor, but more emphasis was given to firepower than maneuver in neutralizing enemy defenders as the main attack pushed through. Though larger forces (such as field armies) might have multiple penetrations underway simultaneously or sequentially, the outcome was still the same, to conduct a pursuit to destroy the enemy.[101]

Given its inherent ability to provide increased firepower through ground attack, the 1941 doctrine provided an expanded role for what was now designated the Army Air Forces (AAF) under the command of Major General Henry (Hap) Harley Arnold. The AAF was to fight hostile aircraft, operate in concert with ground or naval forces in land and sea warfare, and engage designated enemy targets. The doctrine acknowledged that combat aviation was "a powerful means to influence battle," but only with reservations. "Because of the speed and powers of evasion inherent in all aircraft," the doctrine counseled, "air fighting is generally of brief duration and the results are often indecisive. As a result, unless greatly superior, aviation is incapable of controlling the air in the same sense that surface forces can control an area and can therefore reduce hostile air operations only to a limited extent." Thus, the War Department acknowledged that air power enhanced ground operations, but only for a limited period with limited results.[102]

Defensively, army forces sought to gain time so they could develop favorable conditions to undertake the offense. The primary means to defend was the *battle position,* a place to be held at all costs augmented by obstacles and the use of terrain. Covering forces were employed in front of the battle position to hide its true location and to delay and disorganize the enemy advance. Battle positions were organized into a main line of resistance, the line in front of which all units concentrated defensive fire, and the regimental reserve line where units awaited commitment to the battle. Between the two lines, companies and battalions organized their reserves and support troops. To control fire and the maneuver of defending forces, units were designated specific sectors within the battle position, as well as boundaries to coordinate fire support and prevent dispute over which unit was responsible for a key piece of terrain within an enemy's avenue of approach.[103]

The 1941 FM 100-5 revealed that the service was seeking to regulate virtually every aspect of modern warfare. In attempting to anticipate every possible contingency, the manual was both a guide and a reference work, an operational and tactical education stuffed between two covers. For those who grasped the various concepts and nuances within its pages, the 1941 manual was a highly intellectual art form that contained the collective experience and wisdom of the Army, as an institution. Still, many missions were overlooked. They remained informal practice, as World War II demonstrated.

The manual remained in effect for three years. It served as the fundamental philosophical tome for training and operating an army in the field through the early years of World War II. America's industrial capacity for producing war materiel met or exceeded the Army's vision for employing an offensively oriented, combined arms doctrine. Between 1941 and 1945, American industry produced the tanks, trucks, radios, weapons, and other equipment necessary for executing the war that doctrine had envisioned.[104]

The 1941 FM 100-5 in War

Faced with global war after the Japanese attacked Pearl Harbor, Hawaii, on 7 December 1941, the president of the United States and the War Department organized the war effort into a zone of the interior, theaters of war, theaters of operation, and combat zones, as doctrine prescribed. Although the Army's involvement in a two-ocean war proved difficult, doctrine was comprehensive enough to provide sufficient tactical guidance for most operations.

The first significant combat action was in North Africa, preceded by beach landings against a hostile foe. Doctrine did not address amphibious landings *per se,* but Chapter 12 of FM 100-5 addressed an attack on a fortified locality. Such an operation was envisioned to unfold in four phases:

1. Reducing the hostile outpost system and gaining close contact with the main position.
2. Breaking through the fortifications at the most favorable point.
3. Extending the gap by isolating and reducing hostile emplacements on its flanks.
4. Completing the action by moving mobile reserves through the gap to complete the encirclement and isolation of remaining fortifications while continuing the action against them from the front.

In conducting amphibious assaults from North Africa to Normandy, army forces followed this operational and tactical scheme.

Chapter 12 devoted five paragraphs to desert warfare. Variances of desert terrain and its impact upon navigation and mobility were emphasized. Motorized and mechanized units were preferred as they were less dependent upon water sources than infantry and horse cavalry. Camels, though, were recommended. Air observation was deemed improved, since concealment was considered difficult to achieve. Army doctrine writers imagined that the general principles for conducting offensive and defensive operations held true, with special emphasis upon controlling water sources and the use of fast-moving units to encircle or turn opposing forces.[105]

Once army forces were ashore, contact with the Axis armies during the North Africa campaign (1942–1943) proved that the service was capable of fighting an offensive combined-arms war, although tactical inexperience led to difficulties such as an inability to control units. In the defense, many commanders had difficulty executing their plans because of tactical inexperience, such as at Kasserine Pass (1942). Army force commanders were less familiar with directing large units above division and found that controlling them in combat was difficult. Thus, while the doctrine was sound, its execution was often flawed.

In Sicily and Italy (1943), army forces faced a stalwart opponent defending in ruggedly steep terrain. FM 100-5, however, provided guidance in Chapter 12. Mountainous terrain was deemed to offer no insuperable obstacles if troops were properly equipped, clothed, supplied, and trained. The manual acknowledged that terrain often restricted movement to within valleys and narrow winding roads, which required taking the heights to dominate them. Commanders were advised that success often meant adapting their units to the terrain because firepower was diminished and concealment increased. Infantry forces were paramount, cavalry less important, and high-angle artillery necessary. Engineers were required to construct new roads and to improve old ones. Offensive operations were to favor more of a frontal attack because of terrain compartmentalization, while defenses were strengthened by the difficult terrain. The army in Sicily and Italy found that executing combined-arms warfare proved as difficult as the 1941 manual had revealed, for speed and flexibility were often reduced to a halt in the valleys while infantry forces cleared the surrounding high ground assisted by artillery and aerial bombardment. The Battle of Monte Casino (1943) in Italy was one such example.[106]

The Normandy invasion of 6 June 1944 and the subsequent breakout and pursuit across France beginning in July characterized the 1941 doctrine's conceptualization of combined-arms warfare and the attack of fortified positions. After the lodgment was established, troops fought through the French hedgerows in what was a tactical penetration using infantry and support arms. Striking toward Cherbourg and Caen, army forces "wheeled" or enveloped German forces through maneuver supported by artillery and air power. On 25 July, ground and air forces struck the German positions southwest of Caen. Three infantry divisions broke the German line. One motorized and two armored divisions then exploited the gap, with forces turning west, south, and east. American units holding the flanks of the penetration area fought off German counterattacks, as other forces reached Argenten on 21 August. In a classic turning movement, the Americans bypassed German units to create a pocket, from which the enemy extricated themselves at the loss of 50,000 prisoners. By now, other American units, as well as allies, were in pursuit of German forces seeking to cross the Seine River. Ultimately, the pursuit continued across France, became delayed in the Ardennes and Hurtgen Forests in December 1944, and then carried forward across the Rhine River into Germany.[107]

In support of this advance was allied air power, of which the U.S. Army Air Forces contributed more than 2,000 bombers stationed in England plus fighter escort. American daylight bombing of factories had produced no significant reduction in German logistical strength by mid-1944, though later attacks on railways and synthetic petroleum plants proved more effective. German coal shipments were crippled, drastically affecting steel production. The bombing of the Romanian oil fields significantly weakened German fuel supplies for aircraft, motorized, and mechanized forces. The strategic bombing campaign contributed to Germany's defeat, as doctrine noted, because "the mobility, speed, and range of aircraft make possible their rapid intervention at critical points in a theater of operations . . . and deep incursions into enemy territory."[108]

In the Southwest Pacific, army troops encountered a variety of terrain in fighting the Japanese. Invading islands was not addressed, but doctrine did cover jungle operations as the blending of wooded terrain with night combat because of limited visibility and decreased movement. Rail networks and roads were few and maintaining a given direction of advance was difficult. Disease, insects, heat, heavy rains, and other tropical issues meant that control and maneuver was often impossible and aerial observation restricted. Where combined-arms warfare focused upon motorized and mechanized speed to bypass enemy forces, a jungle operation limited ranges for weapons. Close-in fighting performed by infantry was the norm. Pack howitzers, small, compact artillery capable of mule transport, were the primary means of fire support, along with mortars and light bombardment from the air. Surprise and ambushes were paramount, as was security. Army forces came across all of this and more in the Pacific War, with Buna (1942–1943) being one example. There, troops of the 32d Division, along with Australian allies, slugged their way through jungles and mountains in warfare characteristic of what FM 100-5 had described.[109]

Field Service Regulations FM 100-5 Operations 1944

On 14 June 1944, the War Department produced another FM 100-5 *Operations*. In copying the 1941 version, the new doctrine nonetheless included additions that reflected the War Department's views of changes in operational concepts and technology derived from combat experience. That the service could alter its keystone doctrine during wartime indicates that the War Department general staff and the education system continued to gather and analyze reports from the field and react to them, as Pershing had done immediately following the Great War. However, whether or not the 1944 doctrine had any measurable effect upon the World War II army remains debatable since the war ended in August 1945. Japanese officials signed the formal surrender documents on 2 September.[110]

The 1944 FM 100-5 eliminated Chapter 14 (Air Task Forces) and replaced it with a new Chapter 3 (The Air Forces). Relocating the chapter suggested that air power had risen in importance within the Army. Organizationally, Army Air Forces were divided into a strategic air force, a tactical air force, the air defense command, and the air service command. Their mission was to destroy opposing air forces, as the largest aviation tactical unit. While its overall purpose did not change from the 1941 edition, the flexibility of air power was determined to be its greatest asset.[111]

The strategic air force employed heavy bombers, but they were no longer considered to be "flying fortresses." Vulnerability to enemy air and ground attack meant that doctrine now included long-range fighter escort to further bomber effectiveness. A large portion of the strategic air force was to destroy enemy national aircraft factories or conduct long-range reconnaissance of sea approaches to the combat zone. Completing the strategic air force was the tactical air force. Composed of light- to medium-bombardment units, fighters, reconnaissance assets, and aircraft warning units, the tactical air force's mission was achieving air superiority. Air superiority is local and ephemeral, for gaining control of the air is one thing but sustaining it is another. Still, air superiority was intended to prevent the movement of hostile troops and supplies into the theater of operations and especially into the combat zone, thus it enhanced both the ground force commander and other air force assets. Fighters fixated on destroying opposing aircraft, while light and medium bombers destroyed aircraft on the ground and their bases. Air-defense command included fighter aircraft, antiaircraft artillery, searchlights, barrage balloons, and air and aircraft warning devices such as radar. Its mission was to prevent, oppose, and reduce the effectiveness of hostile aircraft through active defensive fire and passive means such as camouflage and dispersion. The air-service command was responsible for the procurement of aircraft. It was also responsible for all logistical functions associated with processes such as repair, maintenance, salvage, and other operations. Conceptually, the 1944 FM 100-5 had given the air force sufficient autonomy and structure to create a separate service air force in the future.[112]

Chapter 13 contained one addition, amphibious operations. Described as a joint land, naval, and air action, the purpose was to establish a beachhead in order then to carry on a normal [conventional] type of war. Long-range meteorological forecasts were essential, as well as secrecy, centralized planning, and decentralized control once the landings commenced. Complexity demanded extensive training and cooperation among the services, to include full-scale rehearsals. The most important phase was the approach to the beach, as hostile fire might break up boat formations and cause numerous casualties. During the landing, effective communications allowed the extensive coordination required for naval gunfire and air support to neutralize enemy fire, while the use of engineers to clear obstacles added to a safe landing for the force. Air superiority over the beaches was essential. Airborne troops were used to capture enemy defenses and airfields, as well as to delay the movement of enemy reinforcements to the landing site. Naval forces protected the beach from the sea, while tanks cleared beach defenses and attacked hostile local reserves. Sustaining the beachhead meant planning for resupply over the beach from off-shore vessels but also securing a port facility as soon as possible.[113]

Airborne troops received a dedicated chapter. Chapter 14 described airborne forces as ground troops specially organized, trained, and equipped to use air transportation to enter a combat zone. An airborne task force consisted of aircraft designed to drop paratroopers plus air-landing gliders.

The basic organization was the division, which constituted regimental combat teams and glider units with intrinsic infantry, artillery, engineers, and service elements. Airborne units were considered to be theater assets under the theater commander. Due to the uniqueness of their mission, they were not to be delegated to lower headquarters. Missions included seizing and holding key localities, attacking the enemy rear area, blocking or delaying enemy reserves, capturing airfields, creating diversions, reinforcing friendly units, or seizing islands not accessible to ground forces. Airborne units were not to remain in combat longer than three to five days, due to their need to resupply and reequip for future missions.[114]

As early as 1941, Lieutenant General Lesley McNair had sought to make divisions more fluid in maneuver warfare by reducing their size. Specialized units required for specific missions only, such as anti-tank or air-defense units, were removed or "streamlined" out of the infantry division and formed into "pools" of specialized nondivisional elements. Higher headquarters, typically corps and field armies, then allocated the pooled units to a division for a particular mission or employed them en masse on the battlefield as the situation warranted. In the 1944 doctrine, following McNair's methodology, tank destroyers and nondivisional armored units (Chapter 16) replaced the "General Headquarters Tank Units" of the 1941 edition. Tank destroyers were organized into brigades, groups, or battalions armed with self-propelled or towed tank-destroyer weapons. Highly dependent upon mobility more than armor for protection, their mission was to destroy enemy tanks through direct fire in both the offense and defense, as well as retrogrades. Tank-destroyer units were also used to defend beaches, to destroy enemy pillboxes (positional defenses generally made out of reinforced concrete), and to support infantry by direct fire in assaults. Tank destroyers were described as best used in mass and held in readiness to strike opposing tanks quickly. Tank-destroyer battalions were used as a unit but could be parceled out in companies. Nondivisional armored units consisted of armored groups, tank battalions, armored infantry battalions, and armored field artillery battalions attached to corps and divisions. They were to be employed as any armored force.[115]

Army Doctrine at War's End

Between 1905 and 1945, the Army and its doctrine underwent considerable change. The realities of the Spanish-American War and subsequent American government engagement in world affairs had led to two global wars in four decades. Army doctrine had adapted to these changes through the vision of Elihu Root and his restructuring of the Army, the War Department, and the army education system. By 1945, doctrine had become an effective means to organize, train, and fight a service that employed combined arms warfare in three dimensions.

By September 1945, however, the army juggernaut that had won the largest war ever in history was in disarray. The dropping of two atomic bombs upon Japan via Army Air Forces strategic bombers fueled the cries of air power advocates for a separate service. Naval warfare supporters soon joined in the discussion, contending that strategic nuclear weapons, not land armies, were now the technology of choice for achieving national foreign policy objectives. In 1946, less than six months after the end of World War II, many policymakers wondered if a land army was even necessary to win future wars.

COLD WAR DOCTRINE: FROM ARMY OPERATIONS TO MULTINATIONAL AND MULTISERVICE OPERATIONS, 1945–1991

Doctrine provides a military organization with a common philosophy, a common language, a common purpose and a unity of effort.—General George H. Decker, Speech to U.S. Army Command and General Staff College, 1960

During World War II, army forces successfully executed land and air operations in Europe, the Mediterranean, and Asia through offensive-minded combined-arms warfare. In reflecting upon what the Americans had accomplished, former British Prime Minister Winston Churchill called the U.S. Army a "prodigy of organization," the envy of the Allied powers. Such high praise was due, in part, to the 1941 and 1944 editions of *Field Service Regulations* (*FSR*) Field Manual (FM) 100-5 *Operations,* the intellectual framework that regulated the Army in war.[1]

Although War Department expertise and the service's educational system had produced a doctrine suitable for defeating the Axis powers, atomic technology heralded a new era. After months of aerial bombing failed to overpower Japan, Colonel Paul W. Tibbets and his B-29 crewmen exploded an atomic bomb over the city of Hiroshima on 6 August 1945. A second detonation over Nagasaki followed on 9 August. Japan capitulated. The American scientific community had weaponized the atom, but it was the Army's strategic bombing doctrine and an appropriate airframe that delivered it.

Changes in the international order soon commenced. On 24 October 1945, the anemic League of Nations passed the challenge of international discourse to the United Nations (UN), an organization dedicated, in principle, to ending wars. Talks between the Soviet Union and the Western allies over the status of postwar Europe only heightened mutual suspicions over motives and security threats. By 1947, the "Cold War" had emerged with its accompanying scramble for allies and markets. Empires fragmented, new nations appeared, and two global superpowers rose to dominance.[2]

The 1944 FM 100-5 did not address the Army's role in a postwar world. War Department leaders initially were not concerned, for demobilization and occupation duties occupied their time. Still, Congress and the general public began to perceive the Army's ground troops as obsolete, believing that nuclear bombs had brought Japan to its knees.

In truth, by early 1945, General Curtis LeMay had already concluded that strategic airpower alone had achieved national-security policy objectives. The firebombing of Japan had been as much a political campaign to assert air ascendancy over American land and naval forces as to compel Japanese submission. Generals Henry H. (Hap) Arnold and Carl A. Spaatz testified before Congress that nuclear weapons delivered by bombers had achieved national policy objectives while reducing American casualties on the ground and at sea. The two prominent generals' testimony reinforced the view that the ground force was obsolete.[3]

Such matters weighed heavily upon General of the Army Dwight D. Eisenhower as he assumed his army chief of staff duties in November 1945. Through President Harry S. Truman's directives and various War Department circulars, Eisenhower began an intensive internal service review not only to contemplate future war, but to justify the service's *raison d'être*. His revisionist efforts led to greater autonomy for the Army Air Forces to deflect the call for a separate service and to create research and development programs to exploit new technology.[4]

In 1946, General Joseph W. Stilwell's test board concluded that the 1941 Pearl Harbor attack was a surprise assault typical of many wars. Strategic surprise had crippled the Pacific Fleet at Pearl Harbor; such an event seemed repeatable in the future. The next time, Stilwell argued, American forces must retaliate with bombing, long-range missiles, and biological weapons.[5]

Given the changing international situation and in light of the new threat posed by nuclear weapons, Congress overhauled the national defense establishment, the first major change since 1905. The National Security Act of 1947 provided for a National Military Establishment headed by the secretary of defense and included the Joint Chiefs of Staff representing the

Departments of the Army, Navy, and, now, a separate Air Force. The War Department general staff became the army staff. Although the Army lost its air forces, it shed the intraservice cultural tension that had long divided infantry advocates and the champions of airpower. Internal debate was now transformed into an external struggle between separate, and presumably equal, service departments over funding and missions.[6]

The 1944 *FSR* FM 100-5 *Operations* in Greece and the Philippines

While service leaders debated the nuances of atomic and conventional warfare, Cold War dynamics sent army advisors to Greece and the Philippines. The 1944 FM 100-5 contained sufficient guidance for army forces to address a number of operational situations. Chapter 13 specifically concerned partisan warfare, small independent or semi-independent forces operating against a superior enemy in small strike forces. The manual said nothing, however, about the Army's role in support of government agencies or the formation of joint planning groups. In that regard, informal practice held sway.

In the Communist-led Greek insurgency (1945–1949), army troops operated in defense of the Truman Doctrine to free populations from armed minorities or outside interventions. Under President Harry Truman's American Mission to Aid Greece (AMAG), Major General William G. Livesay established U.S. Army Group, Greece. Although intended to function only as observers, army personnel soon took control of various Greek ministries. In December 1947, Pentagon officials created the Joint United States Military Advisory and Planning Group, Greece (JUSMAPG) to assist the Greek National Army, the Greek National Defense Corps, and Home Guard. Greeks received training in guerrilla, patrolling, and ambush tactics, as well as conventional operations, all of which had a doctrinal basis contained within FM 100-5.

The Philippines insurgency (1945–1955) originated with a peasant uprising in response to horrific socioeconomic conditions. Communist-led Huks, inspired by American doctrinal infantry manuals, formed into bands of 100 men posing as innocent civilians by day and attacking Filipino government agencies by night. The U.S. government created a Joint United States Military Advisory Group (JUSMAG) in Manila in 1947, but its commander, Major General Leland S. Hobbs, had little authority to provide aggressive assistance until 1950. American advisers expanded the Armed Forces of the Philippines by creating twenty-six battalion combat teams capable of independent operations. Thanks to a combination of American and Filipino military and civic action, the Huk movement lost its support base by 1952. By 1955, the insurgency was all but defeated.[7]

As the Army dealt with Communist-led insurgencies in Greece and the Philippines, the army staff evaluated the service's recent World War II experiences to situate them within the emergent national security environment shaped by the burgeoning Cold War and the presence of atomic weapons. In 1949, the arrival of Chief of Staff General Omar Bradley marked the Army's future role as a force to be used in securing strategic bases for the U.S. Air Force via the use of airborne forces. U.S. Air Force strategic bombers would be responsible for delivering firepower by employing nuclear ordnance. The shocked enemy was expected to succumb to army units subsequently unleashed in World War II–like ground invasions. As far as Bradley was concerned, however, nuclear weapons were just another tool in the toolbox, an additional means for ground forces to achieve national objectives.[8]

The *Field Service Regulations* FM 100-5 *Operations* 1949

On 15 August 1949, Chief of Staff General Omar Bradley approved the latest version of FM 100-5 *Operations,* a doctrinal text reflecting his views. Nuclear weapons were not mentioned anywhere in the manual, as the Army had yet to acquire them. A product of Fort Leavenworth, Kansas, and the army staff, the 1949 FM 100-5 blended the 1944 manual with new Cold War realities. Decisive warfare was to be accomplished through a war of movement akin to the 1914 *FSR* open-warfare concept advocated by John J. Pershing in 1917–1918. But, in a direct affront aimed at the Air Force, doctrine advised that any new technology merely extends existing organization capabilities. In exceptional cases only, "a [technological] development may possess potentialities, which dictate radical review of the conduct of tactical operations. Thus the crossbow, firearm, the machinegun, and the airplane, in turn resulted in major changes in the tactical doctrine of the periods." This qualified statement represented the army leadership's grudging admission that regulating the chaos of war through doctrine required continually adjusting its tactical concepts to ever-changing technology. Such situations, however, were considered to be historically rare.[9]

Technology aside, the 1949 manual assessed that the Army remained the dominant service. Its field units were to be properly organized, trained, and equipped for separate service combat operations but subject to support of the Air Force and the Navy, especially with movement overseas and missions such as airborne drops and amphibious assaults. Although the 1944 manual had indicated that the Army was reliant upon the Army Air Forces and Navy for conducting certain operations such as beach invasions, the 1949 doctrine truly recognized that the Army was strategically dependent upon other services for transportation.[10]

The 1949 FM 100-5 explained the Army's national-security mission as the destruction of opposing armed forces, their military effectiveness, and will to fight. Clearly shaped by the experiences of World War II, the field manual espoused the continuing dominance of infantry. Infantry divisions fixated on maneuver enabled by the fire superiority of artillery, combat aviation, and supporting armored or armored-infantry elements. As with the 1944 manual, army forces were to envelop a flank while artillery fire and air support massed against enemy defenses. Secondary attacks engaged enemy troops along a broad front to prevent their repositioning and affecting the main thrust.

Nineteenth-century Jominian theory was further introduced through a discussion of nine principles of war. The intent was to impose an intellectual framework of lasting truisms upon the Army that no commander could ignore and expect to win. The *objective* (to destroy the enemy and his will to fight) was the first priority. *Simplicity* was also important, thus all plans should be simple and direct. No army could fight without *unity of command*. This meant that the senior officer was in charge unless another was specified. Imposing freedom of action and a commander's will upon the enemy lay in the *offensive. Maneuver* required orchestrating the offensive, mass, economy of forces, and surprise to overcome hostile superior forces. *Mass*, the concentration of forces on the ground, on the sea, and in the air, meant applying resources at the critical place and time in a decisive direction to attain an advantage. In order to concentrate in one place but have resources available for other missions, *economy of forces* meant using just enough of the proper assets to get the job done. *Surprise*, the unexpected arrival of friendly forces upon the enemy, meant denying information or deception plus rapid movement and power in execution. Conversely, *security* prevented the friendly force commander from being surprised.[11]

Although infantry reigned supreme, the doctrine did advocate combined arms warfare. In arguing that no branch could win wars alone, FM 100-5 established that infantry, armor, and armored cavalry (a mobile force capable of reconnaissance, attacking, and defending) should attack the enemy defense, while the artillery, to be equipped with missiles as well as cannons, fired in support of maneuvering units. Commanders were directed to attack "a physical objective such as a body of troops, dominating terrain, a communications center, or lines of communications or other vital areas in the hostile rear." Maneuver remained paramount to offensive operations. Occasionally, however, the manual advised that maneuver and firepower might fail to destroy the enemy. In that case, close combat was required, closing with and fighting the enemy hand-to-hand. Regardless, the offense was the decisive form of war. The defense was but a temporary measure until transition to the offense occurred.[12]

A conventional doctrine, FM 100-5 devoted only two pages to unconventional partisans. The Army's newly resurrected force, Rangers, was not mentioned. Having disappeared from service by the 1830s, Ranger units emerged in World War II. They, and other commando-style units, had contributed to operational success at Normandy and elsewhere. The 1944 doctrine, however, made no mention of them. Ranger operations, as far as keystone doctrine was concerned, were informal practice. Separate guidelines existed for their use.[13]

The 1946 Stilwell panel's finding that future war would be characterized by a surprise attack launched by a foreign power impressed Bradley enough that he ordered the "Lessons of the Pearl Harbor Attack" added as an appendix. Twenty-five principles were listed, including operational and intelligence work, supervisors not taking anything for granted, supplying information, delegation of authority, alertness, and other factors. The intent was that the principles "be studied throughout the Army and that they be explicitly enunciated in appropriate field manuals and other publications."[14]

With the 1949 FM 100-5, the Army now possessed a doctrine for general global war commensurate with the reorganized Department of Defense. The manual conveyed the army leadership's dependency upon the Air Force to prepare the battlefield using nuclear bombs. The Army would then mop up. That thinking was in line with fighting a nuclear battlefield in Cold War Europe in conjunction with the recently formed North Atlantic Treaty Organization (NATO). Despite the Soviet Union's detonation of a nuclear device in September 1949, the army staff remained confident that the current doctrine was appropriate for an offensive-minded, World War II–based force designed to outmaneuver or physically annihilate the enemy. Conflict on the Korean peninsula (1950–1953) would test that notion.[15]

The 1949 FM 100-5 in Korea

As the 1949 doctrine had envisioned, the next war began with a surprise attack. On 25 June 1950, the North Korean Army invaded South Korea. In this Asian civil war, however, the service's doctrine was soon found to be awry. Whereas doctrine had envisioned an immediate nuclear response by air force strategic bombers, no such rejoinder occurred. As the first commanding general of a United Nations (UN) war, General Douglas MacArthur followed doctrine and pressured President Truman to use nuclear bombs against Chinese airbases harboring fleeing North Korean aircraft. In a 30 November 1950 press conference, Truman, who considered Korea to be a police action in the tradition of President Theodore Roosevelt's famous corollary rather than a war, indicated that atomic bomb usage was under consideration. He later recanted amid British misgivings and fear of sparking yet another large-scale Asian conflict that might trigger a Soviet invasion against a fledgling NATO. Truman thus limited the struggle in Korea by refusing to unleash nuclear weapons.[16]

Neither the Army's senior leadership nor the 1949 doctrine had anticipated that a president would circumscribe the use of military ordnance. General Matthew Ridgway remarked that the concept of a limited war had never been considered. By doctrinally imposing upon the service a culture of maneuver and overwhelming firepower based upon air force nuclear weaponry, the army leadership had failed to foresee that diplomatic and political pressures might prevent the use of all weapons available within the arsenal.[17]

In strategic and operational terms, the Korean conflict was not what the 1949 FM 100-5 had envisioned. As a UN war, the Army fought alongside allied armies. But coalition warfare was not mentioned in the field manual, meaning multinational operations were unregulated informal practice. By mid-1950, army forces found themselves on the defensive. Doctrine had envisioned using atomic weapons and fire superiority to allow ground forces to drive into the enemy's country, annihilate the foe, occupy terrain, and destroy the will to resist. Instead, the service fought for three years first retreating, then attacking, then retreating and attacking again until reaching a stalemate with the North Koreans and their Chinese ally. Reminiscent of World War I, three years of warfare ended with entrenched armies facing each other about where the conflict began.

Korea did not resemble the tactical war of movement described within FM 100-5. Unlike Europe, only one-fifth of the Korean peninsula was arable land, most of it in the south and west. Every valley was terraced for cultivation, and the higher rugged terrain dominated fields of fire in the lowlands. Summers were hot, humid, and subject to monsoons. Winters were often brutally cold. The combination of terrain and climate meant a miserable situation for infantry forced to take the high ground. Valleys compartmentalized armored formations that reduced maneuver. During the war, the enemy augmented conventional linear offenses and positional defenses with human wave assaults and infiltration tactics similar to Japanese methods during World War II. Given its European war heritage, the 1949 doctrine had not accounted for this style of fighting.[18]

The Korean conflict also consisted of small- and large-unit actions in which opposing forces engaged one another in linear or positional warfare. On 5 July 1950, a reinforced infantry battalion known as Task Force Smith defended the high ground near Osan-Ni. Characteristic of many small-unit actions in Korea, the battalion occupied high ground with dominating fields of fire. However, the force was no match for two regiments of North Korean infantry accompanied by six tanks, which quickly encircled the battalion's exposed position and drove the Americans off with heavy casualties. In the summer of 1950, the linear Pusan defensive perimeter resembled World War II European tactics. American defenders held a river line supported by mobile reserves. From September until November 1950, the American Eighth Army conducted an offensive penetration of the North Korean defense line along the Naktong River followed by a pursuit that ultimately reached the Yalu River. Chinese intervention turned the tide.[19]

Although Truman's police action eventually ended in stalemate with opposing forces more or less where they had begun, one positive outcome was that the Army had became relevant once again. Although the 1949 doctrine was not designed for a limited war, the Korean conflict demonstrated that army forces were still necessary in the atomic age. Given Korea's rugged terrain, nuclear weapons would have had difficulty producing mass effects. Air force conventional munitions and the Navy's weaponry contributed to ending the conflict but could not bring about a cease-fire by themselves. Infantry-dominated army forces had contained the spread of Communism by holding ground, a concept as old as warfare itself.

Lessons Learned, Future Issues

Even before Truman left office in 1953, the Army began gathering information concerning the lessons of the Korean conflict. It quickly became apparent that Korea had demonstrated that doctrine must account for political decisions that imposed constraints (what you must do) and restraints (what you cannot do) on the battlefield. Instead of a radiating terrain for combat troops to overrun and occupy, the Korean affair was a bloody infantry conflict similar to the war against Japan. When the North Koreans or Chinese could be drawn into the open, American firepower inflicted a heavy toll. However, enemy human wave attacks intermingled with American and allied defenders, thereby reducing the effectiveness of aircraft support and heavy weapons fire. During the stalemate years of 1952 and 1953, the conflict became a firepower slugfest more reminiscent of World War I than a war of maneuver. Korea reminded the service that overwhelming firepower at the decisive point on the battlefield was just as important as maneuver alone.[20]

Although massive UN firepower contributed significantly to the eventual cease-fire, a 1952 U.S. Army 2d Division report praised the fluidity of enemy maneuver through infiltration tactics. Infiltration enabled the enemy to use terrain to attack in multiple directions at once. Flowing like water around obstacles and prepared defenses, Chinese and North Korean troops isolated army units while hugging their own artillery barrages. Enemy artillery rolled over American defensive positions closely followed by the assaulting forces, which struck the defenders immediately. Expert defenders in their own right, the Chinese and North Koreans dug in quickly and in depth to create strong points interlocked by fields of fire. Due to the complexity of their defense and the tenacity of their troops, American offensive operations were incapable of rapid penetrations followed by armored and mechanized exploitation forces, as doctrine had envisioned.[21]

The Korean experience provided food for thought but became a secondary priority when the new president, Dwight D. Eisenhower, and his administration sought a "New Look" in national security policy. In October 1953, the chiefs of the various services received National Security Document 162/2, which now granted nuclear munitions equal status with all weapons. By January 1954, John Foster Dulles was calling for "massive retaliation" if the Soviets went to war against the United States or its allies. Eisenhower's massive-retaliation policy did not guarantee automatic nuclear release, but it clearly established the possibility that it might occur. It was the uncertainty of the American response, backed by the country's known capability, that was intended to prevent a nuclear war. But opposing policy analysts suggested another option, limited nuclear war. One such individual was J. Robert Oppenheimer, who argued that nuclear bombs were too big for modern wars. Instead, technology was capable of producing smaller devices that could be placed within artillery shells, missiles, and rockets.[22]

Oppenheimer's position gave the army leadership a way to compete with the nuclear-armed Air Force and Navy. Small nuclear weapons enhanced the service's firmly established offensively focused warfighting culture without risking a large-scale nuclear exchange or requiring a significant shift in tactics. In 1950, army staff studies examined the future of small, tactical nuclear weapons. Later, in May 1953, a vintage 280mm gun successfully fired a nuclear projectile, paving the way for the service to equip forces with the Honest John rocket in 1954.[23]

Political considerations, however, also continued to shape army doctrine developments. The Eisenhower administration viewed nuclear weapons as more cost-effective than maintaining large conventional military formations. "Cheap" technology therefore substituted for costly manpower, a situation that American labor had understood since the rise of industrialization in the mid-nineteenth century. But it was the deterrent threat that made nuclear weapons popular and conventional forces less so. During the Eisenhower administration, "the mere threat of dropping a few atomic bombs, combined with the knowledge of their destructive potential, would intimidate would-be aggressors and maintain world order." But the so-called "strategy of deterrence," the prevention of war by professing the willingness to wage it, required a powerful capability for it to have any teeth. That capacity existed within the Air Force and the Navy. Under Eisenhower, airpower grew in importance, with the Air Force budget doubling that of the Army by 1955. The Army, the force that had won World War II, had moved from center stage to the periphery with the advent of the atomic age. The service that was largely responsible for having delivered victory in a horrific global war was now considered to be principally useful for constabulary missions such as occupation duty following an offensive nuclear war or maintaining domestic order if enemy nuclear bombs struck the United States.[24]

By 1954, the nuclear-oriented air force leadership had made a powerful argument that airpower was decisive and ground forces unnecessary. Despite Korea proving otherwise, the Army was on the verge of extinction even as it clung to its World War II legacy. The more service leaders struggled to adjust to strategic nuclear warfare, the more disillusioned its members became. In 1954, one army officer remarked: "I do not know what the Army's mission is or how it plans to fulfill its mission. And this, I find, is true of my fellow soldiers. At a time when new weapons and new machines herald a revolution in warfare, we soldiers do not know where the Army is going and how it is to get there."[25]

While professional soldiers were confused about the service's future, the army staff searched for a doctrine that would enable ground forces to function effectively on a nuclear battlefield. Predictably, service analysts concluded that nuclear warfare still mandated a land force capable of destroying the enemy in battle. The Army was not obsolete, it simply had to adapt to the challenges of modern warfare. To adjust, the service drew upon lessons learned from Korea while devising a tactical scheme for its forces not only to survive a nuclear attack, but to fight one offensively using nuclear weapons.[26]

The *Field Service Regulations* FM 100-5 *Operations* 1954

On 27 September 1954, a new doctrine for a nuclear army was released. Created under the auspices of the Command and General Staff College (CGSC) at Fort Leavenworth, Kansas, and the army staff, the manual was a hybrid of previous operational and tactical thought but with an atomic battlefield thrown into the mix.

The focus of the 1954 FM 100-5 was the division and below. It made but a few changes to the format of the 1949 edition. Chapter 3 was renamed "Branches of the Army," while adding a section on military police. "The Exercise of Command," the subject of Chapter 4, now included leadership (formally a stand-alone chapter), as well as electronic warfare, mass-destruction weapons, psychological operations, and command in combat. Chapter 8 (The Offensive) and Chapter 9 (The Defensive) each included a section on fundamentals. Where a separate chapter in the 1949 FM 100-5 had previously been devoted to airborne operations, those units were now included in Chapter 11 (Special Operations). Within the same chapter, the section on partisans was deleted and replaced by a discussion of guerrillas. Two appendices were included, "References" and "Lessons of Pearl Harbor," the latter being a carryover from the 1949 manual. Issues related to atomic warfare permeated the manual's pages; the word "atomic" appeared seventy-one times in the 229-page document. Three later minor changes included the addition of an index.

The 1954 manual was as much a political statement as a guide for regulating war. In the introduction, the doctrine writers clearly were addressing both internal and external audiences when they proclaimed: "Army forces, as land forces, are the decisive component of the military structure by virtue of their unique ability to close with and destroy the organized and irregular forces of an enemy power or coalition of powers." The statement was designed to convince readers that the service was still preeminent in achieving national strategic objectives.[27]

In reference to the National Security Act of 1947, the manual stated that the service's peacetime mission was to organize, train, equip, and indoctrinate "field units capable of performing their wartime mission to attain national policy." Recent Korean conflict experience was reflected by the manual calling for the Army to prepare for "wars of limited objective," as well as general war. Thus, the Army girded for two wartime scenarios, limited wars such as Korea and general war akin to World War II. In recognizing that the "nature of the political situation at any time may require employment of armed forces in wars of limited objective," the army staff also incorporated the ideas of the nineteenth-century Prussian theorist Carl von Clausewitz, no doubt because of input from the Command and General Staff College and the Army War College.[28]

As an infantry-fixated manual, the 1954 doctrine described the purpose of the offense as to "destroy the enemy's armed forces, impose the commander's will on the enemy, or the seizure of territory in order to further operations." In preparing for the attack, as with the 1949 edition, fire support was critical to allow freedom of maneuver. Maneuver would allow atomic and conventional weapons to reposition so they could increase their effectiveness while forcing the enemy to defend in unfavorable terrain and thereby be destroyed. The division and combined-arms operations remained the preeminent means to maximize atomic and conventional fire, although "the integration of atomic weapons into tactical operations does not change tactical doctrine for the employment of firepower. Decisive results are obtained when a maneuvering force promptly exploits the destructive and psychological effects of atomic weapons."[29]

The Army fixated upon a war of movement, one in which maneuver would prove decisive and firepower allow forces to maneuver. In the opening stages of a campaign, if enemy contact had not already occurred, a "covering force" would precede the main body in the offense. The covering force was to operate well in front of the main body to make enemy contact in order to develop the situation and protect the main body. It was deemed crucial because "its initial engagement may determine the entire course of the battle." A covering force must not only gather information but be "a large mobile force, strong in armor and artillery and effectively supported by tactical aircraft." Behind the covering force would be an "advance guard," a force taken from the main body. The advance guard pressed forward in columns on a broad front. Its purpose was to prevent surprise and to conduct ground observation while allowing the main body to deploy into combat formations. Security of the entire force was paramount. Once the enemy was destroyed, pursuit would be initiated to annihilate the hostile main force.[30]

The war of movement advocated in 1954 depended upon infantry, armored, and mechanized units for speed and shock. But the forms of maneuver to secure objectives through offensive operations had changed little from the 1949 FM 100-5. After preliminary preparations, a force crossed a "line of departure" at a prescribed time to coordinate the attack. Within the zone of action, a commander specified a "direction of attack" for the main effort, although the route was very specific and restrictive, thus it was used only when necessary. The main attack occurred on a narrow front to seize objectives or destroy enemy forces while the secondary attack assisted the main effort. A reserve awaited commitment to influence the action and, if committed, was reformed immediately to maintain flexibility. Conventional fire support remained under centralized control, while atomic weapons were retained by higher echelons to be used on a wider front. Concepts regarding envelopments, double envelopments, turning movements, penetrations, and frontal attacks were left unchanged. Penetration, however, was understood to benefit the most from weapons of mass destruction.[31]

In executing the war of movement, atomic weapons were viewed as simply another means of fire support. Although extremely powerful, "the integration of atomic weapons into tactical operations does not change tactical doctrine for the employment of firepower heretofore mentioned." In truth, the army leadership saw no need for innovative tactics to enhance the effectiveness of atomic weapons fire. The service relied upon forms of maneuver that Napoleon would have recognized and employed in the early nineteenth century.[32]

In the defense, the purpose remained to gain time until conditions changed to allow the offense to commence or to preserve forces in one area while concentrating others elsewhere. The 1954 doctrine now employed a covering force normally supplied by the corps headquarters to delay or disrupt the enemy before the latter struck the main defense area. The division organized a "general outpost" of infantry, armor, and engineers sandwiched between the covering force and the main battle area to protect the covering force as it withdrew and prevent the enemy from directly observing the main defense. Local security elements completed this layered concept, positioning themselves behind the combat outpost and in front of the defending battalions.[33]

The division might defend by using a "position defense" or a "mobile defense." A position defense meant that the defending forces maintained their location and held terrain. A mobile defense retained the bulk of the defenders as a striking force, while a portion of the force formed forward defensive positions called "islands of resistance," which might or might not be mutually supporting. Conceptually, the forward defense offered resistance in order to slow an attacker down while the striking force counterattacked to destroy the enemy at a favorable location.[34]

Both types of defenses required extensive preparation of terrain, use of engineers, consideration of depth and width, fields of fire and observation, and other matters. Fire support integrated all available weapons into a plan to bring the enemy under fire as soon as possible (increasing in intensity as the enemy closed) in order to destroy the force if it penetrated the defensive line, and then counterattack. Fire support included friendly air assets being directed first against enemy aircraft to ensure freedom of movement for the defender. Atomic weapons were to be used once an effective target was identified.[35]

In considering how the Army of 1954 intended to fight in the future, it became clear that atomic weapons had not produced a revolution in tactical warfare. The Army clung to fundamental beliefs it had held since the early 1900s; war was mobile and infantry based. Weapons of mass destruction delivered by air force strategic bombers were powerful but only enhanced ground capabilities. Nothing that atomic weapons offered in destructive capacity had compelled the army leadership to deviate from decades of accepted conventional practice in its effort to regulate the chaos of war.

The Pentomic Era: Changes in Tactical Thinking

From 1955 to 1958, under the direction of Army Chiefs of Staff General Matthew B. Ridgway and General Maxwell D. Taylor, doctrine aimed to refute the Eisenhower administration's policy of massive retaliation via the Air Force's strategic bombers. Given what had happened in Korea,

Ridgway, Taylor, and others (including General James M. Gavin, the deputy chief of staff for plans and research), had concluded that future wars were to be limited politically and a strategic atomic-bomb strike doubtful. What the nation required was a strategically mobile army to fight in places where nuclear bombs were unsuitable.[36]

Army senior leadership soon became bent upon using tactical nuclear technology to prevent sparking a global atomic war. Ridgway, Taylor, and Gavin leveraged their mutual World War II experience and command at higher echelons. Ridgway had commanded UN forces in Korea. As Eighth Army commander in that conflict, Taylor had advised his allies to increase their dispersion and flexibility by reorganizing into five unit formations instead of three. In 1951, Gavin's fascination with rocket and missile technology was honed by his participation in Project VISTA, a study of future atomic warfare in Europe. As VII Corps commander in 1952, he became convinced that existing airborne and infantry divisional forces would not survive an atomic war. Dispersal was necessary to avoid the effects of atomic blasts; mobile combat commands and armored divisions were best suited to do so.[37]

Military innovators require powerful patronage. Steuben had George Washington during the American Revolution, while Emory Upton had been taken under the wing of William Tecumseh Sherman after the American Civil War. In the army staff of the 1950s, Gavin found his benefactor in Taylor. Each had commanded airborne divisions in World War II and both were influential members of the so-called "airborne mafia." As chief of staff, Taylor not only shared Gavin's service culture background, but was in a position of authority from which he could force change.

Doctrine and force structure redesign soon became a means to fend off critics and ensure survival on an atomic battlefield. Taylor advanced the argument that small nuclear devices would make it possible to avoid a larger nuclear exchange. The force best suited to this task, he claimed, was the Army. This meant creating a force with increased air-movement capability and ground mobility, one that could fight in greater depth to survive an atomic detonation while possessing the flexibility to disperse, concentrate, and disperse again virtually at will. Such units had to fight independently, not linearly, and be expendable in the larger scheme if struck by a nuclear blast. The battalion, a unit larger than a company but smaller than a division, a brigade, or a regiment, seemed the proper force structure.

Taylor and Gavin's version of a nuclear-age battalion became known as

Matthew Ridgway, center (Courtesy of the National Archives)

the "battle group." The battle group was organized for "pliability and sustainability." Reminiscent of Emory Upton's nineteenth-century concept of "fours," and the "threes" of the triangular division, each had five companies containing five platoons, as well as a headquarters with a mortar battery, reconnaissance, signal, maintenance, and medical units to allow for independent action. Five battle groups formed one division, with the division headquarters controlling nuclear-capable artillery and Honest John rockets. Given that groups of "five" dominated the force structure, it earned the moniker of the "Pentomic" division, an organization of 11,486 officers and men.[38]

To accommodate its new force structure, the 1954 FM 100-5 was modified in 1956 and 1958. As the Pentomic division matured, operations shifted from a war of movement to an increased emphasis on firepower, as Korea had demonstrated in 1952–1953. In 1958, to satisfy the maneuver advocates, the army leadership had determined that a direct thrust into the heart of an enemy defense was the fastest and cheapest form of attack on the nuclear battlefield. But making that penetration required firepower, so future battle would begin with tactical nuclear fires blasting

James Gavin (U.S. Army Military History Institute, Carlisle Barracks, Pennsylvania)

gaps within an enemy defense long before the field army attacked. Army combat units would lay hidden until the decisive moment, poised to rush forward in columns once the nuclear fires created holes within the enemy positions. The shock of instant nuclear eradication was expected to instill terror within the now-fleeing enemy survivors. American mechanized forces would then drive through the breach, mopping up pockets of resistance and pursuing the fractured enemy to the point of annihilation. The frontal assault, once viewed as a tactical blunder because of its propensity for producing high casualties, now became the supreme form of maneuver, thanks to the destructive power of tactical nuclear weapons.[39]

While radiation would assuredly affect friendly forces in the attack, Taylor and Gavin discounted it. It was far more important to prove that the service was a strategic force capable of winning a tactical nuclear battle without prompting a global nuclear war. In August 1955, a nuclear spectacle at Yucca Flat, Nevada, revealed the Pentomic division's capabilities. As invited dignitaries and reporters watched, a 30-kiloton atomic device detonated on an objective 2 miles from "Task Force Razor" without any apparent adverse effects to the soldiers or their vehicles. Within a minute of the blast, the soldiers opened fire with conventional weapons. Eight minutes after detonation, the task force advanced to its objective, skirting ground zero by about half a mile against the backdrop of a mushroom cloud dramatically rising over 40,000 feet in the atmosphere.

The exercise was as compelling as it was orchestrated. Commanders had waited three days for proper wind conditions while carefully positioning each vehicle during numerous rehearsals. The 1954 doctrine was ignored, as leaders withdrew local security forces, placed the soldiers inside their vehicles, closed the hatches, covered the glass with tape, and turned all weapons to face the rear. In truth, Task Force Razor's precautionary preparations would have made the units vulnerable to enemy attack in war, as the classified after-action report indicated.[40]

For all its promise, the Pentomic division proved a difficult proposition to nurture. Tactical nuclear weapons research and development consumed service funds at an alarming rate and other army equipment suffered for it. In 1957 alone, nearly half of the service's research and development budget went toward missiles and nuclear weapons compared with 4.5 percent for new vehicles, 4.3 percent for artillery, and 4 percent for aircraft. The Pentomic division, dependent upon an expensive armored and mechanized force structure, made little progress. A large portion of the budget was expended to create the Davy Crockett tactical nuclear weapon and delivery system.[41]

By the mid-1950s, some pundits began questioning the Army's direction. S. L. A. Marshall decried the service's infatuation with technology by writing: "One might think that the whole future is to be won through the augmenting of fire power." Marshall noted that the enemy seldom cooperated by becoming lucrative targets and often mingled with civilians and hid in residential areas. Using atomic munitions "would be like hunting fleas with an elephant gun, though it is being proposed as a tenable theory of operations." Marshall cautioned further that nuclear munitions technology was a passing fad; the Army must not deny its cultural heritage, "the primacy of the individual fighting man." To do so, he warned, flew in the face of the American military experience and invited cultural collapse.[42]

Army officers also questioned the Pentomic division as they struggled to understand the nuances of the tactical nuclear battlefield. Prior to the 1957–1958 academic year, the Command and General Staff College at Fort Leavenworth, Kansas, had emphasized a conventional general war in Europe. After the Pentomic concept was approved in 1956, the Leavenworth faculty was now expected to instruct student officers in atomic warfare. Absent any practical experience to speak of, such instruction was pure conjecture. In seeking both to explain and sell the atomic battlefield as the Army's future, the faculty confused not only themselves but also the students who devised tactical operations and calculated the effects of tactical nuclear munitions upon soldiers and equipment. From 1957 through the early 1960s, Leavenworth graduated mid-career officers who were schooled as Pentomic, not conventional, warriors.[43]

Army general officers in the field were as confused as the Leavenworth crowd. In Germany, Lieutenant General Arthur S. Collins Jr. wondered how a 500-kiloton nuclear weapon had any tactical utility. Collins later recalled being brushed aside by his staff when he asked if they understood just how much damage a weapon of that magnitude would cause to the densely populated Stuttgart-Munich area. The detonation had the potential to kill tens of thousands of people, reduce habitats to rubble, and destroy infrastructure. In his opinion, the Army had ignored the ramifications of nuclear weapons in an effort just to make it work somehow.[44]

FM 100-5 in Lebanon

The 1954 FM 100-5 was the first keystone doctrine to discuss wars of limited objective, although crafting a theoretical basis for it would take years. In what became known as "situations short of war," service manuals discussed operations to bolster failing governments, deter aggression, and stabilize regions by functioning as an international police force. When the Army was ordered to intervene in Lebanon (1958), a country fragmented by political, factional, and religious differences, the 1954 doctrine had accounted for limited objective wars to destroy aggressors while restoring the political and territorial integrity of the friendly nation. It was assumed that hostile forces originated from satellite countries and were backed with minimum equipment by a major power.

In 1957, President Eisenhower adopted a policy in which the United States would provide military assistance to any Middle East country seeking to counter Communist incursions. On 14 July 1958, Camille Chamoun, the pro-Western Christian president of Lebanon, accepted Eisenhower's offer. The formation of the United Arab Republic (UAR) (Egypt, Syria, and Yemen) in 1958 had provoked an internal Lebanese debate to join. Rioting and violence erupted. Fearing a coup, Chamoun soon requested American intervention.

Eisenhower immediately assented. The Marines arrived on 15 July. Soldiers from the 187th Airborne Battle Group of the 24th Division entered Lebanon on the 19th. Eventually, 8,000 soldiers and 6,000 Marines plus air force and navy assets assumed a clear but difficult mission: protect American lives and property, discourage UAR and Soviet mischief, and keep Chamoun in power. These nondoctrinal tasks evolved into a military show of force using patrols, roadblocks, and proffered (but politely refused) assistance to the Lebanese Army. Two weeks after intervention, Lebanon held presidential elections. Chamoun did not run for office and unrest waned. Army forces left on 25 October, at the cost of $200 million, millions of additional funds in aid, and one dead American.[45]

From Pentomic to ROAD

By the late 1950s, the Army's debate over the nuclear battlefield became public. In the influential magazine *Army,* Colonel Francis X. Bradley argued that any future war with the Soviet Union must involve nuclear weapons, given the Red Army's superior size. Conventional methods lacked sufficient punch to stop the Soviet juggernaut, thus "we must go nuclear." Outraged, Lieutenant General Collins responded in the next issue that America's allies "would become neutral and tell us to go home" if the Army pursued nuclear war as the first option. Besides, he noted, "We talk about what these weapons can do *for* us, but we seldom discuss what they can do *to* us." Colonel William E. DePuy joined in by offering that the Army had to embrace both conventional *and* nuclear forces, for the Soviets had both capabilities. DePuy believed that technology must match technology because the Soviet Army was too large to defeat conventionally. The nation required both conventional and nuclear capabilities "to maintain a rough symmetry of capabilities with the Communist Bloc in each category of force." Yet, he also acknowledged that the service was now more a deterrent than a warfighting force, existing only to "deny an enemy the chance of victory through tactical nuclear warfare."[46]

As the debates continued to rage into 1960, the commandant of the U.S. Army's Command and General Staff College at Fort Leavenworth, Kansas, urged the faculty to continue developing instructional models that supported changes in doctrine. General George H. Decker, army chief of staff between 1960–1962, along with Generals Hamilton H. Howze, Donald V. Bennett, and Paul L. Freeman, lambasted the Pentomic concept as farcical. Within the pages of *Military Review,* one of the Army's sounding boards for intellectual matters, Colonel (Retired) Henry E. Kelly chastised the Pentomic doctrine for believing "the attacker will employ tactics obviously unfavorable to himself." Besides, Kelly argued, the concept lacked the equipment to make the "doctrinal words" work. Kelly's statement hit at the crux of the matter; doctrine absent means was hollow. Fort Leavenworth faculty members observed that the Pentomic concept was so vague that student officers believed falsely that "a few well-placed nuclear weapons" would allow an army to sweep across the battlefield. Instead of instructing students in technological panacea, instructors asked, "How [does] the Army intend to fight with today's [obsolete] equipment?"[47]

The updated 1954 FM 100-5 had tried to both justify the Army's existence and explain tactical atomic warfare. But Taylor and Gavin had failed to convince the officer corps that tactical nuclear weapons overcame skepticism about land power and massive retaliation policy. The army staff could not convincingly articulate Taylor and Gavin's vision for the Army to the service. This sparked an intellectual debate over detaching the service from its cultural symbol, the American soldier, in favor of trendy technology. Atomic theory also ripped at another cultural conviction— trust in maneuver—and replaced it with the frontal assault and penetration using firepower.

In following their leaders, army staff officers and other like-thinking individuals dutifully "charged off to develop the doctrine, tactics, and organization needed to convert technological promise into combat capability." Units converted to the Pentomic organization even as many officers believed that the warrior ethos was rapidly disappearing. The Army had deviated from being warriors into a deterrent force armed with tactical nuclear weapons. Many officers believed that tactical nuclear weapons served no purpose except to deter Soviet use. Shaken to their core, officers at all echelons of the service refused to accept the concept of the atomic army.[48]

In 1956, when Pentomic divisions arrived within the U.S. Army Europe (USAREUR) Seventh Army, the commanding general, Bruce C. Clarke, was unimpressed. Training exercises revealed flaws in battalion operations; Clarke preferred the regimental system. Divisional artillery was inadequate in numbers and range to support broadly dispersed battle groups with conventional or atomic fires. The mass effectiveness of indirect fire, the hallmark of the Army during World War II and Korea, was gone.

Clarke eventually assumed command of Continental Army Command (CONARC), where he attempted to impose doctrinal innovations of his own. He believed that atomic warfare was likely but would require armor and centralized artillery control to survive. Pentomic forces lacked flexibility and firepower. His solution was the Modern Mobile Army (MOMAR) concept, to be developed and implemented between 1965 and 1970. The MOMAR division was to be either heavy (armor) or medium (armor/infantry) and built upon a triangular force structure and task forces that could be organized by mixing different types of units. The MOMAR structure was flexible enough to conform to changing circumstances such as a war in Europe or Asia.

MOMAR failed, however, because it too closely resembled the Pentomic division. Heavy dependence upon armor and mechanized infantry was similar to the German force structure of World War II. The newly revamped German *Bundeswehr* discounted those ideas. In the Pentagon, Army Vice Chief of Staff General Clyde D. Eddleman also rejected it in favor of infantry, armored, and mechanized divisions. MOMAR was developed conceptually but never fielded.[49]

By 1960, despite army staff attempts to rationalize the strategic and tactical purpose of the Pentomic concept, senior field commanders were convinced that the idea had failed. Five battle groups were difficult to control; the added complexity of divisional artillery and support units made it very problematic to support broadly dispersed forces. While the Army fielded Pentomic units in Europe, the German division focused upon brigades and more flexible battalions. NATO adopted the German division as its standard.

By 1960, lacking both NATO and army support, the service prepared a new division concept for the incoming Kennedy administration. But

Kennedy soon expressed concern over Soviet-sponsored insurgencies, a type of war unsuitable for nuclear weapons. Strategically, he moved away from Eisenhower's massive retaliation policy toward a flexible response where the United States government responded to aggression with "suitable, selective, swift, and effective means." Kennedy's administration directed Chief of Staff General George H. Decker to revamp the Army's divisional structure. Decker was to improve mobility and nonnuclear firepower, as well as "to insure [Army] flexibility to meet any direct or indirect threat, to facilitate coordination with our major allies, and to provide modern mechanized divisions in Europe and new airborne brigades in both the Pacific and Europe."[50]

By mid-1961, the Army discarded the Pentomic concept for the Reorganization Objectives Army Division (ROAD), a divisional organization that revived World War II doctrine. Major General Harold K. Johnson explained at the time, "The basic fighting structure to which we are returning is one with which most of us have a reasonable degree of familiarity." Indeed, the Army had returned to its pre-Pentomic roots with a maneuver-oriented force bent upon physical annihilation of the enemy through the offense and overwhelming firepower.[51]

In July 1961, *Army* magazine described the ROAD division organization as appearing in four forms: airborne, infantry, armor, and mechanized. Three of the four divisions were infantry-dominated units while the fourth, armor, had several battalions of infantry within its structure. The ROAD divisions were capable of fighting an enemy force on either a nuclear or traditional battlefield and were terrain-specific. Mechanized and armored forces went to Europe with its more open terrain, while infantry divisions were sent to Korea because of its more mountainous topography. Airborne forces were primarily located in the United States, as strategic response forces with forward brigades overseas.[52]

The *Field Service Regulations* FM 100-5 *Operations* 1962

On 19 February 1962, the Army replaced the 1954 FM 100-5 and its updates. Published by the army staff, the new manual was written to accommodate joint chiefs of staff (JCS) directives and its publications, which were of a strategic nature. FM 100-5 made numerous references to JCS publications as the "doctrinal basis" for concepts the Army included in the manual. Moreover, the manual reflected the Army War College's curriculum and its effect upon doctrine, as well as other service schools.

Chapter 1 began by defining national objectives (attainment of national interests), national policies (courses of action to achieve objectives), national strategy (the sum of national policies, plans, and programs to support national interests), and the elements of national power (political, economic, military, and psychological). More than a strategic primer, its intent was to demonstrate that military strategy was derived from national strategy and that the Army played a significant role in attaining national strategy objectives.[53]

In recognition of the Kennedy administration's concern over conventional, nuclear, *and* insurgency warfare, the manual addressed a "spectrum of war," a bar scale with Cold War at one pole, limited war in the center, and general war at the opposite pole. Cold War was defined as a power struggle between contrasting nations or coalitions, in the absence of overt conflict. Limited war was "limited in objective, weapons, locale, or participants," but one important enough to commit military force without the unrestricted employment of all available resources. General war was undertaken by opposing nuclear powers or coalitions against the homeland of either side employing all available means.[54]

Given the ongoing Cold War, FM 100-5 argued that land forces maintained stability, law, and order. They provided a credible and visible guarantee of commitment to allies while establishing and maintaining peace. In limited war, land forces had to oppose aggression but minimize the risk of expanding the conflict. National objectives overrode military ones, thus army forces had to be flexible and appropriate in composition and size to meet political ends and operational context. In general war, the nature of a conflict meant the Army must operate in two phases. The first phase was a massive nuclear exchange that would be important but not conclusive. The second phase, the imposition of one belligerent's will over another, as Clausewitz had argued in the 1830s, would prove decisive. While the Army was a vital deterrent force, it must be prepared to operate in all three war situations.[55]

Chapter 2 described the "operational environment" as the strategic "conditions, circumstances, and influences that effect employment of military force elements." The Army recognized that national objectives and supporting policy considerations determined by civilian and military authorities delineate a military objective. The Army must grapple with the physical features of the area of operations (geography), the characteristics and attitudes of its people (cultural awareness), the enemy, and the weapons systems available for use. Clearly, the army staff aimed to convey its legal obligations as a service and its role and purpose as one instrument of national power.[56]

In recognition of the operational environment, the manual was the first to articulate the difference between joint and combined operations. Joint operations involved more than one service, such as the Army and the Navy, operating together in a single mission. Combined operations involved the armed forces of two or more allied nations. Both required an understanding of divergent doctrines, techniques, and customs, as well as differences in political, religious, and cultural backgrounds, including philosophies, procedures, and languages. Commanders and staffs were advised to intermix as joint and combined headquarters to provide balance.[57]

The ROAD concept was articulated in Chapter 4. Large units were defined as the army group, the field army, and the corps. Army groups consisted of multiple field armies. Field armies directed tactical operations and provided administrative support for assigned and attached corps. Divisions were attached to corps. Under certain conditions, a field army might not have a corps headquarters but control divisions directly. Infantry, mechanized infantry, armored, and airborne divisions formed the basis of corps, while the Army also fielded a missile command and logistical commands. Units were now designated as close combat, combat support, and combat service support.[58]

FM 100-5 devoted considerable attention to the conduct of battle. Battle now combined both offensive *and* defensive actions, thus shifting away from emphasis on the offense alone. Combining both aspects of war allowed commanders the flexibility to determine what best fit the situation, nonnuclear or nuclear. Reducing vulnerability to nuclear weapons fire required increased dispersion and, thus, increased physical areas of responsibility. Mobility was essential to survival, as were the combined effects of fire and maneuver, not just one or the other. In the offense, the attacker's purpose was to destroy the enemy, seize terrain, or otherwise contribute to ongoing operations by diverting attention. Fundamental maneuvers such as penetrations and envelopments, atomic or conventional strikes, did not change from the previous doctrine. Pursuits continued to be the recommended way to destroy enemy forces. In the defense, the purpose remained to transition to the offense, to destroy opposing forces, or to hold terrain. Both the mobile and area defense were retained.[59]

The 1962 doctrine also recognized the importance of airmobile operations, which were discussed along with airborne operations in Chapter 7. Airmobile forces had resulted from army experiments with helicopters post–Korean War and involved moving land combat elements and their equipment about the battlefield using rotary aircraft. Airmobile operations were considered of limited mission range and short duration. Although the forces moved rapidly by air, they were recognized as vulnerable once on the ground. Commanders were advised to accomplish their missions rapidly and disperse. Airmobile operations were affected by weather but less so than airborne operations where low wind speed was critical. Still, the primary advantage of airmobility was the ability to fly over obstacles and the rapid deployment and recovery of infantry, artillery, cavalry, and reconnaissance units, which they facilitated. Infantry units were considered to be particularly suited for outflanking enemy positions, raiding, and seizing limited, lightly held objectives.[60]

In meeting the intent of the Kennedy administration, the 1962 doctrine discussed unconventional warfare in Chapter 10. Insurgency was not new to the Army, as the Spanish-American War (1898) and Philippines War (1899–1902) had included either working with or fighting against insurgents. Army units had been involved with guerrillas in Russia (1917–1919), as well as in the Banana Wars era (1898–1933) in the Caribbean. More recent international experience had come from the British in Malaysia (1946–1954). However, the Army did not draw upon its own history *per se* or use the terms insurgency or counterinsurgency in the 1962 edition. Instead, the FM described the use of guerrilla, evasion and escape, and subversion activities.[61]

The 1962 FM 100-5 defined unconventional warfare as war conducted by local personnel and resources to further military, political, or economic objectives. Small clandestine and covert forces attacked targets deep in enemy territory using "unique capabilities that cannot be reckoned with in a conventional manner." In a Cold War situation, the doctrine explained that this consisted of fighting a war for the minds of men but the field manual did not specify how victory in such a war might be obtained.[62]

Commanders involved in unconventional warfare were advised to consider such factors as an ideological struggle, mass communication effectiveness, new weapons, the availability of "special forces type units," psychological operations, and improved transportation. Civil affairs units were also of importance, particularly when calculating the ramifications of sponsoring guerrilla or subversive units. The field manual noted that the consequences of supporting unpopular political groups not compatible with national interests "may be detrimental to long term objectives." This situation was amplified by postwar ambitious leaders with their own agendas. Use of excessive force was to be avoided, as well as the destruction of religious, cultural, agricultural, and humanitarian facilities, which, it was argued, might foment dissent and result in the lack of cooperation or even armed aggression against army forces.[63]

For the first time, special forces teams were mentioned in a keystone doctrine, an indication that the army staff not only recognized their growing importance but also sought to assume control over certain operations that had once been informal practice. Special Forces units had formed in the early 1950s following the demise of the Office of Strategic Services. Their specialties included a capability to organize, equip, train, support, and control guerrillas and conduct guerrilla warfare. Guerrilla missions included assisting airborne or airmobile actions and attacking static yet vital targets. Infiltrated by air, water, or land, special forces units could also act as stay-behind forces during a retrograde.[64]

Where Chapter 10 did not address insurgents, Chapter 11 included them in its guidance for conducting operations against irregular (unconventional) forces. An irregular force was defined as any nonconventional force or operation that included guerrillas, insurgents, subversives, terrorists, revolutionaries, or similar personnel, organizations, and methods. Such forces, the field manual noted, tended to operate in agrarian societies with underdeveloped resources and industry, inadequate transportation, communication, and food distribution systems, and low standards of living. Periodic crop failure led to famine and privation. Nations with high illiteracy rates, lack of educational institutions, and poor medical and sanitation were likely to form and employ irregular forces. Moreover, the immediate and decisive results of operations against such forces were seldom observable. Attacking irregular forces often scattered them to re-form and strike again.

FM 100-5 indicated that irregular forces organized as two primary and mutually supporting elements, a guerrilla element that operated overtly and a subversive element that operated covertly. Both depended upon local support by individuals or small groups that provided intelligence, evasion and escape, and logistics. The subversive element was recognized as the major force in an area where overt use of guerrillas is precluded by friendly irregular forces but both types of activities are hard to detect. Often, mission success was considered to depend more upon changing people's attitudes than the physical destruction of an enemy force.[65]

With its publication, the Army had a doctrine that explained the service's role strategically, operationally, and tactically. It was the intellectual product of the army staff and educational system. It contained both theoretical and practical approaches to future atomic, conventional, and unconventional warfare. And it soon drew considerable attention from critics both external and internal to the service.

Reaction to the 1962 FM 100-5

In March 1962, *Army* magazine devoted extensive coverage to the noticeable shift from conventional to unconventional warfare in the service's doctrine and philosophy. "Through the President's prompting," wrote the authors, "the Army has established an office of Special Warfare whose chief will report directly to the Chief of Staff." In truth, the conventional-minded army leadership wanted little to do with irregular warfare, although the service was allegedly "inaugurating a stepped-up program of education and training of officers and men in guerrilla and counter-insurgency methods." General Herbert B. Powell, commanding general of CONARC, vowed that the Army was "making excellent progress in Special Warfare, that there's a lot of life in the program and it is going well." Despite Powell's cheerful words, President Kennedy remained skeptical. He appointed a committee consisting of Maxwell Taylor, McGeorge Bundy, Roswell L. Gilpatric, U. Alexis Johnson, Robert Kennedy, and Lyman L. Lemnitzer to ensure that Powell's words rang true.[66]

Kennedy had good reason to doubt the Army's sincerity, for the president and Army Chief of Staff General Decker disagreed over the proper conduct of special warfare operations. Kennedy believed that it took a guerrilla to fight a guerrilla, but Decker held to a service cultural truism dating back to 1794 that any conventional soldier could handle guerrillas or unconventional forces when given proper leadership and additional training. As Decker saw it, the service required "a proper balance among the various forms of ground power" to meet particular situations, but conventional soldiers should dominate. In 1962, army leaders prepared for general or limited wars that required a universal soldier, one possessing the fundamental combat skills to devastate any nation that an army was unleashed against. In creating chaos through the annihilation of opposing armies, governments, or societies, conventional troops cared little for the culture and traditions of the local people, let alone guerrillas.[67]

Conventional thinking ran counter to army missions, for, in 1962, Kennedy deployed army forces to Thailand. The impact of the 1962 doctrine on the mission is difficult to discern. Special Forces units, which had previously trained the Laotian military in 1957 to counter Communist aggression, operated with the 25th Infantry Division and allies. Their mission was to bolster Thai confidence so that the deteriorating situation in Laos would not spread across the border (although a crisis did not materialize). Instead, Major General James L. Richardson of the 25th Division focused on infrastructural repair and public relations until the Army withdrew in August.[68]

While a Cold War fixation on conventional and atomic warfare dominated the army leadership's thinking, the educational system was transforming to some degree. Beginning in 1958, army schools such as the Command and General Staff College offered instruction in irregular warfare, though the bedrock of the curriculum remained Cold War force-on-force instruction. Astute students nevertheless came to appreciate that their service was in effect culturally composed of two armies, a conventional one regulated to value conformity, maneuver, and firepower and the special forces governed by boldness of action, cultural awareness, and innovation. Doctrinal and cultural differences in approaches to war between the conventional army and the special forces community bedeviled the service then and now.[69]

Doctrine and the Escalation in Vietnam

With the publication of the 1962 FM 100-5, the Army came to possess a doctrine capable of meeting conventional or atomic war expectations in Europe and unconventional wars wherever they might occur. But given the degree to which the conventional army leadership dominated the service, irregular warfare rested primarily upon the shoulders of the special forces. Two armies entered Vietnam in 1965, a conventional army trained and equipped to fight a future war in Europe and the special forces organized to operate independently as guerrillas.

In 1962, the Army was considering deployments to Vietnam using ROAD units. The force of choice was a ROAD infantry division of 13,500 troops composed of seven infantry, two tank, and two mechanized infantry battalions plus additional supporting troops. It was assumed that Vietnam would be an infantry-heavy war, an idea fomented from the Korean experience and misunderstood French after-action reports concerning the use of armor. French commanders had failed to concentrate their obsolete tanks to support infantry attacks, thus they had used them with little benefit. For an American Army doctrine fixated on Europe, using armor extensively in Vietnam seemed absurd.[70]

Still, the army staff understood that Vietnam, with no front lines to crack or flanks to assail, was a different war than imagined in Europe. It was what the 1962 FM 100-5 had described as a limited war, a war that demanded understanding political reality in order not to expand the conflict but to "defend land, people, and resources without destroying them, to meet aggression in a manner compatible with Free World and U.S. security interests" without total nuclear war or interaction. Such a situation developed in the Caribbean.[71]

The Dominican Republic

In 1965, as America became more involved in Vietnam's civil war, President Lyndon B. Johnson directed an intervention to the Dominican Republic. Reflecting a situation that FM 100-5 described as "short of war," the American military operation responded to a leftist coup against Donald Reid Cabral, a military-backed conservative who had supplanted the leftist elected President Juan Bosch. Cabral's downfall resulted in a provisional government under leftist Jose Rafael Molina Urena. In-fighting ensued as both the left and right jockeyed for power. Urena fled the country. Civil war loomed. So-called Loyalist (anti-Bosch) forces were unable to secure power and asked for U.S. intervention in exchange for promises to hold free elections. After weighing overblown U.S. intelligence reports of a potential Communist-led takeover, Johnson agreed.

On 27 April 1965, many foreigners began fleeing Santo Domingo. Five hundred Marines arrived twenty-four hours later. On 29 April, 1,500 Marines reinforced the first contingent, followed by elements of the 82d Airborne Division joined still later by special forces. For more than a year, troops adjusted to the whims of civilian policymakers and diplomatic issues under circumstances that doctrine had not accounted for. The field manual had made no mention of rules of engagement, orders that not only restricted weapons fire but changed so often that the soldiers became confused. Conventional army troops reacted to sniper fire, while special-forces units attacked television towers to prevent the dissemination of anti-American rhetoric. It was impossible to determine with any precision who was a leftist or not, although Lieutenant General Bruce Palmer Jr., commander of forces, managed to bottle up 80 percent of the left-leaning Constitutionalists in the southeastern portion of the capital.

Eventually, elections were held in June 1966. Joaquin Balaguer defeated Bosch in free elections, and the last American troops departed on 21 September 1966, at the cost of 27 soldiers dead and 172 wounded.[72]

The 1962 Doctrine in Vietnam

Although army doctrine depended upon massed fire support to allow maneuver to occur, the nature of the Vietnam War and fighting both conventional and unconventional forces meant dispersing artillery into fire bases to support discrete operations. As doctrine had explained, dispersion increased due to the helicopter's ability to transport men and equipment rapidly about the battlefield, as demonstrated in November 1965. The 1st Battalion, 7th Cavalry, in reality an infantry battalion, was transported via helicopters to the Ia Drang Valley for the first battle employing the air-mobility concept. American infantrymen numbering 450 troops swooped down from the sky into a landing zone where, within an hour, Lieutenant Colonel Hal Moore's battalion was attacked by approximately 2,000 North Vietnamese soldiers. Unable to maneuver as doctrine prescribed, Moore's unit slugged it out for two days under the protection of artillery and air support. The action was the first test for FM 100-5's airmobility doctrine; the North Vietnamese gained respect for American firepower and army troopers discovered an enemy that knew how to fight.[73]

By 1966, the Army had determined that using tanks and armored cavalry vehicles in Vietnam was possible, but the doctrine for employing them was developed on the ground, not by the army staff. Armored personnel carriers were upgraded to possess two machine guns for suppressing fire while the infantrymen dismounted from within. Armored vehicles were also recognized as useful for perimeter defense, as the North Vietnamese troops and Viet Cong insurgents often resorted to infiltration of perimeters or human wave attacks.[74]

Vietnam was an infantry war that saw the deployment of five infantry divisions, one air cavalry division (airmobile), and one airborne division, as well as a number of separate light infantry brigades, a mechanized infantry brigade and two airborne infantry brigades. From 1965 until 1968, these units followed FM 100-5 precepts in the offense by clinging to the fundamental principle that decisive results could be obtained by strong, mobile exploiting forces fixated on terrain objectives or the enemy. Warfare shifted from the envisioned attack against Soviet defenders to search-and-destroy operations against North Vietnamese regulars and Viet Cong guerrillas that required hunting down an elusive foe and, upon discovery, rushing mobile forces to the scene. In truth, search-and-destroy operations were nothing more than a movement to contact based upon time-sensitive intelligence. But army officers found that making doctrine work required managing long-range fires, close air support, helicopters and airmobile operations, ground attacks, armor, mechanized infantry and cavalry, satellite communications, logistics, and a host of other requirements.[75]

While the conventional army in Vietnam gained an appreciation for the complexity of search-and-destroy operations, other forces conducted the irregular war that FM 100-5 had also anticipated. However, two schools of thought emerged concerning how to execute it. Small-unit warfare advocates believed in covering an area with patrols that pursued an enemy until destruction. A second group fixated on destroying the guerrilla base or the source of power. General William C. Westmoreland, commanding general of Military Assistance Command Vietnam (MACV) accepted both concepts, since they each achieved a similar purpose, the destruction of enemy forces.[76]

Various doctrinal and nondoctrinal army units were used to counter guerrillas. By 1967, every division and separate brigade had provisional long-range reconnaissance patrols (LRRPS). Army Rangers were used as strike units. Vietnamese defectors joined with army units to become Kit Carson Scouts. Special-forces units worked with Civilian Irregular Defense Groups, Mobile Guerrilla Forces, and various elements designated Delta, Sigma, Omega, Gamma, and Apache. These units were made up of Americans plus South Vietnamese troops or mercenaries. They fulfilled a variety of missions, from border patrol to population security, deep reconnaissance to special attack missions.[77]

In July 1965, Army Chief of Staff Harold K. Johnson directed an executive study of search-and-destroy methods to measure their effectiveness in Vietnam. The study was known as the Program for the Pacification and Long-Term Development of South Vietnam (PROVN). A group of mid-level officers was to explore new courses of action for the United States and its allies that led to the accomplishment of national-security objectives. Completed in March 1966 and with the concurrence of Vice Chief of Staff of the Army Major General Creighton W. Abrams Jr., the study revealed that the creation of an Army of the Republic of Vietnam (ARVN) under a U.S. Army model in the late 1950s was flawed. Transferring American doctrine to the Vietnamese was hampered by translation and cultural differences, as well the Vietnamese general dislike for a Western-fixated conventional approach. Interspersing security forces within the local population, not search and destroy, was deemed more appropriate. Westmoreland predictably discounted the study. Yet, it indicated that the Army was a learning institution that attempted to change as situations dictated.[78]

By 1968, ROAD divisions and other units had adjusted doctrine to conduct numerous operations in Vietnam against conventional and unconventional forces. Americans fought well and usually prevailed in a war of attrition in which body counts measured success. Operation CEDAR FALLS in 1967 saw 30,000 troops engaged in the Iron Triangle region just north of Saigon. Air force B-52 aircraft intensely bombed the area, which was then followed by airmobile and ground troops who removed local villagers. Heavy equipment leveled forested areas, denying the Viet Cong areas to hide. The region was then burned to eradicate miles of enemy underground tunnels. By late 1967, the American Army had made a difference. It appeared that Vietnam might be saved.[79]

Field Manual 100-5 *Operations of Army Forces in the Field* 1968

On 6 September 1968, the army staff and the Combat Developments Command Institute at Fort Leavenworth, Kansas, published a new keystone doctrine. Its object was more to reassure NATO allies that Europe remained the priority during the Vietnam War than to alter how the service intended to wage future conflict. Units within USAREUR had become manpower and replacement pools for Vietnam, with commensurate shortages in munitions and personnel. The new doctrine shunned most of the Vietnam experience and was filled instead with references to NATO standard agreements (STANAG), international agreements, and the joint service community. The title changed from "Field Service Regulations," which had dated to 1905, to "Operations of Army Forces in the Field," a shift from a more authoritarian perspective (regulations) to the conduct of missions (operations). The FM 100-5 designation was retained and the manual's intellectual thrust and tactical preconceptions remained virtually unaltered since the 1962 edition.[80]

There were some modifications, however, such as the Army's view of the Cold War. The Cold War now included stability operations (Chapter 13) that were "internal defense and internal development operations and assistance provided by the Armed Forces to maintain, restore, or establish a climate of order within which responsible government can function effectively and without which progress cannot be achieved." Progress to what end was not defined. Moreover, within the Army, the terms "stability operations" or "internal defense and internal development" were preferred to "counterinsurgency," for the service viewed counterinsurgency as a government or joint-services program dependent upon national resources. In the 1968 FM 100-5, the army leadership recognized that defeating a counterinsurgency was beyond the scope of the service alone, though army forces were recognized as a contributor to the "overall national program."

The Vietnam War did have some impact upon the 1968 doctrine. Although the Army had defeated communist conventional and unconventional forces during the Tet Offensive of January 1968, the subsequent political and public outcry had taken its toll. Service reaction within the 1968 FM 100-5 included redefining the Army's role in stability operations as "small, mobile training teams administered by the U.S. Army attaché, through the in-country establishment and operation of a U.S. military assistance advisory group or mission, to the involvement of U.S. combat, combat support, and combat service support troops (described in Chapter 4). The effectiveness of U.S. Army participation in stability operations depends on how well this assistance is integrated into the internal defense and internal developments of the host country." After more than a decade of various operations in Greece, Turkey, Lebanon, Asia, and the Caribbean and Central America, the Army of the 1960s had become less independent-minded than its World War II predecessor. The army leadership carefully articulated that the service was part of a maturing joint service system, while further acknowledging civilian leadership in foreign interventions.[81]

Although Chapter 5 further articulated principles of war and operational concepts through phasing, the addition of the fire support coordination line (FSCL) (a permissive measure to allow for rapidly acquiring and destroying targets), and the importance of civil affairs and psychological operations, the most significant alteration was the inclusion of electronic-warfare operations. As technology advanced, electronics became capable of emitting overpowering electromagnetic energy. Commanders were advised to both use this asset and protect against its use, particularly in gathering intelligence, where the benefits of "friendly electronic devices may outweigh the immediate tactical advantages of jamming enemy communications."[82]

Chapter 6 (Conduct of Battle) articulated that nuclear warfare was subject to "scaling" or "a range of operational environments wherein the employment of nuclear weapons, in both quantity and yield, is selective." Nuclear warfare was certainly possible, but the effects of a scaled war meant that maneuver might not be sufficiently degraded. Combat units were to plan accordingly. Moreover, Ranger unit operations were no longer informal practice but specifically mentioned under Section VII as "overt operations by highly trained units to any depth into enemy-held areas for the purpose of reconnaissance, raids, and general disruption of enemy operations."[83]

The Vietnam War's influence was further evident in two chapters, Airmobile Operations (Chapter 8) and Unconventional Warfare (Chapter 11). Airmobile operations had become such a significant factor in Vietnam that it now merited its own chapter apart from airborne operations. Airmobility concepts had evolved into five functions: intelligence; mobility; firepower; command, control, and communications; and combat service support. Air mobile units were placed under the control of the land force commander to engage in ground combat. Unconventional warfare contained a section on support of conventional military operations to direct that "commanders in the field insure that UW operations support conventional military operations" and not vice versa. Given the tension between the conventional and special-forces communities in Vietnam, the army leadership took decisive action through its keystone manual to resolve that situation.[84]

Chapter 12 (Cold War Operations) recognized that direct threats to U.S. national security existed through the "illegal occupation, subversion, or coercion of friendly countries; a show of force; or the establishment of hostile military forces near U.S. territory" and warranted military action. Missions included a show of force, truce enforcement, international police action, legal occupation, and stability operations, the latter receiving a separate chapter, Chapter 13. Army forces were expected to engage in a range of activities from "parades, maneuvers, demonstrations, police and patrol duty, operations against irregular forces, or reinforcement of a threatened area."[85]

In further articulating the purpose and role of army forces along a spectrum of conflict, the 1968 FM 100-5 represented the army staff's efforts to gain more control over army operations. More than just emphasizing the Army's role in international and European defense, the army staff regulated the behavior of special forces and Ranger units by establishing the dominance of the conventional force commander. The rise of airmobility became part of that dominance, for this unique asset was also controlled by conventional doctrinal precepts.

The 1968 FM 100-5 in War and Peace

As the Army in Vietnam began to draw down after 1968, the nature of its missions changed. In 1969, the new administration announced the "Nixon Doctrine," which sought to extract the United States from foreign entanglements unless they were tied to international treaties such as NATO or other vital interests. Europe was considered the first priority followed by the Middle East with its strategic oil reserves and Israeli security concerns. Vietnam soon languished.[86]

The Nixon administration's shift in policy from major combat operations to Vietnamization, the handing over of missions to the South Vietnamese, was anticipated to some degree by doctrine. Civic actions and humanitarian relief operations increased and such missions, although not specifically mentioned in FM 100-5, fell under the rubric of unconventional warfare. Civic Action Teams were designed to assist paramilitary units, with team members acting as advisers. Many of them lived in Vietnamese hamlets not only to train local inhabitants in defense but also to improve infrastructure.

With the withdrawal of army forces from Vietnam in 1973, service leaders found very few lessons that applied to Cold War Europe. Earlier, in 1971, the Army experimented with a TRICAP (triple capability) division a Fort Hood, Texas. This concept blended infantry, armor, and airmobile forces into one unit, but it was never fully fielded throughout the force. In Vietnam, the Army had strength of numbers but in Central Europe, Warsaw Pact forces overmatched NATO numerically. The Vietnam War did not involve the realistic possibility of a nuclear, biological, or chemical attack, but the Red Army had the capability for using those means in a conflict with NATO forces. More important, the Army's image had been tarnished, for the war was lost strategically. Internal problems from the draft, ill-discipline, and drug abuse had leaked into the service from an increasingly permissive American society. The Army was the product of its Vietnam experience yet "coincidentally relevant to conventional war in Europe."[87]

By 1973, it was apparent to many pundits that the Vietnam War had degraded training for conventional war. In Vietnam, the service had focused too much upon small-unit action instead of large-scale operations. To prepare the Army to support the Nixon doctrine in Europe and the Middle

East, the service required a new manual that proffered a philosophy not only for fighting a future war, but one that reestablished confidence in the force's capabilities to deter aggression. The task of designing that doctrine fell to General William E. DePuy.[88]

In the twentieth century, one man rarely affected army doctrine. DePuy's opportunity to do so resulted from his appointment as the first commander of Training and Doctrine Command (TRADOC), which was created in July 1973 to develop combat doctrine. DePuy worked with now Army Chief of Staff Creighton W. Abrams Jr. to return the service to "its basics," an organizational culture where leaders led by example and soldiers executed precisely what they were told to do.[89]

Battalion command in Europe during World War II influenced DePuy to adopt German tactical methods in training his companies and platoons. Speaking at Fort Polk, Louisiana, in 1973, he argued that his methods generated combat power at the battalion and lower levels. Methodological failure while a commander in Vietnam had resulted from high personnel turnover rates and mentally limited men just "going about their business." Army warfighting was "an intellectual exercise being performed by nonintellectuals and scholarly leaders must help them." To DePuy, war was simple, as what worked in one worked in the next. Soldiers resist authority and lack initiative in battle; doctrine should educate the Army to grasp that a leader leads by example and soldiers do as they are told.[90]

Strong-willed, competent leaders meant that "[all the soldier] has to do is go where the squad leader tells him to go. If [the squad leader] goes to the right, they go to the right. If he goes to the left, then they go to the left; if he crawls, they crawl. If he shoots to the right, they shoot, if he runs in behind the barn, they all get behind the barn." To DePuy, dynamic leaders led disciplined, motivated soldiers in battle, but leaders and soldiers were far from ideal. War was an imperfect human affair, but the flawed American soldier was capable of obedience. Society and the service produced substandard infantrymen, for "our system does not put the smartest people in the Army in rifle squads in the best of wars." American soldiers were "people who have not had the advantage and privileges of education and perhaps a family and culture, etc., they are the ones who end up in the rifle squad. They are great guys but not articulate. They find it difficult to express themselves and cannot write [articulately]. Furthermore, they are not intellectuals."[91]

In 1778, Baron von Steuben had discovered that American soldiers were ideal-driven rabble, so he wrote tactical doctrine suitable for those in

William DePuy (U.S. Army Military History Institute, Carlisle Barracks, Pennsylvania)

the ranks to understand. DePuy, in Steuben-like fashion, wrote doctrine for the type of soldier that he was familiar with. His directions to the doctrine-writing team were clear: "Wars are won by draftees and reserve officers. Write so they can understand."[92]

For DePuy, doctrine was the means to regulate the Army's behavior and to compel conformity through relevance. Examples of contemporary warfare helped soldiers to understand his views, especially the use of mechanized forces during the 1973 Arab-Israeli War. The Egyptian and Syrian armies had caught the Israelis by surprise and the Egyptian armed forces performed far better than they had during their stunning defeat in 1967. While Western pundits had expected a decisive Israeli victory, rejuvenated Arab ground and air forces revealed Israel to be vulnerable.[93]

Egyptian Army methods became DePuy's model for success. In a report entitled "Implications of the Middle East War on U.S. Army Tactics, Doctrine and Systems," he noted that the war held significant lessons for the Army. Modern weapons were more lethal than any to date, requiring combined-arms warfare and improved tactics to accommodate them. Army doctrine long advocated combined-arms warfare, but the service would need to focus upon routine individual and unit training to guarantee success.[94]

Although Israel won the 1973 war, the Arab forces had used Soviet technology to easily destroy American armor. The ramifications of this for the defense of Europe were not lost upon DePuy, who directed TRADOC's combat developers to ensure that doctrine prodded Congress to procure modern equipment. In the early 1960s, Secretary of Defense Robert M. McNamara had introduced the Pentagon to a planning, programming, and budgeting system approach to obtain new equipage. DePuy, having recently left the Pentagon for Fort Monroe, Virginia, understood McNamara's methods and together with Generals Donn A. Starry and Paul F. Gorman used doctrine to justify modernization.[95]

DePuy sought to change the Army tactically but his success depended upon political buy-in. Correspondence and conferences with senior army leaders such as U.S. Army Europe (USAREUR) Commander General George S. Blanchard provided input. Considerable time was spent to ensure that the U.S. Air Force's Tactical Air Command vetted the draft manual, for the Army required tactical air support to defeat the Soviets. West German Army endorsement was essential since a Soviet invasion of Central Europe involved transiting Germany. Thus, the field manual complied with German manual 100/100, *Command in Battle*. Between 1974 and 1976, TRADOC officers not only produced a new doctrine but accepted Air Force-influenced precepts of electronic warfare and the German mechanized doctrine previously embraced by NATO. The blending of American and German ideas gave rise to the "AirLand Battle."[96]

AirLand Battle evolved not only from interservice and international ideas, but also from historical data from the Arab-Israeli War. The Israelis mobilized reservists and threw them into the 1973 conflict quickly because the battlefield was close to the homeland. In a future war with the Soviets, however, the American Army had to mobilize reserves and send them across the Atlantic Ocean to Europe. Army doctrine had to defend Europe with on-hand forces before the Soviets won outright. An immediate offensive was deemed out of the question; the Warsaw Pact outgunned NATO.

As part of NATO, USAREUR was expected to destroy the initial Soviet echelons in a conventional battle. Although outnumbered, concentrating forces and firepower at the proper time and location would prove decisive, provided commanders at the brigade and battalion levels were competent enough to do so. Combined-arms units, an amalgamation of various forces, used surprise and deception, as well as cover and concealment to minimize the effects of enemy combat power. Army forces employed an "active defense" or an "elastic but not brittle" tactical scheme, first resisting then withdrawing one element while another engaged the enemy. This "overwatch" method was intended to cause the attrition of enemy forces while grudgingly trading ground to buy time for reinforcements to arrive from the United States. Commanders were to conduct limited attacks against enemy flanks only when the outcome was expected to result in decisively greater enemy losses or to capture objectives crucial to larger battle outcomes. European realities required that priority be given to the defense, although the Army had historically maintained that victory lies in the offense.[97]

The 1976 Field Manual 100-5 *Operations*

On 1 July 1976, Army Chief of Staff General Bernard W. Rogers approved a new FM 100-5 *Operations,* the title of which changed from "*Operations of Army Forces in the Field*" to signify a "capstone" doctrine for the entire service, not just deployed army forces. The 1976 version resembled nineteenth-century army drill manuals in form and content. The doctrine's visual appearance was the first indicator of transformation; a cover pattern of green and black camouflage replaced the typical beige or brown one. The manual was placed in a three-ring plastic binder to signify its tactical focus, facilitate revisions, and make it usable in the field. Pages contained historical vignettes, pictures, charts, and insets printed in two colors with varying fonts to draw the reader's eye to critical passages. The authors included graphs and tables filled with data aimed at convincing readers this doctrine must be accepted as truth.

Within the table of contents page, the manual announced its purpose as providing foundation for service schools' educational curriculum guiding training and combat developments throughout the Army. The Army's primary mission was clearly stated: winning the land battle. Winning meant destroying enemy forces by massive and violent firepower, the chief ingredient of combat power. Modern technology's ability to generate firepower took priority over maneuver. Technology and men combined to form a weapon system, the blending of armor, a main gun, fire control, automotives, communications, and a crew capable of effectively employing it. But doing so meant that the Army had relegated people to being components of a machine, a concept advanced by Elihu Root now come to fruition. In promoting technology, the Army had replaced the soldier as its cultural icon.[98]

Chapter 1 (U.S. Army Objectives) fixated upon Central Europe and indirectly the Middle East, since Arab forces were trained and equipped by Warsaw Pact advisers. Unlike past wars in which army forces could conceivably lose the first battle and then counterattack with mobile reserves, the Warsaw Pact's numerical superiority in weapons and personnel made that situation implausible. West Germany was about the size of Oregon; therefore its defense lacked depth, making large-unit counterattacks problematic. Given the Israeli experience, Soviet weapons were recognized as so lethal that the first battle might be the last one because of high defender losses in a brief time. Europe meant fighting outnumbered but winning the first battle outright.[99]

To emphasize the significance of technological advancements in modern weapon systems, Chapter 2 (Modern Weapons on the Modern Battlefield) described the lethality of contemporary warfare. Where the nineteenth-century naval advocate Alfred Thayer Mahan had celebrated the battleship, DePuy championed the tank. In his mind, all great armies based their forces on the tank; few great states were without armored forces. Modern armored forces destroyed opposing forces faster and at greater ranges with fewer projectiles than ever before. Mechanized infantry supported tanks, not vice versa as previous doctrine had taught. Modern infantry, properly trained and equipped with antitank missiles, could destroy armor in both the offense and the defense at ranges out to 3,000 meters. But doing so meant taking advantage of mobility, the crossing of ground. Armor was much faster and more capable than infantry. To keep up, infantry had to be mechanized, not motorized or operating on foot. When armor and mechanized infantry were amalgamated with artillery, air defense, aviation, engineers, and electronic warfare, they formed an armored combined-arms team with the firepower to defeat a Soviet opponent on either a conventional or nuclear battlefield.[100]

The third chapter (How to Fight) was the crux of the 1976 doctrine. DePuy expected the doctrine to tutor division and lower command and staff officers. Generals at the corps and division levels concentrated forces, colonels and lieutenant colonels at brigades and battalions controlled and directed the battle, and captains within companies, cavalry troops, and artillery batteries fought the battle. Competent leaders understood battlefield dynamics, how to concentrate forces at critical time and places (Jomini's decisive point), that battle must be controlled and directed (managed), to use cover and concealment and combined-arms teamwork to maximize effectiveness, and to train teams and crews to use a weapon's full capability. The 1973 Arab-Israeli War had provided evidence that a unit commander could destroy a force three times the size of his own by substituting firepower for manpower when possible.[101]

Given the manual's defensive focus, Chapter 4 (Offense) placed restrictions upon offensive operations. Since 1905, the Army had advocated an aggressive offense to break an opposing line and then pursue the enemy to destruction. Attacking was expected and normal. With the 1976 doctrine, commanders were advised to carefully weigh the situation and to attack only when maneuver and fires could inflict disabling losses. In a European scenario, an aggressive commander might lose more than he gained, thus the offense was risky. In the offense, commanders had to understand (what DePuy considered to "see") the battlefield and know enemy intent, concentrate combat power, suppress enemy defensive fire, use shock (combined arms), attack deep into the enemy rear area, and support mobility. Commanders organized their resources as "task forces," mixing armor and infantry or remaining "pure" according to the situation. In open ground, tanks dominated. Restricted terrain required more infantry. Air defense, artillery, engineers, attack helicopters, and tactical air supported the task force. The forms of the attack, however, were retained from previous doctrine: movement to contact, hasty and deliberate attacks, exploitation, and pursuit.[102]

The importance of the defense was emphasized in Chapter 5. Eight purposes were identified: cause an enemy attack to fail; preserve forces, facilities, installations, activities; retain tactical, strategic, or political objectives; gain time; concentrate forces elsewhere; wear down enemy forces as a prelude to offensive operations; control essential terrain; and force the enemy to mass so that he is more vulnerable to firepower. Of note was that a prelude to offensive operations was mentioned sixth; army doctrine since 1905 had made that the top priority of the defense. Defensive fundamentals included understanding the enemy (i.e., Soviet doctrine), seeing the battlefield, concentrating forces at the critical times and places (or the decisive point), fighting as a combined-arms team, and exploiting the advantages of the defender through preparation and firepower.[103]

Defenses changed but little from previous manuals. A covering force was still expected to delay and disrupt the attacker while protecting the main battle area. Within the main battle area, the defender was to fight an elastic defense using terrain advantages to employ tanks and antitank weapons effectively at maximum ranges. Mines and obstacles would be used to slow the attacker to allow more time for an engagement. The rear area was directly behind the main battle area, a place for logistical units and communications nodes, as well as surveillance units. Rear units must be capable of self-defense.[104]

Chapter 8 (AirLand Battle) described DePuy's view of modern war: land and air forces working together. Both services had redundant capabilities of intelligence collection, tank killing, air defense, logistical capacity, and electronic warfare. But neither could do everything effectively, therefore the Army could not win the land war without the Air Force. In AirLand Battle, suppressing enemy air defenses was critical, for it made ground attacks more effective. Intelligence assets fed information to the Army and Air Force to enable the targeting and destruction of enemy air defense weapons, radar, communications links, and control centers. Blinded and defenseless ground forces would become vulnerable to close air support and other fires. In sum, the Air Force was to serve as an enabler that allowed army forces to fight outnumbered and win.[105]

Swift Reaction

By late 1976, the manual had become known as the "DePuy Doctrine" and percolated through the service as its primary author had envisioned. To expose soldier-students to the manual, every army school taught the doctrine as a matter of course. Senior and mid-career officers at the Army War College and the U.S. Army Command and General Staff College received tutelage in active defense and destroying the enemy through firepower. Lower-ranking officers and enlisted men at Fort Benning, Georgia, and at Fort Knox, Kentucky, practiced moving rearward while simulating the destruction of an advancing enemy.

To ensure that every soldier complied with doctrine, DePuy revised how the service trained and evaluated unit and individual performance. While unit training evaluations were not a new concept, by 1978 TRADOC produced the Army Training and Evaluation Program (ARTEP), which prescribed doctrinal missions for specific units. For example, the ARTEP manual for the mechanized infantry battalion contained a checklist of required tasks to conduct an attack. Commanders would train their units to perform the tasks and evaluators would determine if the unit met army standards by awarding a "go" or "no go" rating upon task completion. Commanders who failed to meet ARTEP standards retrained and retested their units until they passed. Individual soldiers undertook a periodic Skill Qualification Test (SQT). Every individual was issued an SQT manual that explained his or her job skills. Officers trained each other, and noncommissioned officers instructed soldiers in the various tasks. A supervisor administered a written or hands-on test to ensure that the skill had been mastered. TRADOC's creation had allowed DePuy to achieve a historical first: the service had approved a doctrine that affected every unit and soldier because the bureaucratic architecture was in place to force compliance.[106]

Not surprisingly, the 1976 manual drew mixed reactions. In October 1976, civilian academics Philip A. Karber and Archer Jones wrote that the manual signified a "doctrinal renaissance." Many senior army leaders, however, were less than enthusiastic. General Alexander M. Haig, supreme allied commander, Europe, preferred a doctrine stressing the defense as a temporary measure, as he believed only the offense shattered the opponent's will to fight. Lieutenant General Donn Starry, a former member of DePuy's writing team, found fault after becoming a corps commander in Europe. Starry later recalled that the doctrine he had labored over failed to apply to any unit larger than a division.[107]

The 1976 doctrine unleashed an intellectual tempest within the Army. Military analysts argued that while the Army learned from past wars, so did the enemy. Doctrine had addressed lessons from the 1973 Arab-Israeli War, but the Soviets had also changed their precepts. Soon after the Middle East conflict, the Soviets discarded a concentrated breakthrough technique in favor of a multipronged attack to probe for weaknesses. Once a weak spot was detected, the Soviets expected to create a penetration in the NATO defenses and then commit operational maneuver groups to strike deep into NATO rear areas. The 1976 doctrine, wedded to Europe, had not contemplated a change in Soviet thinking. For many critics, the manual lacked the flexibility to adjust to a new style of Warsaw Pact of-fensive operations.[108]

Many individuals who had championed the 1976 manual were puzzled by the negative reaction. Given recent war outcomes in the Middle East and the overwhelming odds facing NATO, a defensive-focused manual that positioned technology over the soldier seemed both appropriate and logical. The Army simply could not win a tactical war by itself; the Air Force had to play a critical role. But doctrine acceptance has as much to do with comfort as reason. Since the 1779 *Regulations,* the American sol-dier had been upheld as a cultural icon. Despite changes in technology since the American Revolution, no manual advocated subjugating people to technology, armor and mechanized infantry included. Although the 1855 rifle-musket and the post–World War II tactical nuclear weapon had affected army doctrine, the *Regulations* and field manuals of those eras accommodated technology as a tool for enhancing soldier performance in battle. In seeking to regulate the chaos of war through technology, DePuy's 1976 doctrine tore at the very fabric of the Army's cultural be-liefs. By 1981, the strain became readily apparent. Senior officers decided that the 1976 manual had to go.[109]

Discarding the 1976 manual was not an easy process, but rank and clout came to the fore. Lieutenant General Richard E. Cavazos, commander of the III Corps at Fort Hood, Texas, believed that the Army required a doc-trine to prepare soldiers psychologically for combat. Lieutenant General William R. Richardson desired to educate officers in the principles of war, truisms that every officer must study and apply to succeed in battle.[110]

It was General Donn Starry, commanding general of TRADOC (1977–1981), who weathered the torrent. Applying his experiences in Europe, Starry began work on Division 86 in 1978, a reorganization of the heavy division primarily for fighting in Europe against the Soviets. Designed to be more flexible and mobile, a division increased to 20,000 troops. Still, in finding an appropriate doctrine for the Army, Starry believed it essen-tial to address global missions, not just Europe. Where DePuy admired systems analysis and mathematics, Starry believed statistics to be antisep-tic. Instead, he, as with Arthur Wagner in the 1890s, believed historical study allowed those who wrote doctrine to be better informed. In 1979, to further the study of military history as an essential part of doctrine development and officer education, Starry approved the creation of the Combat Studies Institute (CSI) within the Command and General Staff College (CGSC). CSI's mission was to conduct research on historical

Donn Starry (U.S. Army Military History Institute, Carlisle Barracks, Pennsylvania)

topics pertinent to doctrinal concerns of the Army and to produce publications related to doctrinal developments.[111]

Given his European experience as a general officer, Starry focused TRADOC's attention on changes within the Soviet military. To prevent the Warsaw Pact from effectively employing operational maneuver groups, army analysts soon developed the concept of the extended battlefield based upon "deep battle" or "deep attack." Rather than defend a front, Starry saw the battlefield in three dimensions, frontage, depth, and altitude. The covering force battle was critical, as it set the conditions for eventual victory in the main defense. Assets such as satellites, sensors, and long-range reconnaissance units would locate enemy forces and then electronically pass information to units capable of employing aircraft, rockets, and helicopters to strike Soviet forces. The deep attack method focused upon the Soviets second-echelon forces, while defending army ground units destroyed the attacking Soviet first echelon. Once the first- and second-echelon forces were defeated, army forces would assume the offensive. Starry borrowed the term "AirLand Battle" from the 1976 manual to describe a new doctrine that balanced the importance of the offense and the defense.[112]

The 1982 Field Manual 100-5 *Operations*

Resplendent in its paper camouflage cover, the newest FM 100-5 *Operations* appeared on 20 August 1982. The manual was written primarily at Fort Leavenworth, Kansas. Starry had recognized that much of the resistance to the 1976 version was because army schools had little input. Their buy-in, he believed, was essential. Still, although army leaders were consulted during manual production, TRADOC maintained overall control as Lieutenant General William R. Richardson was both the deputy TRADOC commander and the commanding general of the Combined Arms Center (CAC) in Kansas.

The manual acknowledged that the Army must be prepared to fight anywhere in the world in combating Soviet aggression, thus removing the 1976 fixation upon West Germany. It was also the first to address levels of war and specifically the operational level. This concept had been discovered by Soviet thinkers prior to World War II as a means to strike an opponent's reserves and rear area using deep attack assets while simultaneously using mechanized forces to attack front-line enemy troops. Although the initial theorists had been purged after 1937, the Soviets rediscovered this theory in the 1960s and 1970s and applied it to an echelon style of war in which waves of forces, separated by time and space, would fall upon a defending enemy. American AirLand battle was designed to counter Soviet multi-echelon attacks.[113]

The operational level of war filled a gap in how wars had been conceived of in the past. Traditionally, in the twentieth century, wars were hierarchical, the strategic or national level being the highest and the tactical or combat level being the lowest. The 1982 manual inserted the operational level between strategic and tactical, defined as the theory of large-unit operations through the planning and execution of campaigns. Although campaigns had been discussed in previous manuals, the 1982 version was the first to associate them with the operational level of war. In sum, a military campaign had become the link between a national strategy to achieve a political objective and the tactical battles fought to achieve that intent.[114]

In Chapter 1, the manual shifted away from winning the land battle (primarily on a conventional battlefield) to the bolded assertion that "the fundamental mission of the Army is to deter war." Where the 1976 edition contained numerous mathematical graphs, the 1982 edition had but one table. Starry's promotion of historical inquiry permeated the manual, for descriptive historical vignettes assisted the reader in comprehending its contents—the 1863 Vicksburg campaign to describe the offense and the World War I battle at Tannenberg for the defense.[115]

The 1982 manual changed terminology from being a "capstone manual" to "the Army's keystone How to Fight manual." Where the 1976 manual was firepower driven and defensive, the 1982 version focused upon "maneuver, firepower, and movement," though in an offensive spirit. Removed was the emphasis on technology, for future war required "good people, soldiers with character and resolve who will win because they will not accept losing." Within but a few sentences, the 1982 doctrine reestablished maneuver over firepower, upheld the offense over the defense, and restored the soldier to prominence.[116]

Chapter 2 was a theoretical tutorial. The Army was introduced to the operational tenets of initiative (offensive spirit), depth (time, distance, and resources), agility (flexible organizations and quick-minded leaders), and synchronization (coordinated action and unity of effort). Also introduced were three "levels of war." The highest level, strategic, was an abstraction, where the nation's armed forces collectively achieved government policy objectives. The Army fought wars at the operational level that had corps or higher units planning and executing campaigns, with other armed forces and allies (specifically within NATO and the Pacific region) achieving strategic goals. At the tactical level, the Army fought battles to destroy the enemy and accomplish operational purposes. Linking the tactical to the highest echelon of government was "operational art," the idea that national political decisions governing war directly influenced tactical battles through the conduct of operational campaigns. Orchestrating campaigns was an art form, something that had to be mastered. In a way, a successfully designed and executed campaign was a masterpiece. The dynamics of battle, or the interaction of factors that decide a battle, were also discussed. This included combat power by combining maneuver, firepower, and protection orchestrated by skillful leadership. Applying combat power at the decisive place and time would decide battles. To do so meant ensuring unity of effort, putting friendly strength against enemy weakness, designating a main effort, moving fast, striking hard, and finishing rapidly, using terrain and weather, and protecting the force.[117]

Organizationally, the manual combined both operational and tactical chapters designed to elevate the service's intellectual view of war. Chapter 4 (Battlefield Environments) discussed nuclear and chemical weapons, as well as electronic warfare and special environments such as smoke. Chapter 7 (Concept of Operations) articulated the fundamentals of AirLand battle, which included indirect approaches, speed and violence, flexibility and initiative of junior leaders, rapid decision-making, clearly defined objectives, a designated main effort, and a deep attack. Army troops had to be agile, to act faster than the enemy to exploit weaknesses and frustrate plans. An agile commander would concentrate forces using maneuver to exploit advantages quickly. The 1982 manual represented the maneuverist school of thought, as opposed to the firepower camp championed by DePuy.[118]

Chapter 7 (Conduct of Operations) changed the term "Air-Land Battle" to "AirLand Battle." Officers were to pay attention to their *area of influence,* a designated area where organic weapons could affect the current battle, and an *area of interest,* a geographic location extending beyond the area of influence that would affect future battles. In terms of time, a brigade's area of influence considered what could be affected within twelve hours, while the area of interest was twenty-four hours out. A division's area of influence was what could happen within twenty-four hours and area of interest was seventy-two hours ahead. Thus, brigade and division areas of interest and influence overlapped. The senior headquarters (division) collected intelligence and interdicted enemy forces that would affect the subordinate brigades. Subsections described combined-arms warfare, battle command and control, types of units, maneuver and fire support to include offensive air support, joint (multiservice) suppression of enemy air defenses (JSEAD), deep battle, psychological operations, unconventional warfare, Ranger operations, and civil-military operations.

Chapter 9 (Offensive Operations) reestablished the offense as the decisive form of war. Drawing upon the insights of British theorist Basil Liddell Hart, commanders were advised to allow subordinates to take aggressive initiative to bring about the deepest, most rapid destruction of the enemy possible. The ideal attack was an expanding torrent characterized by the main attack following reconnaissance units through gaps in the enemy defenses, shifting forces rapidly to widen the penetration by reinforcing success, then carrying the battle deep into the enemy rear area.[119]

A rapid attack meant selecting from five types of major offensive operations: movement to contact, hasty attack, deliberate attack, exploitation, and pursuit. Although these forms of the offensive were considered to be sequential, the fluidity of modern combat meant that operations might begin at any place in the sequence or shift back and forth as the situation changes. Movement to contact meant using combined-arms formations to seek out the enemy, resulting in a meeting engagement. From that point, a hasty attack developed, an improvised affair with little preliminary planning requiring surprise, speed, flexibility, and audacity to wrest the initiative from the enemy. Such attacks were to be practiced through training battle drills at company level and lower, although battalions developed hasty attack plans ahead of time. Deliberate attacks required careful planning with special emphasis upon "synchronized execution," bringing together a force in time and space to maximize effect upon the enemy. A successful attack must be followed by an exploitation, essentially the continuation of the attack to destroy enemy forces before they organize a defense or to deal a psychological blow to demoralize an opponent. A pursuit followed the exploitation, a form of attack where direct pressure causes an enemy main force to be cut off while in retreat and then summarily annihilated.[120]

The fundamentals of the defense were discussed in Chapter 10, not as a stand-alone option but where offensive combat "is as much a part of defensive operations as strongpoint defenses or delaying actions." The defense thus changed in purpose from previous army doctrine as the less decisive form of war to one that removed the initiative from the enemy through a series of strong defenses and successful counterattacks. The basis of a strong defense was the infantry, a designated "pivot for maneuver" that would hold a strongpoint while more mobile forces anchored themselves upon it. Defenses were conducted to cause an enemy attack to fail, gain time, concentrate forces elsewhere, control essential terrain, wear down enemy forces as a prelude to offensive operations, and retain tactical, strategic, or political objectives.[121]

The 1982 manual recognized that military theorists had determined the defense to be the stronger form of war because it was much easier to deny success than to achieve it. Defensive advantages, well articulated in army doctrine since 1905, continued to include knowing the ground and preparing well in advance for an anticipated attack. But the manual stressed that it was crucial to wrest initiative from the attacker. Attackers chose when and where to strike, therefore the defender had to slow the tempo of the offense until the main attack was discovered and then shatter it. Deep attack assets struck enemy follow on forces to disrupt any momentum gained from an initial attack or, as the manual pointed out, using Napoleon's concept that a well-planned defense was followed by a rapid, audacious attack.[122]

Chapter 11 (The Defense) emphasized destroying the attacker's ability to continue fighting rather than eliminating every vehicle and soldier. Doing this meant employing a defensive framework, an intellectual method to organize available forces and terrain. The framework consisted of deep operations ahead of the forward line of own troops (FLOT) to disrupt enemy attackers and take away initiative, a covering force to delay and disrupt attacking forces, a main battle area where the defender intended to destroy the enemy, rear area protection forces to counter infiltrators and other opposing forces, and reserve operations designed to support the main effort through flexibility. Army doctrine did not prescribe a single technique but a defensive spectrum. At one end of the spectrum was a static or strongpoint defense that used firepower to defend terrain; at the other end was a dynamic defense that emphasized maneuver to destroy enemy forces. Large-unit operations, corps and divisions, combined both forms while smaller units typically chose but one, as mission and the situation dictated. Regardless, combined-arms operations were considered critical for any successful defense.[123]

The 1982 doctrine writers had made significant adjustments to the previous manual in an effort to instill a more offensive mindedness while avoiding fixation upon European warfare. Still, out of seventeen chapters, the manual devoted fewer than four pages to "contingency operations," the deployment of army forces during a crisis. In 1983, even as the army school system shifted instruction from the 1976 manual to the latest doctrine, contingency operations or "low-intensity conflict" (LIC) became reality in Grenada.[124]

Operation Urgent Fury and the 1982 AirLand Battle Doctrine

The 1982 doctrine had barely been issued when the Army was called upon to invade Grenada. Many officers had been trained under the 1976 doctrine. Regardless, Operation Urgent Fury (the codename for the 1983 Grenada invasion) was not fought according to AirLand Battle precepts. President Ronald Reagan's willingness to challenge communist-sponsored insurgencies meant countering not only the Soviets but the Cubans and others who were constructing a naval base and airfield on the Caribbean island of Grenada.

Securing Grenada meant taking 119 square miles of island, an area roughly the size of an average American city. Instead of Soviet echelons, American forces faced small groups of lethal rabble armed with mostly light weapons. A hastily planned affair, the multiservice operation involved army Special Forces who infiltrated the island while Rangers parachuted onto the airfield at Port Salines. Soon thereafter, air force aircraft carrying elements of the 82d Airborne Division landed on the airstrip in broad daylight. Under scattered enemy fire, dismounted infantrymen moved forward in columns and then deployed into battle formation when necessary. Firepower replaced maneuver; a lack of mechanized forces turned Grenada into a slugfest.[125]

The Grenada invasion was far more a LIC operation than AirLand Battle. But the army leadership had paid little attention to LIC theory within the 1982 doctrine because of a fixation upon a mid- to high-intensity war probably in Europe. In garrison, most army troops spent little if any time training for LIC missions. The Army was not alone, however. Subsequent postinvasion congressional hearings revealed numerous problems among the services regarding joint planning, fire support, communications, and logistics. Poor cooperation among the services and special operations forces had not only killed American service personnel through fratricide, but also delayed the timely occupation of the island.[126]

Refining AirLand Battle Doctrine

Although Grenada had revealed problems with the 1982 AirLand Battle doctrine, bigger issues demanded that the manual be revised. In 1983, army forces were deemed insufficiently mobile, giving the leadership cause to examine the Army of Excellence (AOE). AOE included the concept of light infantry divisions, as units with less heavy equipment could be moved faster by air transportation. In shifting from the 1976 active defense to AirLand Battle in 1982, the Army had also ruffled the feathers of the German government. The 1982 manual had acknowledged that army doctrine supported allies, but its offensive spirit disturbed certain German officials who believed that NATO procedures took precedence over U.S. Army principles.

A 1983 German White Paper noted misgivings with the Army's doctrine as being overtly aggressive and possibly capable of provoking war. The White Paper revealed that "the Alliance forces must be capable of conducting joint defense operations in the Federal Republic of Germany and the calls for new doctrines must offer no obstacles to such operations." American critics joined the Germans in arguing that the 1982 doctrine was flawed, for pundits believed deep-battle emphasis came at the expense of the close battle along the front lines. Others commented that the 1982 doctrine reduced the Army-Air Force cooperation that was evident with the 1976 edition.[127]

In 1984, General Donn Starry discussed the concept of AirLand Battle III, an idea that sought to realize the potential from acquisition assets, targeting, intelligence fusion, and weapons systems. The battlefield had become deeper than previously imagined because technology allowed enemy forces to be detected at greater distances than before. Where soldiers continued to be primary in defeating enemy forces, it was imperative to fight on an integrated battlefield. An integrated battlefield used sensor and surveillance systems to prevent surprise. It included sophisticated command-control methods using timely intelligence to direct modern weapon-delivery systems possessing increased range, accuracy, and lethality. This "operative tactics" approach did not reject AirLand Battle but altered it so that enemy echelons were attacked earlier than before and sufficiently delayed and disrupted in order to prevent them from accomplishing their objectives.[128]

In 1986, General William R. Richardson, the commanding general of TRADOC, attempted to end the doctrinal criticism by arguing that the 1982 doctrine exposed the Army to the operational level of war and stressed initiative, attacker momentum, flexibility, violent execution, and surprise and shock effect. While the offense was primary, he did not advocate offensive operations alone. Richardson further asserted that the 1982 manual continued Army-Air Force cooperation in war but properly avoided procedural issues. To Richardson, the 1982 doctrine was "on target."[129]

Richardson's statements masked strong criticism, as early as 1979. In preparing a draft manual to replace the 1976 edition, General Donn Starry had opened a dialogue with the German military by mid-1979. Dubbed "AirLand Battle 2000," Starry brokered an arrangement in which the U.S. and German armies agreed to design an organizational concept for a future NATO battlefield. The German press accused the German signatory, Lieutenant General Meinhard Glanz, of kowtowing to American demands. After a lengthy brouhaha, General Richardson, upon assuming command of TRADOC, terminated the AirLand Battle 2000 concept. The German presses not only contributed to the end of AirLand Battle 2000 but guaranteed that the German people would hold sway in future army doctrine if it involved defending their territory.[130]

The army leadership's belief in AirLand Battle had to be balanced with both foreign and domestic agendas. Still, Lieutenant Colonel Leonard D. Holder, a member of the doctrine revision team, announced that the next version would retain the precepts of the 1982 doctrine while expanding the operational level of war. Holder argued that the new version "gives greater emphasis to low-intensity conflict and the use of light forces, and removes some detailed material on the employment of units in favor of broader guidance." As a refined manual only, "It will," he noted, "be a second edition of current doctrine."[131]

The 1986 Field Manual 100-5 *Operations*

On 5 May 1986, the Army published a revised FM 100-5, announcing that "this edition reaffirms the Army's doctrinal thrust introduced in 1982." Produced by a Fort Leavenworth, Kansas, writing team, the preface established that the manual was the Army's keystone publication and source for explaining how army forces executed the operational level of war through planning and conducting campaigns, major operations, battles, and engagements in conjunction with other services and allies. In acknowledging German concerns while also addressing army global missions, the authors stipulated that "FM 100-5 is compatible with and will serve as the U.S. implementing document for NATO land forces tactical doctrine (Allied Tactical Publication 35A), but is both more theoretical and more general so as to meet U.S. needs in other theaters. U.S. troops operating in the framework of FM 100-5 will execute NATO's forward defense plans in compliance with ATP 35A."[132]

In seeking to appease numerous constituents, the 1986 doctrine played to air force sensitivities over tactical air support by discussing the role of air force and army assets separately. Air force missions were to be flown against targets that the Army nominated but would be "executed by the air component commander as an integral part of a total air interdiction effort." The air force leadership agreed to that concept as a means to alleviate tension between the services.[133]

The 1976 doctrine had promoted firepower while the 1982 doctrine upheld maneuver. The 1986 version aimed to balance the two by advocating that maneuver and firepower were essential components of warfare regardless of attacking or defending. The manual reinforced the concept that the offense was "the decisive form of war—the commander's ultimate means of imposing his will upon the enemy." Meanwhile, in the defense, "reactive and offensive elements [worked] together to deprive the enemy of the initiative." The doctrine advised that maneuver and firepower complemented each other in attaining battlefield victory.[134]

The 1986 manual expanded upon the operational level of war by refining the 1982 concept of operational art. Operational art was a skill acquired through years of study, one that enabled an individual to consider national and coalition strategic objectives, translate them into military action through a campaign at the operational level of war, and then use successful tactics to fight battles and engagements that ultimately achieved the strategic purpose. To further assist commanders in understanding the concept of operational art, an appendix addressed "Key Concepts of Operational Design," including centers of gravity, lines of operation, and culminating points, theoretical concepts that borrowed heavily from the nineteenth-century military intellectuals Carl von Clausewitz and Antoine Henri de Jomini. Each operational concept was accompanied by a historical example that explained how commanders used them in campaigns.[135]

The Army's intent for the 1986 manual was announced in Chapter 1, "The Army and How It Fights." The service was to prepare for "strategic challenges across the full range of threats from terrorism through low- and mid-intensity operations to high-intensity and nuclear operations." Most of the manual, however, continued the 1982 version's devotion to conventional operations; terrorists were mentioned as a threat in a single sentence on page four. The army leadership's interest in LIC warranted only seven paragraphs out of 200 pages. In virtually every way, the 1986 manual was what Lieutenant Colonel Holder had described, an updated version of the 1982 doctrine.[136]

Congressional Action and Doctrine

Soon after the 1986 FM 100-5 was published, Congress passed legislation that eventually shaped future army doctrine. The Goldwater-Nichols Department of Defense Reorganization Act of 1986 (henceforth called Goldwater-Nichols) was the most radical government defense reorganization since the National Security Act of 1947. Frustrations over operational failures among the services from the Vietnam War through the Grenada intervention had resulted in legislation that altered how the armed forces responded to civil authority and redesigned the Executive Branch's methodology for creating operational plans.[137]

In 1987, additional legislative action revamped the Department of Defense by establishing the U.S. Special Operations Command (US-SOCOM). Congress had expanded its interest in special operations in the early 1980s at a time when the Department of Defense fixated more upon a conventional approach for war against the Soviet Union than LIC. The legislation placed certain army, air force, and navy assets under US-SOCOM without separating the personnel from their parent services. Doctrinally, the Army lost control of Special Forces and Rangers, which later developed their own doctrinal precepts under their new command headquarters.[138]

Further Steps For Doctrinal Compliance

By the 1980s, the Army had established a long history of written keystone doctrine seeking to regulate the chaos of war. However, both the War Department and the army staff lacked the means to enforce compliance. The establishment of TRADOC changed that situation, for it gained greater control over the Army's doctrine production and education system. In 1976, it accrued additional control when Major General Paul F. Gorman developed the concept for a national training center, a place where brigade and lower commanders could train their units to doctrinal principles. By the mid-1980s, the service had established national training centers in California, Arkansas, and Germany. Each center contained doctrinally educated observer/controllers who observed a unit while training, then led group discussions about the unit's strengths and weaknesses in executing doctrine. In 1987, the army chief of staff, General Carl E. Vuono, created the Battle Command Training Program (BCTP), an organization composed of staff procedure doctrinal experts at the corps, division, and brigade levels. BCTP conducted seminars and computer simulation exercises to prepare general officers and their staffs for mid- to high-intensity combat in Europe, Asia, and the Middle East. In May 1987, BCTP joined with the training centers to form the Combat Training Centers, or CTCs. The CTCs capped decades of army effort to regulate the chaos of war through doctrinal compliance, merging with the school system, the ARTEP, and the SQT to affect every officer and soldier in the service.[139]

By 1989, the Army had established its 1986 AirLand Battle doctrine as the basis for guiding preparations for future war. Army schools taught its principles and units trained to its standards in garrison and at CTCs. Most forces prepared for a European conflict, but others applied AirLand Battle doctrine to Korea or the Middle East. The 82d Airborne Division used the 1986 manual to prepare for contingency operations anywhere in the world. The first test of a contingency operation under the new doctrine came in 1989, a LIC operation into Panama.

AirLand Battle and Operation Just Cause

While the 1986 AirLand Battle doctrine had discussed LIC in but a few paragraphs, it had not envisioned a scenario like Operation Just Cause, the invasion of Panama. Instead of a contingency operation into a foreign and isolated hostile area, army personnel stationed within Panama seized designated target areas and fought both the Panamanian Defense Forces and President Manuel Noriega's thugs while the families of American soldiers watched from their quarters.[140]

The invasion was designed to take down and capture Noriega, who for years had fostered internal terror that included horrific actions against American citizens. In a well-planned and coordinated multiservice effort, special operations forces (SOF) descended on Panama to search for Noriega while army Rangers parachuted onto Panama's Tocumen military airfield to prevent his escape by air. Elements of the 82d Airborne Division parachuted onto Panama and fanned out into Panama City to secure government buildings and other targets. The 7th Infantry Division from Fort Ord, California, moved to Panama virtually overnight and patrolled the streets while mechanized infantry forces from Fort Polk, Louisiana, joined in the invasion.[141]

Panama was an operational and tactical success, although it was far from being AirLand Battle. The invasion was in actuality a simultaneous attack by multiservice forces, which doctrine had described as "special force composition and task organization, rapid deployment, and restraint in execution of military operation." The 1986 doctrine had advised that

LIC "must be fully coordinated with national strategy and fused at the operational level into a coherent effort which usually include economic and political actions as well as military activities." To some degree, that is exactly what happened: U.S. government strategic policy aimed to remove Noriega, secure the Panama Canal, and protect U.S. citizens. Operationally and tactically, the Panamanian operation, while not in complete accordance with the 1986 AirLand Battle doctrine, demonstrated that the Army was capable of operating under LIC circumstances.[142]

The Army's success in Panama received mixed reviews from those in uniform. Marine Major Robert L. Click wrote that "the courage and sacrifice they [the Rangers] exhibited during Operation Just Cause will undoubtedly be added as yet another chapter in the U.S. Army's illustrious history of airborne operations." The Army's senior leadership agreed. Two general officers attributed success to "a ten-year training revolution that involved clear warfighting doctrine, clear training doctrine, and a generation of training-center experienced leaders." But Steven N. Collins, a company commander during the invasion, argued that Panama was the "first experience with rules of engagement and peacetime contingency operations." Although not exactly true (rules of engagement were in use during the 1965 Dominican Republic incursion), Collins explained that LIC operations and the rules of engagement in Panama were problematic. His soldiers had trained as warriors to use extreme violence but were soon relegated to "constables and were to perform the unaccustomed function of maintaining order." His concern was that warriors had trouble adjusting to LIC, precisely the opposite of what several army generals believed to be true. Collins requested that army doctrine should look at Panama to reconsider roles for light-infantry forces and realize that more work needs to be done. He warned that failing to do so might cause future problems, for "as the Army's doctrine of Airland Battle evolves, it is essential to thoroughly assess the recent Panamanian experience and revise our attitude concerning low-intensity conflicts."[143]

Collins had identified a critical service flaw: the army leadership's desire to avoid unconventional operations. Although the 1986 army doctrine had stressed that army forces must prepare to fight a variety of operations across the spectrum of war to include LIC, the overwhelming amount of time spent in BCTP seminars and doctrinally based training at the national training centers was devoted to mid- to high-intensity conventional combat. That was hardly a surprise given that conventional officers dominated the service and that army keystone doctrine had avoided discussing such missions in any detail since the first manual was produced in 1779.[144]

Desert Storm: The War Doctrine Imagined

By 1990, the Army was in peak form, the recipient of new weapons systems and a training program "designed to provide tough, realistic combined arms and services training in accordance with AirLand Battle doctrine, for units from squad through corps." Soldiers had trained hard under doctrinal principles conveyed and enforced through the school house, BCTP and unit training at the CTCs, the ARTEP, and the individual SQT. The conventional Army had shed its Vietnam image of a force that was doctrinally bankrupt. Special Forces and Rangers had developed unique capabilities under USSOCOM in support of the 1986 AirLand Battle doctrine.[145]

AirLand Battle doctrine had been written to face Soviet forces across the globe. In August 1990, a threat materialized that tested its preparedness. When Soviet-equipped Iraqi forces entered the sovereign state of Kuwait, President George H. W. Bush undertook steps to pressure Saddam Hussein to withdraw his forces. Under Operation Desert Shield, the Army used air force assets to deploy the 82d Airborne Division to Saudi Arabia on 6 August 1990. When Hussein proved recalcitrant, additional military forces from the United States, West Germany, and other locations joined coalition partners in conducting combat operations during Operation Desert Storm.

A commander's personality often shapes the employment of doctrine, and Desert Storm was such a case. General H. Norman Schwarzkopf, an AirLand Battle advocate, was the senior American commander in charge of the operation. He understood that Southwest Asia was the right environment to employ the concepts prescribed within the 1986 edition of FM 100-5. A student of history, Schwarzkopf recognized that doctrine was not perfect and it often required modifications. The 1986 AirLand Battle doctrine was no exception. AirLand Battle doctrine was written primarily to fight a war against the Warsaw Pact in Central Europe, where the Army and NATO were initially on the defensive. In Kuwait, however, the Iraqi Army was defensively entrenched and required an offensive to break its line. Following doctrine, Schwarzkopf conducted deep attack operations using bombers, ground attack aircraft, and helicopters to strike Iraqi communications networks, air-defense systems, and other vital targets before the ground war commenced.[146]

Coalition forces required assets capable of locating Iraqi Army reserves and other targets deep in Iraq. The SOF community had prepared itself for just such a mission. Schwarzkopf, however, had disallowed any SOF activity without personally approving their missions to ensure that they met the campaign objectives. Some might argue that his decisions reflected an anti-SOF bias, but army doctrine supported him. A 1986 Air-Land Battle doctrine tenet—synchronization—allowed Schwarzkopf to coordinate all activities that affected the campaign.[147]

Desert Storm was what the Army's 1986 doctrine had prophesized: a fight against a symmetrical enemy using maneuver and firepower. The ground war became a rolling offensive, a series of loosely related independent battles that collectively and methodically destroyed the Iraqi forces in the Kuwait Theater of Operations. After 100 hours of ground operations, the Army and its allies had crushed Iraqi forces and, for some, validated AirLand Battle doctrine.[148]

The 1986 Doctrine As the Cold War Ends

In 1989, after the Berlin Wall fell, it became apparent to military analysts that the Soviet Union was also on the verge of collapse. By 1990, American foreign policy makers produced strategic policy papers that shifted the threat to United States interests from bi-polarity to regional crises. For strategists, the emerging "New World Order" was murky at best. The *National Security Strategy of the United States* predicted a period of uncertainty and uproarious transformation. The Army was to continue to deter aggression through military alliances and a limited forward presence while reducing active-duty force structure and increasing dependency upon reserve units that were deemed less costly than active forces. TRADOC analysts understood that the imminent breakup of the Soviet Union meant a shift in geopolitical realities from Cold War deterrence to the threat of regional contingencies.[149]

Regional contingencies meant the increased possibility of LIC. The army leadership searched for a doctrine that would address those issues. In 1990, Army Chief of Staff General Carl E. Vuono believed that the service needed to become more strategic in its thinking and to operate across "the operational continuum of war, conflict, and peacetime competition." The use of continuums in doctrine was nothing new, but Vuono offered that the United States was involved in a daily struggle with nation states over diplomatic and economic competition. Competition might lead to state-versus-state conflict that escalated along a scale containing two opposing poles: the lesser pole of diplomatic competition and the greater pole of total war. The Army required a doctrine to fight any war along the scale, thus linking the strategic level of war to the Army's operational and tactical missions.[150]

In April 1990, Vouno directed TRADOC commander General John W. Foss to produce an updated army doctrine. With a strategic tie to ground operations in mind, the new manual was to reflect the 1986 Goldwater-Nichols Defense Reorganization Act and be coordinated with the Air Force Tactical Air Command. TRADOC analysts produced a hierarchical model that resembled a ladder, with the top or strategic rung representing a military strategy tied to a national-security policy that achieved national objectives. The strategic rung was linked like a chain to the operational level, the next level downward where multiple services fought as one force. The operational level was further linked to the tactical level where army units fought the battles that attained national policy objectives. The model stressed the accomplishment of strategic objectives through war-fighting at both the operational and tactical levels of war.[151]

In August 1990, Foss directed the faculty at the School of Advanced Military Studies (SAMS), a department within the Command and General Staff College at Fort Leavenworth, to form a doctrine-writing team. Foss told the writing team to examine Panama as a case study for LIC and future wars. His goal was to publish a new doctrine by late 1991. However, many of the Army's more senior general officers disagreed that LIC was important enough to incorporate within FM 100-5, the Army's premier warfighting doctrine. Some senior officers believed that AirLand Battle had proven itself in the 1991 Gulf War and was best left alone. As the Army's senior leadership fumed over the future of its keystone doctrine, the 1986 manual was tested within a different war.[152]

The Drug War

Although the U.S. government was involved in a drug war in 1990, the 1986 AirLand Battle warfighting doctrine had never mentioned service support for a government agency-led operation. In truth, the Department of Defense became involved in supporting the national counterdrug strategy in 1989 after the 1986 manual was published. Still, although the 1986 doctrine had not envisioned interagency operations and the development of relationships among disparate bureaucratic groups, U.S. Southern Command (USSOUTHCOM) provided military support to counterdrug enforcement units by 1990.

The drug war was as complex to conduct as any combat operation. Disconcerted USSOUTHCOM personnel tackled security assistance, counterdrug training, intelligence collection, nation assistance, and operational support. Army officers and forces trained to 1986 AirLand Battle parameters now supported operations along the United States-Mexico border, Bolivia, Colombia, Peru, and other locations. One individual called the drug war a "soup sandwich," a jumble of ideas lacking a key decision-maker. By 1992, however, several multiagency briefing teams had traveled throughout South America to share information about ongoing counterdrug operations. Without appropriate interagency and army keystone doctrine, however, army forces consumed time and resources working out procedures.[153]

The drug war had some impact upon the drafting of a new FM 100-5 but the Army remained fixated upon its new role after the Cold War. Major General Stephen Silvasy Jr. wrote that the 1986 AirLand Battle doctrine was sound but required updating to include the discussion of the deployment of army forces from the United States to various theaters of war. Retired Lieutenant General Frederic J. Brown contended that AirLand Battle doctrine was obsolete, as national strategy had reduced the number of forward-deployed army units and placed greater faith in reserve component forces. By late March 1991, the SAMS writing team had addressed many key issues over doctrine and produced one draft chapter. As various reviewers throughout the service examined its contents, army forces responded to a LIC operation in northern Iraq.[154]

Operation Provide Comfort

In April 1991, the Army deployed forces in support of United Nations Security Council Resolution (UNSCR) 688 under Operation Provide Comfort. The mission reflected certain precepts espoused within the 1986 doctrine, for conventional army units and SOF established a command-and-control headquarters, as prescribed for any military operation. However, doctrine had not envisioned army forces feeding a half-million refugees. Along with Marines and NATO partners, army forces established security zones in northern Iraq, where soldiers patrolled, manned checkpoints, and kept tabs on Iraqi forces. By mid-July, the situation stabilized, American forces withdrew, and the UN assumed control.[155]

The northern Iraq operation demonstrated that AirLand Battle was inadequate for considering all facets of LIC operations, especially, in this case, the care of refugees. Drawing upon doctrinal precepts such as flexibility, however, allowed army forces to adjust to circumstances by employing routine procedures such as establishing a command-and-control headquarters and conducting well-practiced tactical missions. Several officers suggested that the service develop a manual specifically for humanitarian operations. The army staff balked, however, arguing that service doctrine was focused on war, an activity deemed much more complex than LIC. If the Army had to perform such a mission, then strong leadership would ensure that the soldiers were adequately prepared.[156]

As the Army entered the post–Cold War world, the 1986 AirLand Battle doctrine represented the service's transition from a World War II army to a modern fighting force. Written to deter war as much as to wage it, the manual was the culmination of forty-five years of army intellectual energy that grappled with congressional changes, the Soviet threat, challenges from other services and allies, and internal criticism over various concepts. With the dissolution of the Soviet Union in December 1991, the Cold War ended. The Army's leaders now struggled to conceive of what the future might hold as they continued to seek ways to regulate the service through keystone doctrine.

DOCTRINE FOR A POST–COLD WAR WORLD: MULTINATIONAL AND MULTISERVICE OPERATIONS, 1991–2008

Doctrine continues to be the engine of change. Thus, as a doctrine-based Army, change begins by changing our body of ideas—changing how we think about warfighting and operations other than war.—General Frederick M. Franks Jr., *Military Review* (December 1993)

On 25 December 1991, the dissolution of the Soviet Union ended a significant chapter in the history of U.S. Army doctrine. Since 1949, various permutations of keystone manuals had fixated upon countering Soviet aggression, primarily in Europe but also globally. Operational and tactical doctrine had been written for conventional and unconventional warfare under nuclear and nonnuclear conditions. Now, with the Soviet Union gone, service leaders searched for a credible national-security threat and a doctrine to defeat it.

Doctrinal revision was already underway under General Carl E. Vuono, the army chief of staff, before the Soviet Union's downfall. When General Gordon R. Sullivan replaced Vuono on 21 June 1991, he viewed the global strategic situation as awash with "partially-developed democratic institutions and emerging free markets" in countries with low standards of living. Sullivan believed that the service must adjust its doctrine to face emerging threats to the national interest.[1]

An intellectual and avid reader, Sullivan envisioned a doctrine that would guide the service into a postindustrial era. He agreed with Alvin and Heidi Toffler's book *War and Anti-War, Survival at the Dawn of the 21st Century*, which argued that the United States had reached a historical turning point owing to computer technology. If computers could determine the outcome of future wars, the Army had to understand and

Gordon Sullivan, upper left (U.S. Army Military History Institute, Carlisle Barracks, Pennsylvania)

exploit their capabilities as an operational imperative, especially in an era of uncertainty for the international situation and the service.[2]

In August 1991, Sullivan selected General Frederick M. Franks Jr., the former commander of VII Corps during Operation Desert Storm, to replace General John W. Foss at TRADOC. An armored cavalry officer, Franks had a long association with army education, having previously served as deputy commandant of the U.S. Army Command and General Staff College at Fort Leavenworth, Kansas, in the early 1980s. Franks believed the 1986 AirLand Battle doctrine had been appropriate for the Cold War and the Gulf but considered its precepts unsatisfactory for projecting army forces from one region of the world to another against foes with dissimilar equipment, size, and tactics. As the Cold War was ending, the Army partially drew down forces in Europe and returned them to the United States. As forces returned to the homeland, the Army required a doctrine that addressed how to project power from the United States in response to global threats.

Producing a postindustrial-age doctrine at the end of the Cold War was difficult but not impossible. But doing so meant convincing many senior officers that there was ample reason to do it. General officers had risen to the top because of their leadership and management ability within a Cold

War bipolar world. They had taken comfort in facing a clearly identifiable threat. Others believed that the 1986 AirLand Battle doctrine was ideal, the culmination of progressive ideas about warfare that had reached their natural outcome. Some understood that times had changed but were unsure of what doctrine was needed.[3]

Forging consensus required a patient salesman with a respected service pedigree, attributes that Franks possessed. Even if it delayed matters, patience involved sending draft chapter outlines to senior army leaders for feedback and revision, a technique with late-nineteenth century roots. Together, Sullivan and Franks designed and implemented what amounted to a psychological operations campaign aimed at prying the Army out of its Cold War comfort zone. Targeting the decision makers, senior army leaders, Sullivan's objective was to attain service consensus in order to transform the Army. To Sullivan, doctrine was an instrument of change, not simply precepts for regulating operations.

Sullivan selected Colonel James R. McDonough, director of the School of Advanced Military Studies (SAMS), to head a writing team at Fort Leavenworth. McDonough was a published army intellectual with an outwardly calm demeanor. He was, however, mission focused with the willpower to get the job done. In his capacity as Franks's point man, McDonough was fully committed to writing and selling the new doctrine. Not only was he personally involved in putting pen to paper, but he also logged thousands of miles flying to domestic and overseas headquarters briefing senior officers and receiving feedback.

In October 1991, McDonough, with Franks's approval, published an article in *Military Review*. It described the current 1986 AirLand Battle manual as the product of a doctrinal evolutionary process undertaken over the course of the service's history. He argued that doctrine progression was periodically necessary to meet changing national security requirements, future warfare challenges, and, in the current situation, the complexities of low-intensity conflict (LIC). Instead of feelings of trepidation, army leaders were to find comfort in the new doctrine, for it was yet another manual incorporating incremental, not radical, change.[4]

The concept of doctrine as an engine of change possessed historical validity. Since 1779, army keystone doctrine had been altered periodically to accommodate new or modified forms of organization, technology, or warfare. At times, doctrine also served to reinforce long-established truisms or to force change.

Many senior army leaders responded to McDonough's journal article favorably. They agreed that doctrine must confront national-security concerns across a full spectrum of military operations ranging from peacetime competition to total war. However, most army general officers also held that Field Manual (FM) 100-5 *Operations* was a poor forum for engaging in national strategy issues. General Crosbie Saint, the commanding general of U.S. Army Europe (USAREUR), believed that FM 100-5 was best left to operations and tactics. Strategic issues and LIC diluted the text. General Edwin Burba Jr. in Korea concurred with Saint. General Carl Stiner, the commander of U.S. Special Operations Command (US-SOCOM), urged Sullivan to integrate the role of special operations forces (SOF) and LIC throughout the manual's contents. General George Joulwan in Panama, concerned with LIC in his region of the world, argued that army doctrine had to focus more in that direction.[5]

By early March 1992, McDonough's SAMS writing team had resolved dozens of conflicting comments and prepared revised draft chapters for a second round of dialogue with senior army leaders. Meanwhile, Sullivan continued his campaign by reaching out to the service at large. In various forums, he described army doctrine as a guide to the future, "how the Army expects to conduct operations." He stressed that recent combat experiences in Panama (1989) and the Gulf War (1990–1991) represented new realities posing different challenges than the Cold War.[6]

As had been true of Elihu Root in 1899, both Sullivan and Franks appreciated that the Army's role had been altered by global events and that the organizational mind-set lagged behind realities. In 1991, the two generals had to determine how best to describe the nation's geopolitical concerns in acceptable terms to the service. To demonstrate that change was warranted, both officers used the Constitution and U.S. Code Title 10, Section 3062, to build their case. Law directed the Army to fight and win wars but also to perform every mission that the civilian leadership assigned, which included LIC.[7]

Sullivan and Franks spent the end of March 1991 pouring over additional comments from the field. Most were favorable. Yet, some army leaders continued to believe that the draft manual had shifted too far away from warfare to nonwar operations. Franks, however, was convinced that his Persian Gulf experience had revealed modern warfare, conventional or not, to be subject to five common battlefield dynamics. He told the writing team to develop the manual accordingly.[8]

Still, Sullivan and Franks understood that anxiety existed over LIC being addressed within the service's warfighting doctrine. From that concern arose the term "operations other than war," or OOTW. Where LIC contained the term "conflict," a nebulous word for war-oriented army leaders, OOTW contained the word "war," something they understood. In delineating war and OOTW as two separate categories, the new doctrine became palatable to warriors.[9]

On 21 August 1992, General Sullivan received the completed draft of the SAMS-produced warfighting manual. As he contemplated its contents, army forces deployed to two LIC operations, Hurricanes Andrew and Iniki. Sullivan, however, portrayed the disaster-relief operations in operational warfighting terms. Believing in the long-established service conviction that a conventional soldier can fulfill any mission given proper leadership and training, Sullivan concluded that the service possessed the proper skills for directing disaster relief in the wake of Hurricane Andrew. He stated, "The Army's warfighting focus and robust doctrine provide a sound basis for disaster relief operations." During Hurricane Iniki, a logistics expert revealed, "This war's gonna be won by engineers and logisticians," a testament to how the Army understood LIC in warfighting terms.[10]

In truth, army operations in response to Hurricanes Andrew and Iniki did reflect certain aspects of 1986 AirLand Battle doctrine. Army forces commanded a Joint Task Force (JTF) headquarters made up of personnel from multiple services. In Florida, the 82d Airborne Division and the 10th Mountain Division formed the nexus of the JTF. In Hawaii, Lieutenant General Johnnie H. Corns, the commander of U.S. Army Pacific (USARPAC), commanded JTF Garden Isle. Combat troops from the 3d Brigade, 25th Infantry Division (Light) used doctrinal procedures to establish a command-and-control headquarters capable of supporting civilian authorities. From a command-and-control perspective, Hurricanes Andrew and Iniki provided opportunities for Sullivan to convince the service that the 1986 Cold War doctrine was suitable for OOTW and that the next manual should continue to do the same.[11]

To achieve service consensus, Sullivan held a Senior Leader Warfighter Conference at Fort Leavenworth from 3–4 November 1992. The conference emphasized the strategic ramifications of a post–Cold War world, a situation fraught with uncertainty and demanding a U.S.-based force-projection army capable of meeting global challenges. Doctrine, Sullivan argued, was the vehicle for institutional change, for the ideas contained in the manual percolated throughout the force through education and training. The assembled officers were told to consider the new doctrine in terms of recent operations in Kuwait and Panama. Kuwait required a power-projection army, one that deployed from the United States and Europe and then used maneuver and firepower to destroy the enemy. Desert Storm showed that the Army must prepare to fight wars over large areas, thus those operations must retain a doctrinal basis. Panama was a simultaneous attack, more OOTW than a war. OOTW was important enough to receive its own chapter within the new manual, for those types of missions required the same depth of planning and execution as combat operations, though they differed intellectually in nature and in scope from war. After a thorough discussion, the attendees agreed that the new doctrine was a significant but not radical shift from the 1986 AirLand Battle manual. With consensus achieved, Sullivan selected 14 June 1993, the 218th anniversary of the Army's birthday, as the official release date for the new FM 100-5.[12]

The 1993 Field Manual 100-5 *Operations*

The army leadership introduced the 1993 FM 100-5 as the authoritative guide for how the service thinks about both war and OOTW. Dynamic, never static, doctrine was understood to capture lessons of the past while anticipating the future. To remind the Army of its recent doctrinal past, the preface summarized the 1976 Air-Land Battle doctrine and the Active Defense as primarily fixated upon Europe against a quantitatively superior Warsaw Pact. The doctrine had accepted force ratios as the deciding factor in war based upon armor and the combined-arms team. The 1986 manual had refined AirLand Battle and extended the deep battle while placing increased emphasis upon operational art.

The 1993 manual was viewed not as radical change but a continuation of ideas that reflected a new strategic era for army forces, one brought about by the 1986 Goldwater-Nichols Act and the end of the Cold War. The four tenets of AirLand Battle remained: initiative, depth, agility, and synchronization. A fifth tenet was added, versatility, reflecting not only the Army's continued faith in the previous tenets but that AirLand Battle had evolved into "full-dimension operations" in a force-projection environment, for which versatility was essential.[13]

Organized into fourteen chapters, the doctrine aimed to educate as well as to provide guidance. To sell the service that it must prepare for both war and OOTW, the readership was apprised of the Army's legal obligations. By law, the mission of the U.S. Army "is to protect and defend the Constitution of the United States of America." Thus, the service must prepare for any contingency that threatened the nation.

In explaining the Army's post–Cold War role in national defense, the 1993 text endorsed three levels of war—strategic, operational, and tactical—as applicable across a range of operations. This range of operations in turn consisted of three environmental states: war, in which the Army attacked or defended; OOTW, including missions such as strikes and raids, peace enforcement, and insurgency support; and peacetime support such as counterdrug, disaster relief, and civil-support missions. War preparation remained the Army's top priority, however, for "winning wars is the primary purpose of the doctrine in this manual." War and OOTW were placed within separate categories but the overall emphasis remained war, for the chapter on OOTW consisted of but nine of the manual's 173 pages.[14]

Strategic objectives were linked to tactical battles and engagements when "translated by the practice of operational art." To accomplish strategic purposes, the United States-based Army must project forces to theaters of operations. Operations Just Cause and Desert Storm were used as historical case studies to clarify the power-projection concepts of predeployment and deployment activities, theater of operations entry either opposed or unopposed, postoperation duties, and reconstitution of forces upon return to home station. The 1993 manual emphasized planning *and* execution, a significant shift from the service cultural mind-set that placed more emphasis upon "operators" than "thinkers."[15]

The 1993 FM 100-5 was organized like no manual before it. Chapter 1 ("Challenges for the U.S. Army") explained the role of doctrine, the American view of war, strategic contexts, and the training and readiness challenge for war and OOTW. The manual noted that doctrine compliance was not optional given that it was how the service communicated ideas, shared its culture, and formed the basis for school curriculum, leader development, and training. While comprehensive, it remained flexible enough to address diverse situations across the globe. It expressed ideals found within the Constitution and reflected the Clausewitzian understanding of the relationship between the government, the people, and the military. It was a product of the national strategy and national military strategy, which together shaped military operations. As a strategic force, the Army was recognized as a key member of a joint team, not a stand-alone service. To be a deployable and expansible national asset, the Army must train with other services, the United Nations (UN), and the government interagency. When the government ordered military action, the American people expected decisive results quickly and without excessive casualties.[16]

Chapter 2 ("Fundamentals of Army Operations") established the range of military operations for which the service must prepare. Three hierarchical "states of the environment" encompassed war, conflict, and peacetime. War, as a military operation to achieve national objectives, spanned the categories of war and conflict while OOTW spanned all three areas, the most heavily being weighted in conflict and peacetime. Examples of war included large-scale operations, attacking, and defending, while conflict meant strikes and raids, peace enforcement, insurgency support, antiterrorism, peacekeeping, and noncombatant evacuations of American citizens from unstable foreign countries. Peacetime operations included counterdrug, disaster relief, civil support, peace building, and nation assistance. Joint operations, combined arms, the principles of war, the Total Army (regulars, reserves, and National Guard forces), branches of the service, tenets, combat power, battle command, and other factors allowed army forces to project power globally. The chapter also described various types of joint and army forces, as well as space assets to familiarize the readers with the capabilities of each.[17]

Force projection was the primary concern of Chapter 3. Using Panama and Desert Storm as historical examples, FM 100-5 explained how army forces mobilized, organized, deployed, entered into a theater of operations, conducted missions, terminated war or conflict, redeployed and reconstituted for future operations, and demobilized. Readers were exposed to the sophistication required to project power, as well as the simultaneous or sequential actions necessary to do so.[18]

While Chapter 4 explained the details of joint operations, Chapter 5 ("Combined Operations") devoted considerable attention to describing the nuances of working with foreign powers. The U.S. government pursued national objectives through coalitions and alliances, thus relationships with foreign powers advanced American agendas. Foreign powers did the same but it was a common purpose that brought various nations together until those purposes were no longer served. Nonetheless, the Army had to grasp that strategic aims varied, doctrines did not often mesh, cultural differences had to be considered, and trust was essential.[19]

In the 1970s and 1980s, the Army had been criticized for deficiencies in strategic and operational planning. This was one of the chief reasons for creating SAMS as a selective second-year follow-on course of study to the Command and General Staff College. In what constituted an abbreviated Army War College and SAMS education in twenty-six pages, Chapter 6 was devoted to the theory and practicality of planning and executing operations. The manual explained how national-security strategy became national-military strategy, how coercive force might accomplish national objectives, the three levels of war, campaign planning, and joint and combined operations. Theoretical constructs taken from Clausewitz and Jomini described elements of theater and operational design to include centers of gravity (the hub of all power and movement upon which everything depends), lines of operation (interior and exterior), and decisive points (geographic areas that when controlled help to bring down the opposing center of gravity). The reader learned of sequencing campaign actions through phases, as well as preparing branches (contingency plans built into the basic plan) and sequels (subsequent operations as possible outcomes to the current plan: victory, defeat, or stalemate). Weapons of mass destruction (nuclear, biological, and chemical) had also to be considered. Retained from the 1986 version was the battlefield framework of deep, close, and rear operations. Historical examples were employed to help illustrate these concepts.[20]

In considering offensive operations, Chapter 8 not only provided guidance for their conduct but also how to plan them. Planning began with the already established concept of METT-T, an acronym for mission, enemy, terrain and weather, troops, and time available. The mission was based upon numerous factors, one key ingredient being the commander's intent. The intent described what was to be done and the purpose behind it, as well as an indication of future action while keeping in mind the superior headquarters mission and intentions. Enemy force dispositions, equipment, doctrine, capabilities, and probable courses of action had to be considered. Terrain considered avenues for rapid advance and protection and focused on key terrain, ground that was essential to control by physical occupation or by weapons fire. Weather affected operations and had to be considered in respect to air support, concealment, and mobility, as well as logistical support. Troop considerations involved not only what was available but the selection of the appropriate types of units for a specific mission given capabilities. Time was of the essence, for the longer one planned, the more opportunity the enemy had to prepare a defense or conduct an attack.[21]

In the defense (Chapter 10), the concept of area and mobile forms was retained from the 1986 edition. As with the chapter on the offense, emphasis was placed upon both planning and execution. In the 1993 edition, however, the reader was advised when to terminate a defense. Defense ended when "an attacking enemy, through its own maneuvers, losses, errors, exhaustion, skillful friendly defense, or other causes, may be in such an unfavorable position that the initiative passes to the defender." Once the defender had wrested away the initiative from the attacker, the prospect of a successful counterattack then allowed transition to the offensive. Without a compelling reason to defend, however, the defender attacked. Knowing when the initiative had passed to the defender was a critical component of command by understanding the situation.[22]

No army can sustain itself without proper logistics. This issue was addressed in Chapter 12 (Logistics), which described planning and execution considerations for a power-projection force during war and OOTW at the strategic, operational, and tactical levels of war. Logisticians had their own tenets, consisting of anticipation (foreseeing requirements), integration (matching operations to logistical support), continuity (uninterrupted supply flow), responsiveness (rapid reaction to crisis), and improvisation (make, invent, arrange, or fabricate whatever was needed). The 1993 manual added considerations for multiservice and multinational forces, as well as host-nation support during contingencies. A host country would potentially supply infrastructure such as buildings, airfields, seaports, and railroads, as well as police, labor, and materials that reduced the cost of the U.S. government. A newly introduced concept was split-based operations in which most logistical support assets remained in the United States while only essential goods and services were deployed to support the mission in the field.[23]

Chapter 13 ("Operations Other than War") began by restating that the Army's primary focus was to fight and win the nation's wars. It immediately qualified that assertion by noting that army forces must be prepared to operate globally in any environment. The writing team provided historical evidence to support this point, noting that the frontier army had protected an expanding nation, built roads, canals, and bridges, assisted nations abroad, and performed numerous other missions in service to the nation. OOTW missions were not new but had grown more frequent, including such things as providing "support to U.S., state, and local governments, disaster relief, nation assistance, and drug interdiction to peacekeeping, support for insurgencies and counterinsurgencies, noncombatant evacuation, and peace enforcement." Justification rested with the legal basis for the Army, the Constitution, as well as various laws that governed the service. The Army was part of a joint team of other services likewise legally obligated to conduct OOTW.[24]

OOTW operations were considered unique. As they were not war, not all of the principles of war applied. OOTW operations had their own principles, which consisted of objective (a clearly defined, decisive, and attainable purpose), unity of effort (endeavoring toward a common goal), legitimacy (supporting the host government's right to govern), perseverance (commitment for the duration), restraint (following rules of engagement to avoid excessive force), and security (force protection and an ability to shift to combat operations, if needed). OOTW operations included antiterrorism, attacks and raids, drug interdiction, insurgency and counterinsurgency support, nation assistance, noncombatant evacuation operations, peace enforcement and peacekeeping, and shows of force. These operations lacked a unifying theoretical basis for determining how to plan and execute them; it was far easier to describe them in broad terms.[25]

The final chapter ("The Environment of Combat") was a tutorial regarding the human and physical dimensions of modern war. The human dimension approach to war emerged from psychological studies of the effects of combat upon soldiers. It consisted of both the psychology of understanding the human condition in war that Clausewitz had espoused in the 1830s, as well as ethics and a strong respect for law, human dignity, and individual rights. Officers and soldiers must be taught that war has its limits concerning acceptable levels of brutality and collateral damage. They should do what was right based upon American constitutional ideals. Descriptions of the physical dimension changed little from previous manuals and included cold weather, desert, jungle, and mountain operations, as well as urban terrain.[26]

Soon after the manual's release, the Army began a publicity campaign to sell it. In the December 1993 edition of *Military Review*, virtually the entire publication was devoted to FM 100-5. A number of general officers submitted articles discussing its impact upon future war and OOTW. General Franks wrote that "the [1993 FM 100-5] goes beyond AirLand Battle to full-dimensional operations, with the Army at the center of the joint team addressing the fundamentals and inherent requirements for a force-projection Army." In articulating a new doctrine for a new strategic environment, the manual incorporated "five warning lights" of change that had directed the service to enter a new era. The new doctrine had made "bold adjustments in how we think about warfare, warfighting, and the conduct of operations other than war [OOTW]." In Somalia, Rwanda, and Haiti, bold adjustments were needed as army forces struggled to apply the 1993 doctrine under diverse and often perilous circumstances.[27]

Straddling Doctrine: Operation Restore Hope

In December 1992, army forces deployed to Somalia in support of the UN Operation Restore Hope. Given that the 1993 manual was still in draft form at the time, army forces operated under the 1986 AirLand Battle doctrine. As a Cold War doctrine, the UN had not been mentioned. AirLand Battle doctrine addressed combined headquarters in broad terms, stressing that army forces must "achieve unity of effort and mutual trust," especially in the areas of "interoperability, well understood [command and control] structures, liaison, and interpreters." Doctrine shed little light upon the bewildering maze of UN bureaucracy in war-torn Mogadishu.[28]

In May 1992, army forces discovered the incompatibility of UN procedures and the 1986 AirLand Battle doctrine during the UN Operation Somalia (UNOSOM) II. No Soviet echelons existed to be struck by deep attack. Unlike North Atlantic Treaty Organization (NATO) command structure in Europe, the UN in Somalia had no government apparatus with which to operate. Instead, the UN headquarters was compelled to engage a variety of entities simultaneously: renegade warlords, a military JTF headquarters, and numerous independent relief agencies. Without any predeployment training to create a cohesive multinational headquarters, the UN could not manage the complexity of daily operations. Army personnel became frustrated, for the 1986 AirLand Battle doctrine did not match Somalia's realities.

Still, the 1986 FM 100-5 had acknowledged that combined operations would be complicated as many nations lacked doctrine. The same was true for the UN, which was without operational doctrine and common procedures. Each national force was expected to implement its national military doctrine, if any existed. During UNOSOM II, most Western nations had some form of doctrinal precepts, but others did not and operated solely by word of mouth. Army personnel attempting to implement AirLand Battle doctrine found their manual incompatible due to "significant differences in doctrine, training, equipment, capability, and size of national contingents" that made interoperability laughable. Army personnel spoke their own doctrinal language, a conglomeration of abstractions such as "attached," "operational control," and "tactical control."

Those terms meant specific things to the Army, but each one had different meanings in UN parlance. Since most nations were not familiar with U.S. Army doctrine, many of these terms were meaningless. Life within the UN headquarters in Somalia came down to individuals trying to do something that was familiar to them but not necessarily to their coalition partners.[29]

The American JTF headquarters in UNOSOM II also proved a troubling place to work. The 1986 AirLand Battle doctrine had devoted less than two pages to the inner workings of a JTF and in the broadest of terms. The doctrine advised that within a joint headquarters "each service's doctrine and applicable joint doctrine will guide employment." Thus, AirLand Battle, a doctrine unsuitable for Somalia, made day-to-day operations incompatible with the other national services. Although the manual was available in the headquarters, and 80 percent of the JTF staff during UNOSOM II consisted of army personnel, most of the augmentees lacked previous joint experience since it had not been a prerequisite for being nominated for the headquarters. Many of the soldiers, regardless of nation, were not familiar with their doctrine or, in the case of the 10th Mountain Division, had trained for tactical combat operations only. Those troops possessed but an embryonic understanding of multiservice considerations and in rudimentary terms.[30]

When the 1993 FM 100-5 replaced the 1986 AirLand Battle doctrine in June 1993, army forces in Somalia under UNOSOM II had little if any exposure to it. Some newly arrived field-grade officers had just graduated from Fort Leavenworth, where they had received instruction regarding the new doctrine. But their knowledge was limited and, given the tempo of operations, they had little chance to explain it to their compatriots. In truth, the 1993 doctrine had little effect upon army operations in Somalia. Few individuals had a copy of the manual or the time to train to its precepts.

When open warfare between UN forces and warlord-controlled thugs began in October 1993, army tactical units used combat procedures that, according to one commander, transcended any doctrine. In an infantry action reminiscent of the Army's experiences in Russia (1919), Grenada (1982), and Panama (1989), army forces engaged Mohammed Fahar Aideed's followers by emphasizing firepower and limited maneuver within the urban confines of Mogadishu. As the soldiers were fighting in urban terrain, the operation turned into a bloody slugfest. Army soldiers killed an estimated 1,000 Somalis at a cost of 19 Americans dead and

84 wounded. An after-action report later concluded, "The commander of U.S. forces did not have the combat-capable staff that a combatant commander would normally have to plan and execute combat operations."[31]

Army operations in Somalia prompted some senior general officers to reconsider the long-held service perception that war was more complex than OOTW. Major General Steven L. Arnold, the commander of army forces in Somalia, concluded that OOTW operations were in fact more convoluted than war "due to different doctrines and cultures." Others within the service still believed that poor leadership, not doctrine, had been the root cause of the Army's Somalia woes. "Leaders," noted one report, "just need to adjust the mindset of soldiers when executing operations other than war." When things turned sour, leaders, not doctrine, were to be blamed.[32]

Rwanda and the 1993 Doctrine

As the 1993 doctrine had been issued in June 1993, its impact upon the 1994 Rwanda crisis was minor. The nature of the mission was primarily political in the form of logistical support. On 22 July 1994, after the U.S. government acknowledged that the situation in Rwanda was genocide, President William J. Clinton ordered European Command (EUCOM) to provide humanitarian support to Rwanda and Zaire. The U.S. armed forces were less than enthusiastic. Stung by the casualties suffered under UNOSOM II, Chairman of the Joint Chiefs of Staff General John M. Shalikashvili believed that the armed services could help in easing human suffering. However, Shalikashvili also believed that the UN was better suited for such work. He, like many within the Department of Defense, cringed "at the notion of [the armed forces] becoming a super, muscle bound Red Cross or Salvation Army." Humanitarian operations such as Rwanda "cut into training exercises, [tied] up equipment and personnel and [took] scarce defense dollars away from other operations."[33]

The 1993 army doctrine had dedicated but two paragraphs to humanitarian assistance and disaster-relief operations. Humanitarian operations involved "[using Department of Defense] personnel, equipment, and supplies to promote human welfare, to reduce pain and suffering, to prevent loss of life or destruction of property from the aftermath of natural or man-made disasters." A disaster-relief operation fell "within the overall context of humanitarian assistance" and was conducted in extreme circumstances to prevent loss of life and property. The Army's role was to "provide logistics support to move supplies to remote areas, extract or evacuate victims, establish emergency communications, conduct direct medical support operations, and render emergency repairs to vital facilities."[34]

In July 1994, army forces joined a multinational military coalition and international relief agencies to support a UN humanitarian assistance effort in Rwanda. Army forces served as staff members in support of the UN High Commissioner For Refugees. At the tactical level, the 21st Theater Army Area Command followed doctrinal precepts in operating water distribution and purification systems at Goma, Zaire, and a cargo distribution center at Entebbe, Uganda. Conventional army forces provided airfield terminal operations services and logistical support for UN forces and army Special Forces personnel coordinated relief efforts with civil authorities.[35]

Although the 1993 doctrine contained a broad framework for providing humanitarian assistance, it had not described the foibles of working amid a sea of international relief agencies whose members had no experience with military procedures. The relief agencies had but one purpose; to provide relief for the suffering while protecting human rights. Lacking transportation and security, relief agency representatives flooded army force commanders with requests for vehicles, aircraft, and security patrols to help in gaining access to dying people. Army leaders lacked sufficient resources to honor such requests and at least one relief agency representative was told to "take a hike."[36]

At the operational level of OOTW, the 1993 Field Manual 100-5 provided insufficient guidance for helping army personnel to understand how to establish support to civilian agencies. Friction arose between army personnel and relief-agency personnel because of unfamiliarity with dissimilar bureaucracies. In its effort to meet legal obligations and ensure a place within the joint community, the 1993 doctrine had announced the service's capacity to conduct humanitarian operations and disaster relief by decreeing that "the Army's global reach, its ability to deploy rapidly, and its capability to operate in the most austere environments make it ideally suited for these missions." Yet, few within the service had any experience working under austere conditions, which showed in Rwanda.[37]

Although doctrine established that humanitarian operations were constitutionally-based OOTW missions, some army personnel found the Rwanda situation to be repugnant. Warrior-trained soldiers were torn between their ingrained warfighting mentality and calls for them to assist international agencies and refugees. Occasionally, soldiers complained in front of relief agency representatives, who concluded that the Army's mission was to leave at the first opportunity. Right or wrong, that perception threatened to undermine U.S. government policy in supporting the UN. As the doctrine had stated, strategic issues were linked to tactical actions and, in this case, vice versa. To end relief-worker allegations that the Army and therefore the U.S. government did not care, army force commanders worked hard to establish liaison with the civilian relief agencies. Establishing liaison was a doctrinal requirement in any military operation. In this case it did ease tensions somewhat between army forces and the relief workers. Still, some civilian agencies refused to cooperate with the military bureaucracy and commanders became frustrated when their liaison efforts were rebuffed. The Rwanda operation lasted but seventy-seven days before army forces handed the operation over to the UN.[38]

In October 1994, the Army Peacekeeping Institute at Carlisle Barracks, Pennsylvania, invited civilian and military representatives to a conference to discuss the Rwanda operation. The attendees concluded that a common doctrine might resolve differences between relief agencies and the military. However, Lieutenant General Daniel Schroeder, the commander of the JTF in Rwanda, disagreed. "The armed forces," he observed, "exist to fight and win the nation's wars." The Rwanda operation was, in Schroeder's opinion, a success not due to OOTW doctrine but because of a very clear and precise mission statement. He added, "I think we need to stay focused on the purpose of the Armed Forces." Schroeder's comments were one indication that the 1994 army had not fully embraced the 1993 doctrinal principles that the Army must conduct both war and OOTW operations due to its legal obligations.[39]

Haiti and the 1993 Doctrine

Although the Rwanda operation had been executed under the auspices of the 1993 FM 100-5, the 1994 Haiti affair constituted the doctrine's first significant test due to its scope and circumstances. Involving thousands of troops from all the services and Caribbean nations, the Haiti operation to restore junta-deposed Haitian President Jean Bertrand Aristide to power was a doctrinal nightmare. Where FM 100-5 had envisioned the possibility that an OOTW might escalate to war and perhaps back again, the warfighting portion of the manual had not considered the opposite prospect. The operation was planned as a war, not OOTW. Only last-minute negotiations between former President Jimmy Carter and junta leader Raoul Cedras prevented an invasion to seize the country by force.[40]

Because the 1993 FM 100-5 contained lessons from the 1989 Panama invasion, army planners had used certain passages in planning the invasion. Given Haiti's island location, size, and terrain, army officers working within headquarters such as U.S. Atlantic Command at Norfolk, Virginia, and XVIIIth Airborne Corps at Fort Bragg, North Carolina, used the Panama invasion concept of simultaneous attack to secure Haiti in one night.[41]

The 1993 doctrine had envisioned a simultaneous attack from the homeland against a hostile force anywhere in the world. In Haiti, this involved a forcible-entry operation launched from the United States against a junta. Major ground forces included the XVIIIth Airborne Corps acting as the Joint Task Force (JTF) 180 headquarters, Army Forces (ARFOR) consisting of the 82d Airborne Division, and a Joint Special Operations Task Force (JSOTF) composed of Special Forces, army Rangers, and navy SEALS, along with Marines. The ARFOR would establish three lodgments, protect U.S. citizens, property, and designated foreign nationals, and neutralize the Haitian armed forces and police. The JSOTF would secure Port-au-Prince International Airport, the National Palace, Dessalines Barracks, the Haitian 4th Police Company Headquarters, and Camp d'Application. Delivering the forces fell to the 12th Air Force and naval forces offshore. Logistical support included deploying with five days of supply on hand, three days of emergency backup, plus four airheads and five intermediate staging bases in the United States and the Caribbean. In anticipation of turning the operation over to the UN at some point, the Army's 10th Mountain Division was to be prepositioned offshore to release the 82d Airborne Division for future contingencies, if they arose.[42]

On the evening of 18 September 1994, as the invasion unfolded, the negotiations team headed by former President Jimmy Carter reached agreement with the Cedras-led junta. The combat invasion was terminated; the invasion force had to shift from a war situation to an OOTW operation within hours. The 82d Airborne was spared the chaotic transformation when the division's air force aircraft returned to home station with its paratroopers aboard once the invasion was aborted.

The 1993 FM 100-5 had cautioned that OOTW operations "often undergo a number of shifts in direction during their course." And, as Chief of Staff General Gordon R. Sullivan had noted previously, troops were expected to be fully capable of OOTW operations when given proper leadership and training. Both of those factors emerged in Port-au-Prince on the morning of 19 September, as soldiers attempted to shift from a war mentality to OOTW. JTF 180, under the command of Lieutenant General Henry H. Shelton, adapted quickly despite Shelton's being placed in the nondoctrinal role of chief negotiator with the Haitian junta. The 10th Mountain Division, led by Major General David C. Meade, had adjustment difficulties. At Fort Drum, New York, the division had prepared for combat, not OOTW. When the soldiers arrived at the Port-au-Prince airfield via helicopter, they dropped to the tarmac in combat gear and brandished their weapons. An officer from the American Embassy walked out to greet them in summer uniform, shined shoes, and sunglasses. Despite assurances that the Haitian junta had capitulated, the 10th Mountain Division troops soon fortified themselves within Port-au-Prince's light-industrial complex behind sandbags and barbed wire.[43]

In truth, because their operational approaches did not mesh, two doctrinally different American armies arrived in Haiti. Conventional forces followed the warfare aspects of FM 100-5, not OOTW. Army Special Forces units under Brigadier General Richard Potter followed the OOTW precepts established within FM 100-5 and FM 100-25, *Doctrine for Army Special Operations Forces*. The 1993 FM 100-5 recognized that SOF, not conventional forces, were more suitable for "insurgencies and counterinsurgencies, contingency operations, peace operations, and counterterrorism operations." SOF were employed at the operational level in OOTW "as part of the theater commander's joint special operations effort," thus were not under ARFOR control and thus not subject to its rules. As the 1991 edition of FM 100-25 stressed, although it was compatible with FM 100-5 (the 1986 AirLand Battle version), special operations were different than conventional ones.[44]

The doctrinal differences between conventional and Special Forces units were readily apparent. The 10th Mountain Division soldiers lived in seclusion from the populace, while the Special Forces resided among the locals, "giving them what they want, food, water, electric power, everything." Conventional troops wore helmets and body armor, and flaunted weapons as they interacted with the Haitians. Special Forces personnel discarded helmets and body armor, slung weapons over their back, and worked among the populace. When a freelance journalist reported two American armies in Haiti, he had inadvertently identified two divergent and incompatible doctrines. Philosophical differences affected the mission but also reflected a cultural rift within the service that doctrine alone did not overcome.[45]

Operation Uphold Democracy illustrated that while the 1993 FM 100-5 listed OOTW principles that must be considered, the chapter lacked sufficient detail to be more than just a general overview. Partially to blame was that the Army lacked a theoretical basis to properly prepare Cold War-minded conventional army leaders and soldiers for OOTW. Although warriors were supposed to make mental adjustments from war to OOTW, army leaders remained wedded to the Cold War warfighting principles of AirLand Battle. In the case of the 10th Mountain Division, home-station training had fixated upon weapons qualification, firepower, and maneuver. In 1996, although some analysts called for training two armies, one for war and the other for OOTW, the Army's Peacekeeping Institute at Carlisle Barracks, Pennsylvania, nonetheless concluded that combat forces were adaptable to nonwar circumstances given "the resources, international and domestic support, time to prepare, and leaders who are mentally agile." Finding mentally agile leaders who could cope with the post–Cold War OOTW of the early 1990s, however, was a rare occurrence.[46]

Still, General Gordon R. Sullivan demonstrated mental agility by seeking to influence the UN mission handover from U.S. forces. In fall 1994, as U.S. forces invaded Haiti, Sullivan directed that a U.S. training team meet with members of the UN Mission in Haiti (UNMIH) to instruct them in U.S. decision-making procedures. Given that UNMIH was to assume responsibility for the mission at some point, his intent was to avoid repeating the disastrous results of Somalia where the UN headquarters proved to be dysfunctional due to variances in military doctrine or the absence thereof. Prior to UNMIH mission assumption, the training team held meetings at Fort Leavenworth, Kansas, and in Port-au-Prince, Haiti. Headquarters members not only received briefings from doctrinal experts but also participated in team-building exercises that simulated potential missions and crises. The result was a cohesive headquarters, as acknowledged by UNMIH members. The UN, however, did not adopt mandatory pre-mission headquarters training for subsequent operations, for reasons that are unclear.[47]

The Search for Appropriate Doctrine

By late 1994, in light of the recent Somalia and Haiti experiences, army leaders raised disconcerting questions about the 1993 version of FM 100-5 *Operations*. The manual had confused readers who saw similarities between war and OOTW. Analysts and others struggled to categorize Somalia and Haiti as war, OOTW, or something else in order to explain what happened and draw lessons from them. They concluded that warfare and OOTW (less natural disasters) shared common ground in that they both attempted to alter an adversary's policy. While some believed that war and OOTW were distinct intellectual constructs, others saw too much commonality among Desert Storm, northern Iraq, Somalia, Rwanda, and Haiti, as all five operations attempted to change an adversary's behavior.[48]

By 1995, the distinctions between war and OOTW had blurred to the point where the 1993 doctrine no longer served its purpose in changing the Army. Army Chief of Staff General Gordon R. Sullivan realized that the service required a new doctrine to rectify the divergence between war and OOTW, to reconcile emerging joint service doctrine with army views, and to further move the service into the information age. However, Sullivan retired in June 1995 and left doctrine development to his replacement, General Dennis J. Reimer. Later, in fall 1995, Reimer ordered General William W. Hartzog, now the commander of TRADOC, to develop a new army doctrine in accordance with Force XXI initiatives, a program to transform the Army for the twenty-first century. On 27 October 1995, Hartzog directed Lieutenant General Leonard D. Holder Jr. at Fort Leavenworth to form a writing team.[49]

Hartzog's guidance to Holder was an attempt to reconcile the debate over war and OOTW. He told Holder that the service was spending too much effort engaged in fruitless arguments when accomplishing missions mattered more. Holder was to "not treat [war and OOTW] as separate and special subsets," and that "[the term] OOTW should not appear in this update of 100-5." Instead, Hartzog wanted the new doctrine to stress the "joint, interagency, and combined aspects of warfare," and to move the service into the national strategic level more than the case previously.[50]

As Holder received Hartzog's guidance, the Clinton administration published the 1996 version of the *National Security Strategy of the United States*. That document called for engaging foreign governments through "military strength, our dynamic economy, our powerful ideals and, above all, our people." The Army and the other services were to deter aggression through a credible overseas presence, countering weapons of mass destruction, and contributing to peace operations while supporting counterterrorism policies, the drug war, and other missions, as directed. Holder's writing team, located within SAMS at Fort Leavenworth, began with an analysis of the *National Security Strategy* and Hartzog's guidance as the Army deployed forces to Bosnia.[51]

The 1993 FM 100-5 in the Balkans

In December 1995, President Clinton ordered military forces to Bosnia and Herzegovina under the provisions of the Dayton Accords. Operation Joint Endeavor (as the mission was called) was a force-projection operation, as described in Chapter 3 of the 1993 FM 100-5. Army forces deployed via air, water, rail, and highway to Bosnia, where they disarmed the Former Warring Factions along a zone of separation or ZOS.[52]

Although the 1993 doctrine had envisioned a force-projection operation from the United States, it had not anticipated the complexities that arose in the Balkans. Deploying the forces became problematic, for although the 1993 doctrine counseled that soldiers must prepare for the possible use of diverse multinational information systems, many army plans officers had not fully realized what that implied. In an attempt to access computer databases for information about unit readiness, officers discovered that computer programs used by the U.S. military and NATO to identify and prioritize troop unit movement and capabilities failed to interface properly. One army transportation officer used a pencil and paper to record information from one computer system and then manually typed the information into two other systems, a routine that consumed hours.[53]

Bosnia was a former war zone and the majority of the Army's missions constituted postconflict operations. The Army's 1993 manual had devoted only a single paragraph to postconflict considerations, which advised commanders to plan for those activities prior to deployment. The manual had foreseen that army forces would need to transition from war to peace but expected this to occur only after fighting a conflict where units were already located within the theater of operations. The Bosnian Civil War had ended with army forces outside of the theater of operations, a circumstance that doctrine had not anticipated.

Doctrine had not considered the transfer of an ongoing operation from one coalition to another. When the Dayton Accord-sanctioned Implementation Force (IFOR) came into being, the UN was already conducting a mission in Bosnia under the auspices of the UN Protection Force (UN-PROFOR). Once the Dayton Accords went into effect in December 1995, numerous UNPROFOR units fell under NATO command with some forces positioned within the Army's designated positions near Tuzla. Falling back upon informal practice, army commanders had to gain control over the former UNPROFOR units while simultaneously deploying troops from the United States and the Federal Republic of Germany. Moreover, some IFOR units that were placed under army control were from former Warsaw Pact members—Russia, Poland, Lithuania, Latvia, and Estonia. The 1993 doctrine, although written for post–Cold War circumstances, had not envisioned cooperating with former adversaries operating under Soviet doctrine or matters such as sharing classified information and intelligence.[54]

Army forces deployed using tactical methods described in the 1993 doctrine. Converging on Bosnia from multiple directions, they conducted an operational force entry and a tactical movement to contact. But the operation was an OOTW, so much so that General William Crouch, commander of USAREUR and Lieutenant General John Abrams, the Vth Corps commander, were concerned that the Bosnia operation would affect combat skills. They used government funds to renovate a Hungarian training area so that army units from Bosnia could conduct gunnery practice and then recuperate before returning to their Balkan mission. One American general stated that continued training in conventional warfighting methods during OOTW reduced decay of conventional warfighting proficiency.[55]

Operationally and tactically, army forces either adapted 1993 doctrinal principles to suit Bosnia's realities or simply worked through the void. But doctrine proved less adequate at the operational level than the tactical. Although the manual provided certain precepts for conceiving campaigns, major operations, battles, engagements, and OOTW, Bosnia proved more complicated than doctrine had imagined. The Balkan mission confirmed that doctrine cannot anticipate every possible contingency.

The Search Continues

In 1996, Holder directed the Fort Leavenworth writing team to consider recent missions in Haiti and Bosnia when merging war and OOTW within one doctrinal concept that unified all army missions. After many spirited debates, the writing team settled upon "land dominance." Land dominance implied that army forces would aim to impose their will upon any adversary regardless of mission. Borrowing from Carl von Clausewitz, doctrine writers decided that imposing the commander's will upon the enemy was valid for every possible operation.[56]

While land dominance seemed an appropriate construct for intellectually blending war and OOTW, Holder rejected it as too vague and raising more questions than it answered. Holder reminded the writers that the TRADOC commander wanted to avoid intellectual debate. The writing team then analyzed the service's experience from the 1980s to the mid-1990s to find common patterns among the missions. After several weeks, the team members determined that each contained some form of the offense and the defense, as well as requirements to stabilize areas under service control and to support U.S. government policies. The Cold War had been a defense with offensive potential in support of government policy. In Desert Storm, the primary emphasis was the offense, but coalition forces initially defended and later stabilized Kuwait in support of government policy. In northern Iraq, army forces supported a government policy to intervene by stabilizing the refugee camps and defending against Iraqi incursions while preparing to attack if necessary. Somalia, Haiti, and Bosnia all shared degrees of defense and offense commonalities, as well as aspects of stability and support of government policy. By identifying mutual elements shared by the diverse missions, the writing team had merged war and OOTW. The concepts of the offense, defense, stability, and support were shortened into the acronym "ODSS."[57]

In August 1997, a final draft of the revised FM 100-5 emerged. Labeled the "Holder Version," the draft "confirmed that the nation will, as it has for over 200 years, call on the Army to conduct a wide array of operations beyond the scope of all-out war." It was clear that the Army had performed a variety of operations over its history, but the draft maintained that FM 100-5 was first and foremost a warfighting manual. Regardless of war or OOTW, modern army operations were joint and multinational actions involving four instruments of national power—diplomatic, informational, military, and economic—to achieve national-policy objectives. The American soldier was the linchpin who executed the doctrinal principles that sought to achieve policy objectives. Five core service functions dominated army operations: see, shape, shield, strike, and move. These five functions were "inseparable parts of the whole" and essential when executing the ODSS aspects of army missions. To explain the five functions, the writers included historical vignettes from the Gulf War, northern Iraq, Haiti, and Bosnia.[58]

Although members of the writing team periodically rotated due to service assignment obligations, the 1997 draft maintained the 1982, 1986, and 1993 doctrines' emphasis upon operational art. Operational art described how strategy, army operations, and tactics were interrelated concepts in the execution of a campaign. In a change from the 1993 doctrine, the text devoted greater detail describing interacting with other services and allied coalitions. The draft also expanded the concept of power projection through further elaboration upon the procedures for mobilizing forces, deploying and employing army troops, conflict termination considerations, and demobilization activities. The authors intended to furnish sufficient details to guide future operations while leaving some room for commanders to modify the precepts as necessary.[59]

While the 1997 draft calmed much of the intellectual debate concerning war and OOTW that the 1993 doctrine had generated, it failed to reconcile the doctrinal quirks between conventional forces and SOF. The 1997 authors believed the conventional army dominated the service and that the special operations community needed to adjust its doctrine to comply with FM 100-5. Interestingly, SOF were listed separately from the conventional Army as equal partners with the Air Force, Navy, Marine Corps, multinational coalitions, and various government and nongovernment agencies because "many senior Army officers viewed them that way anyway." As with the 1993 doctrine, SOF remained under the operational control of a JSOTF.[60]

The 1997 draft FM 100-5 was a significant effort to blend war and OOTW into one theoretical approach. Writing team members were confident it would meet with approval. One officer, Lieutenant Colonel David Fastabend, published several journal articles announcing the arrival of the latest army doctrine. Events, however, soon made those articles moot.[61]

In the summer of 1997, Lieutenant General Holder retired. Lieutenant General Montgomery C. Meigs replaced him as commanding general of the Combined Arms Center. Meigs was an army intellectual with a keen interest in doctrine. Shortly after assuming command, he asked for a draft manual concept briefing. Part way through the presentation, Meigs ordered the briefer to stop. He then announced that something was amiss. He would personally review the draft. In September 1997, Meigs returned the manual with extensive edits and the written comment: "I don't like it." In his opinion, the 1997 version neither spurred intellectual debate nor adequately reflected modern warfare.[62]

Before arriving at Fort Leavenworth, Meigs had commanded the 1st Infantry Division in Bosnia. There he had found himself caught between the competing requirements of national strategy, operations, and tactics. If doctrine was to regulate the chaos of modern conflict, then it should instruct general officers in how to extract themselves from predicaments by practicing operational art. He ordered the writing team to create "a literary narrative to engage general officers in discourse" about operational art and contemporary military operations. Not surprisingly, the writing team was livid. Meigs had tossed out months of hard work.[63]

Facing a deadline of Christmas 1997, the writers immediately began to convert the manual from a text to guide the service in preparing for future operations to one more suited for educating general officers. After several additional revisions, Meigs approved the revised final draft in June 1998. The "Meigs Version," as the 1998 draft became known, stressed that warfighting remained the Army's primary focus and most dangerous enterprise. Drawing upon Title 10 U.S. Code for guidance, the manual acknowledged that the service's purpose was to organize, train, and equip forces for prompt and sustained combat incident to land warfare. Preparing for war was the manual's primary aim, but the draft also created a discussion over the full range of violent and nonviolent operations that army forces might face in the future.[64]

Where the "Holder Version" had attempted to quell controversy, as Hartzog had demanded, the "Meigs Version" deliberately sought to generate debate. The 1998 draft was written to "explain a theory of operational art, which links tactical means to strategic ends," and to "describe how the Army fights." The draft "[established] the Army's doctrine for operational-level warfighting and contingency operations" in attaining national policy objectives. Historical examples throughout the draft were designed to foment professional discussion over the complex situations faced by general officers when they endeavored to achieve national policy aims.[65]

As a military historian, Meigs was interested in the Army's past as a means to justify his doctrinal precepts. He inserted a section into the 1998 draft that provided an overview of how army keystone doctrine had evolved. It explained why it had changed over 200 years. His intent was to educate army generals regarding how foreign military intellectuals had influenced doctrine and the service's role in achieving national policy aims. However, the manual contained its own American views of warfare.[66]

The 1998 version elaborated more upon the strategic level of war than the 1997 draft. To assist general officers in fathoming the intricacies of operational art, the manual included four historical case studies. The reader was expected to gain an appreciation for how senior army commanders had used theoretical constructs in the past to design military operations that attained policy aims. Meigs had achieved his purpose; the manual was ready to stir intellectual debate and to guide general officers in overcoming difficulties when positioned between strategic national-policy goals and military-operational realities.[67]

In June 1998, Meigs released the draft manual to a selected audience for comment. Within days, SAMS received mostly negative criticism. Many reviewers saw the manual as an intellectual exercise suited for the academic "ivory tower." Others claimed that the manual lacked tactical substance and wasted time philosophizing about the vagaries of national strategy. The manual was allegedly a doctrine for operational warfighting, but the reviewers were not impressed.[68]

In the summer of 1998, General Hartzog retired as the TRADOC commander and was replaced by General John Abrams. Lieutenant General Meigs received a promotion and left Fort Leavenworth to assume command of USAREUR. His replacement, Lieutenant General William M. (Mike) Steele, was handed the 1998 draft. He soon discovered what one officer described as "a tempest in a teapot that threatened to boil over." After reviewing the 1998 draft, Steele agreed with the reviewers and concluded that the service would never accept it. Between September and December 1998, the writing team waited for guidance while many members were reassigned and replacements brought onboard.[69]

In January 1999, Steele told the new writing team leader, Lieutenant Colonel Steve Rotkoff, to resurrect the 1997 "Holder Version." He directed Rotkoff to discover where the reviewers had disagreed with the doctrinal concepts and to "fix them." Steele's review of both the 1997 and 1998 drafts had also revealed that they paid little attention to SOF and the reserve components. He wanted to ensure that those elements were woven throughout the text. To assist in that effort, officers from the SOF community and the army reserve components were assigned to the writing team. By 1 October 1999, a series of draft concept papers had appeared. Steele vetted each of them before sending them to TRADOC. While Steele and Rotkoff awaited approval, a modern war was being waged in the Balkans.[70]

Doctrine and Kosovo

Between 24 March and 9 June 1999, NATO forces led by the United States conducted an air war against Yugoslavia in an effort to stop human-rights abuses against the citizens of Kosovo. Operation Allied Force was a bombing campaign that struck targets in Serbia while attacking Serbian troops and supply routes in Kosovo. General Wesley K. Clark, supreme allied commander, Europe, waged what he called an undeclared "modern war," the first fought by NATO. Unlike past wars, it did not involve national or democratic systems' survival. Populations and conscript armies never mobilized. Economies did not turn to wartime production. Modern war meant that it was limited and constrained in geography, scope, weaponry, and effects. Escalation was excruciatingly weighed. Diplomacy continued with numerous actors both in support of and opposed to military action. Confidence building and conflict prevention came to the fore. Technology played a key but restrained role. Casualty aversion and accidental civilian property damage was carefully considered. Victory was carefully defined.[71]

In Kosovo, a doctrinal force-entry option was not the primary effort, for the Army and combined ground forces operated in support of the air component. Still, after months of deliberations over the use of ground forces, the UN passed Security Council Resolution 1244 on 10 June 1999 creating the Kosovo Force (KFOR). A combined NATO and non-NATO ground force, KFOR's initial mission included deterring future hostility among Kosovo, Yugoslav, and Serb forces, maintaining a secure environment, demilitarizing the Kosovo Liberation Army, supporting an international humanitarian effort, and working with an international presence.

Kosovo was a peace-enforcement OOTW operation as defined by the 1993 manual. The manual provided little guidance, however, other than noting how coercive or implied coercive means to cause factions from fighting each other while remaining neutral might be difficult. In Kosovo, army troops established Camps Bondsteel and Monteith as part of Multinational Battle Group East, coordinated efforts with partner troops from Greece, Lithuania, Poland, and Ukraine, provided security, and interacted with the local populace. As doctrine had indicated, gathering information to be processed into operational intelligence was essential. American soldiers initially responded to a significant number of violent incidents directed at troops and civilians to include bombings and shootings, as well as criminal activity and protest. Although the American sector saw far more incidents in the early months than any other, army forces ultimately succeeded in quieting the region. Troop rotations continued, however.[72]

Toward a New FM 100-5

In February 2000, the writing team received TRADOC approval of the Steele-directed concept papers. A content summary booklet, a chapter-by-chapter executive précis, contained an intellectual discussion of the "big ideas." The big ideas established "full-spectrum operations" as the manual's theme. Full-spectrum operations addressed "the range of operations Army forces conduct in war" and introduced a new concept, military operations other than war (MOOTW). In 1997, the Holder Version of the draft had attempted to merge war and OOTW into ODSS. Steele again separated war and OOTW into two intellectual categories, as the 1993 manual had done, but changed OOTW to MOOTW to "reflect that we are talking about military operations in this doctrine." As Steele acknowledged, "the Army keeps those terms separated anyway."[73]

Between March and April of 2000, the writing team revised the manual and solicited comments from over fifty agencies and organizations. In May, the team again revised the 2000 draft and sent a comment copy to all senior army headquarters, other services, and numerous government agencies. By June, the SAMS writing team had received more than 1,000 individual responses. TRADOC also directed Steele to change FM 100-5 to FM 3-0, in accordance with the joint doctrinal manual numbering system.

By July 2000, Steele reached consensus with the senior army leaders and the manual was sent up the chain of command for approval. The manual's publication was delayed because the army chief of staff, General Eric Shinseki, had other priorities. While awaiting approval, the TRADOC commander agreed to publish the draft manual on 1 October 2000 as U.S. Army Command and General Staff College Student Text 3-0, *Operations*. On 14 June 2001, Shinseki approved the manual and the 1993 edition was replaced.[74]

The 2001 FM 3-0 *Operations*

The 2001 version of FM 3-0 *Operations* was written to be a transitional doctrine. Its intent was to reorganize the service from a Cold War "heavy or legacy force" dominated by mechanization to an "interim force" containing both mechanization and more modernized forces by 2007. Significant modification of service organization was expected to commence in 2005, which would result in another doctrine revision by 2007. The 2007 manual was expected to prepare the Army for missions until 2010, when the "objective force," a fully modernized army, would emerge.[75]

The manual was organized differently than any previous edition. Full-spectrum operations now spanned war and MOOTW. The four ODSS concepts cut across all operations. Historical vignettes, spanning not only long-past events such as the American Civil War but also more recent ones such as Haiti, explained key concepts.

In seeking to avoid the intellectual furor set off by the 1993 manual, the 2001 version returned to the spectrum of conflict idea offered in 1962. The latest theoretical model, however, placed at one end of the spectrum the less violent peacetime competition that occurred daily between nation-states. Army forces participated in MOOTW at this theoretical pole by providing forces to promote peace through foreign internal defense, combined exercises with allies, and liaison teams. In the middle of the spectrum, MOOTW escalated to smaller-scale contingencies short of actual war, such as Bosnia, where the service sought to deter war and resolve conflict. War was at the opposite pole.[76]

Unlike the 1997 or 1998 drafts, the 2001 version devoted one chapter each to the nuances of offense, defense, stability, and support operations to ensure that commanders viewed ODSS operations as inextricably linked. While all four ODSS operations were regarded equal in importance, the offense was "the first among equals." But each element of ODSS had to be taken seriously, thus the doctrine was the first to acknowledge that preparing for stability and support duties were equal priorities to war preparation. For senior army leaders to accept this as fact was no small matter as the service had avoided acknowledging such roles in its keystone doctrine since 1779. Indeed, upon reading the chapters on ODSS, one retired four-star general stated: "This manual takes the service down a slippery slope." While he understood that the Army must perform every task assigned by civilian authority, he did not want to read about it in the service's keystone doctrine.[77]

The preface revealed that the army leadership had changed its views of what a keystone manual should be. It established officially that doctrine was a principal tool for professional education in the art and science of war, not just a guide for how to conduct it. As a warfighting-focused manual, it did not express a specific intent to educate the Army in MOOTW. Yet, it did stress that warfighting meant dominating land warfare and land warfare meant dominating any MOOTW situation. FM 3-0 was the first doctrine to conform to political correctness pieties, noting that, "Unless stated otherwise, masculine nouns or pronouns do not refer exclusively to men." The manual also acknowledged that it contained copyrighted material and excerpts taken from secondary sources. Doctrine was now subject to the same conventions and legal rules as scholarly publications.

FM 3-0 was organized into four parts, each with a specific theme. Part One ("The Environment of Operations") discussed the Army's role in peace, conflict, and war. It identified warfighting as the Army's primary purpose, while the ability to dominate land warfare was seen as the service's ultimate function. Part One contained three chapters. Chapter 1 ("The Army and the Role of Land Power") was a tutorial regarding the Army's role in national defense. It described the six dimensions of the operational environment: the threat (both traditional and nontraditional); political (the military as an instrument of civilian-controlled national power); unified action (multiservice or "joint," multinational, and interagency missions); land combat operations (destroying or defeating enemies or seizing land objectives); information (aggregation of information collection, processing, storage, displaying, and dissemination, as well as the media); and technology (as an enhancement of leader, soldier, and unit performance). It also discussed the importance of doctrine as a means to facilitate a shared culture, and introduced the service to ODSS, as those operations applied to war and MOOTW, as well as guidance for training individuals and units.[78]

The second chapter ("Unified Action") described the relationship between the Army, joint, multinational, and interagency, as it applied to full-spectrum operations. It began with an explanation of how the National Defense establishment functioned, as well as public law and joint service publication justifications. It then shifted to a discussion of the strategic, operational, and tactical levels of war, illustrated by a diagram that explained their relationship to one another. Unified action required an understanding of the roles and purpose of the other services. A separate section, approved by representatives of the Air Force, Navy and Marine Corps, Coast Guard, and SOF, addressed these issues. Army forces or ARFOR were explained, as well as the chain of command in unified missions and ODSS considerations.[79]

The last chapter in Part 1, Chapter 3 ("Strategic Responsiveness"), explained that the Army was a strategic force, one capable of projecting military power globally, albeit with the support of other services. A strategic army had seven attributes. It had to be *responsive* through planning, preparation, execution, and assessment. Being *deployable* meant envisioning the process of moving people and assets from start to finish, not just physical movement. *Agile* forces were mobile enough to move rapidly but were also sustainable for long durations. A strategic force must also be *versatile,* capable of reorganizing and adapting to changing situations and missions. The force must be *lethal,* with enough combat power to overcome any adversary in war or MOOTW. No strategic force could accomplish its mission without being *survivable,* which meant an inherent ability to protect itself. Last, a *sustainable* force was necessary, one that generated and sustained combat power by reconciling national and host-nation assets with competing requirements.[80]

Part II ("Foundations of Full Spectrum Operations") explained the theoretical underpinnings for the new doctrine: fundamentals, battle command, and conduct. While the manual acknowledged that war was complex, it also recognized that war's essence was simple: win on the offense, initiate combat on army terms, gain and maintain the initiative, build momentum quickly, and win decisively. In sum, the 2001 manual's foundation represented ideas that had evolved over the course of the twentieth century and were reflected in various past army keystone doctrine publications.[81]

Chapter 4 ("Fundamentals of Full Spectrum Operations") covered a broad range of topics. It identified the elements of combat power, maneuver, firepower, leadership, protection, and information. These building blocks combined to generate the combat power for a strategic army. Each element was intertwined with the nine traditional principles of war to create an operational framework of ODSS that was capable of decision. But the fundamentals were worthless without strong leaders to execute them. Therefore, leaders must be competent in their interpersonal, conceptual, technical, and tactical skills to build teamwork and trust. Competent leaders understood and used the principles of war, as well as the tenets of army operations: initiative, agility, depth, synchronization, and versatility, carry-overs from the AirLand Battle era and the 1993 manual. The crux of the chapter, however, was the operational framework, the intellectual means to organize a theater of operations through decisive (mission accomplishment), shaping (creating and preserving conditions for success), and sustaining (enabling methods for decisive and shaping operations such as logistics) techniques.[82]

The ideas contained within Chapter 5 ("Battle Command") were not new but served to tutor senior and middle-grade army leaders in how to think logically about the application of leadership in generating combat power. Battle command was an art that emerged from a commander's experience and judgment acquired through hands-on practices and education. Battle command skill allowed a commander to visualize what must be accomplished, describe what must be done, and then direct the implementation of that vision. Commanders analyzed the mission, the enemy, terrain and weather, troops and support available, time, and civil considerations both internal and external to their area of operations or anything that might affect mission accomplishment.[83]

Chapter 6 ("Conducting Full Spectrum Operations") was an operational and tactical primer. Its purpose was to educate army leaders in how to use battle command skills to think critically through planning, preparing, executing, and assessing the myriad possible missions that might be undertaken. A key intellectual divide was established between operational and tactical planning. Operational planning was deemed long-range and involved major operations of large units, often joint and/or multinational. Tactical planning had a shorter time frame before execution and focused more upon flexibility and options. Branches, a contingency that might affect the current operation, and sequels, future operations based upon current operation success, failure, or stalemate, were given more specificity than previous doctrine primarily due to the influence of SAMS as an educational institution where campaign design was important. An extensive section discussed rules of engagement and their effect upon military operations. The section provided a means to envision how to conduct operations but also what actions must be taken at the termination of hostilities. The Army now had a doctrine that provided a framework for commanders to conceptualize actions to be taken prehostility, hostility, and posthostility.[84]

Part Three ("Conducting Decisive Full Spectrum Operations") provided detail on four types of army operations, as reflected in ODSS. Each operation warranted a separate chapter. Chapter 7 ("Offensive Operations") made clear that the offense was the decisive form of war. The reader was apprised as to the variance between operational and tactical offense. The opposing center of gravity was the target of an operational attack, the intent being to destroy the enemy's decisive points directly or indirectly through the use of joint and/or multinational forces. Tactical offenses were conducted through battles that exploited the effects obtained by operational offenses. Offensive operations, regardless of level of war, depended upon surprise (unexpected actions), concentration (massing overwhelming effects of combat power), tempo (speed of operations), and audacity (bold action). The forms of maneuver changed little from those espoused in the 1905 doctrine onward. Extensive information was provided, however, for planning, preparing, and executing offensive operations.[85]

Chapter 8 ("Defensive Operations") made it clear that the defense was a temporary action to be undertaken until sufficient strength was achieved to attack. Commanders were advised that defenses required preparation to defeat attacks, security, disruption (to upset an attacker's tempo and synchronization of forces), massing effects (overwhelming combat power at the decisive point), and flexibility (ability to shift forces and main effort). The types of defenses did not change from the previous manual. As with Chapter 7, extensive instruction was provided for planning, preparing, and executing defensive operations.[86]

In Chapter 9 ("Stability Operations"), the writing team illuminated an area of military operations that had received little attention within any previous keystone manual. Stability operations were defined as a way to defend and protect national interests through the promotion of peace and stability and, if needed, to defeat enemies. The Army, as part of a joint team, provided forces to combatant commanders to shape the international environment through peacetime military engagement (PME). PME encompassed all military activities involving other nations to afford mutual understanding and cooperation. Army forces had to consider operating in nonlinear and noncontiguous areas of operation across the globe with identifiable or ambiguous opposing forces and potential threats. Potential missions included peace operations, foreign internal defense, security assistance, humanitarian and civic assistance, support to insurgencies, counterdrug operations, combating terrorism, noncombatant evacuations, arms control, and show of force. Stability operations required interagency and multinational cooperation, legitimizing the host nation, understanding the potential for unintended consequences of various actions, using force in a nonthreatening manner, acting decisively to prevent escalation, and using force selectively and discriminately. After two centuries of Army doctrine, commanders had an intellectual basis for understanding the basics of stability operations and the Army's role in them.[87]

Chapter 10 ("Support Operations") addressed army potential assistance to civil authorities both foreign and domestic, in response to crisis. As a strategically responsive service subject to civil authority and law, army forces were expected to provide essential services and resources to augment civilian agencies that lacked such means. Support operations were complicated because of the number of agencies and services involved, domestic, foreign, and international laws, government oversight, and competing agendas. These operations were mandated by various entities ranging from the United States, UN, and NATO, as well as others. They were typically joint and/or multinational affairs. Domestic support missions involved response to disasters or emergencies and involved food, medical, water, electricity, transportation, fire fighting, hazard identification, flood control, and other resources. Foreign humanitarian assistance often meant easing human suffering from both natural and man-made situations to include preventing starvation and rectifying human-rights violations. While the Army had the capability to assist, the intent was to transfer control to civilian authority as quickly as possible to free up assets for other contingencies.[88]

Part Four ("Enabling Operations") examined how information and combat service support affected ODSS at the operational level. An enabling operation either shaped or sustained an operation and, in the case of MOOTW, might even be decisive.

Within Chapter 11 ("Information Superiority"), the Army described operational advantage as an ability to collect, process, and disseminate an uninterrupted flow of information while exploiting or disrupting an opponent's ability to do the same. Information superiority meant being able to see first, understand first, and act first. This involved understanding that information was a part of combat power and to focus as much upon it as fire and maneuver. Without information superiority, seizing and retaining the initiative would be difficult. Attaining information superiority began with understanding that the information environment was the aggregate of individuals, organizations, or systems that collected, processed, and disseminated information and was not under total military control. Yet, army forces had to operate under those conditions to collect information from all sources and then assess its usefulness and timeliness. Due to the limitation of assets in numbers and capability, information gathering had to be prioritized, which was the duty of the commander. Thus, a commander must understand the situation, determine what information was required, and seek ways to gather that information in a timely manner. Denying the enemy information through deception, operational security, and prevention of unauthorized access to installations and systems was key. Technology, a focus of Chief of Staff General Gordon R. Sullivan in the early 1990s, had increased capacity to gather, process, and display information through video and graphics that allowed commanders to develop a common picture shared electronically across various levels of headquarters. Exploiting technology reduced but did not eliminate risk in that the right information, when processed into useful intelligence, allowed for better informed decision-making.[89]

Chapter 12 examined Combat Service Support (CSS). It noted that a strategic force had to have "operational reach," the ability to support any full-spectrum operation anywhere in the world. Commanders were advised to treat operations and logistics as interdependent and thus must be fully integrated into any plan. This was especially important when army CSS assets were based primarily in the United States and required deployment to support operations. Sending the right amount of logistics, adjusting to requirements, and maintaining continuous support in war or MOOTW required detailed planning and preparation. In a world where multiservice and multinational operations were becoming more common than not, CSS concerned far more than simply supporting army forces. In future operations, host-nation support would increase, as would reliance upon contractors to maintain systems, augment logistical capabilities, and supply special skills, goods, and services that the Army lacked.[90]

Far from being a tactical warfighting manual alone, FM 3-0 reflected the realities of the post–Cold War world where the service not only was reliant upon other services for mission support but also allied nations. Its major contribution was formalizing ODSS as the four activities the Army had always done but had hesitated to acknowledge within its keystone manual. By taking responsibility for full-spectrum operations from battalion to corps and higher, the army leadership had not only acknowledged missions that had once been informal practice but were about to be tested in the international turmoil of the early twenty-first century.

The 2001 FM 3-0 in Full-Spectrum Operations

Assessing the impact that FM 3-0 had upon full-spectrum operations during the manual's 2001–2008 lifespan is problematic. As much of the documentation relating to the manual's use remains classified, there is little upon which to base an analysis. Moreover, FM 3-0 had been in publication for only three months when the events of 11 September 2001 occurred. At the time, army faculty members at the various schools had already received briefings from doctrinal mobile training teams to prepare them to teach the doctrine to students. Presenting and explaining the doctrine's concepts was no easy task, for the new manual contained ideas about stability operations and support operations that few army officers had experienced or understood. As with learning Pentomic army principles during the late 1950s, grasping the nuances of full-spectrum operations and ODSS was a challenge for both instructors and the students.

Still, some provisional observations can be made concerning the 2001 army keystone manual and its intellectual underpinnings in regards to combating terrorism (2001 onward), the invasion of Afghanistan (2001), and the liberation of Iraq (2003). In the broadest sense, the 2001 doctrine had accurately envisioned how the Army would need to operate in the early twenty-first century. The manual stressed that army forces would only engage opponents as part of joint and/or multinational teams. At times, the Army might engage in an area of operation with contiguous lines. At other times, operations might be fluid and noncontiguous or even alternate between the two. Regardless of borders, wars and MOOTW might not only be contiguous or noncontiguous but also linear or nonlinear and separated by more than one simultaneously active area of operation, each with its own requirements, unique nature, and foes, identifiable or not.[91]

Although the field manual attempted to address every possible situation that army forces might find themselves in, the doctrine contributed little to understanding the Army's role in planning, preparing, executing, and assessing missions related to terrorism. In truth, the writing team had often referred to terrorism as a national problem, not an army issue, to be addressed by national assets available to the president. In deference to input from the SOF community, paragraph 1–10 described terrorism missions as part of PME, not a war or MOOTW. In paragraph 9–37, terrorism was defined as the calculated use of unlawful violence to coerce or intimidate governments or societies. Paragraphs 9–38 and 9–39 viewed combating terrorism as counterterrorism (offensive measures) to prevent, deter, or respond to terrorist acts and antiterrorism (defensive means) to reduce vulnerability. FM 3-0 acknowledged that army forces, SOF specifically, conducted strikes and raids in counterterrorism operations under the direct control of the president and the secretary of defense. It was made clear that commanders employing conventional army forces within their areas of responsibility were conducting conventional, not counterterrorist operations. Conventional forces were not sufficiently trained or equipped to fight terrorists.[92]

FM 3-0 made a significant contribution to the Afghan War or Operation Enduring Freedom (commencing October 2001) by assisting planners conceptualize contiguous-noncontiguous and linear-nonlinear operational design. Operation Enduring Freedom included four distinct operations: Afghanistan, the Philippines, the Horn of Africa (primarily a Navy/Marine Corps operation), and the Trans-Sahara (EUCOM's naval component leading Joint Task Force Aztec Silence with SOF support). The relative austerity of these four noncontiguous areas of operation matched doctrinal precepts in which multiple areas of operations were concurrently active and linearity was at times existent and then nonexistent, a situation that posed challenges for operational reach and sustainment.

In the Philippines, in what the 2001 doctrine described as unified action, 1,200 members of U.S. Special Operations Command Pacific (SOCPAC) deployed as part of Joint Task Force 510. Beginning in January 2002, army Special Forces (SF) assisted the Armed Forces of the Philippines in defeating al-Qaeda, Jemaah Islamiyah, and Abu Sayyaf, militant terrorist organizations with radical Islamic beliefs. SOCPAC troops also conducted humanitarian operations in Basilan, a stronghold of radical support. In accordance with the 2001 doctrine and foreign internal defense missions, SF troops trained and equipped local units and created a Light Reaction Company, which then deployed along with their advisers to the island. Army SF subsequently assisted the Philippine forces in denying the enemy sanctuary and routes of operation, observed villages and key leaders for rebel support sympathies, trained locals to defend themselves and villages, and conducted strike operations when needed.[93]

The Afghanistan invasion was also a unified action. On 7 October 2001, the war began with an air component operation to seize air supremacy, the complete domination of air space over the country. By 15 October, SOF had entered the country as part of a U.S.-British coalition and established contact with Afghani rebel forces. In what became a war fought on horseback using laser designators for targeting precision munitions, army and joint forces, along with Afghani allies, methodically dismantled the Taliban régime.[94]

As described in full-spectrum operations doctrine, the Afghan War reflected ODSS. Interagency assets provided forces and intelligence, and army forces supported that effort. Offensively, army SOF units engaged in combat actions beginning on 21 October in support of the North Alliance under General Abdul Rashid Dostum, while Marines, B-52 airstrikes, and coalition aircraft pounded Taliban positions. The Taliban leadership fled Kandahar on 6 December. Defensively, the force failed to prevent coalition partners from allowing the Taliban to flee the Tora Bora area from 1–17 December; key leaders escaped thus damaging U.S. strategic objectives. Still, a tenuous stability was established even as the 10th Mountain Division, the 101st Airborne Division, and the 82d Airborne Division arrived in sequence after most combat operations had ended. As stability operations continued, conventional army forces conducted offensive operations and defended against remaining Taliban forces during the month-long Operation Anaconda (March 2002). In January 2006, NATO assumed control of operations through its International Security Assistance Force. Army forces provided support to NATO Operation Mountain Thrust when Taliban forces reorganized in May and attempted to regain control over Afghanistan. Afghanistan was a true joint, multinational, and interagency operation, as the 2001 doctrine had described.[95]

The liberation of Iraq was designed operationally according to FM 3-0. In conducting unified planning and considering contiguous and noncontiguous operations, the primary headquarters, U.S. Central Command (CENTCOM), worked with the interagency, EUCOM, U.S. Transportation Command (TRANSCOM), and U.S. Space Command (SPACECOM). Planning transpired during the winter of 2001–2002 and focused on toppling the Saddam Hussein régime. In using digital information assets, the plan was not top-driven but emerged in parallel with the numerous headquarters involved. While CENTCOM had established an operational joint headquarters Combined Forces Land Component Command (CFLCC), the U.S. Army's Third Army commanded by Lieutenant General David D. McKiernan, planning was interactive between the CFLCC, the Army's V Corps, and the 1 Marine Expeditionary Force (MEF), among other entities.[96]

The initial stage of Operation Iraqi Freedom, which ended with the fall of Saddam Hussein's government, required the deployment of large army and Marine Corps units supported by limited strategic airlift and sea assets plus a no-notice deployment to achieve strategic surprise. Three possible courses of action emerged, each with a simultaneous attack conducted by air and ground assets. The first was a methodical deployment similar to Desert Storm (1991), labeled "generated start," which created tremendous combat assets but with a price paid in time and resources while forfeiting strategic surprise. The second, a "running start," began with the war commencing using limited troops in Kuwait and fed more into the fight as they deployed and arrived. A third option was a hybrid that combined both.[97]

The final, approved plan blended the hybrid and running start option to provide more troops than planned in the running start but fewer than the hybrid course of action. While the objective remained clear, removing the Iraqi government, debate focused over how to accomplish the task with the available forces. In balancing ways to accomplish the mission with the means at hand to do it, CENTCOM and the CFLCC eventually conducted a successful operation that was doctrinally sound.[98]

The operational scheme of maneuver followed the campaign design principles established in FM 3-0. CENTCOM's focus became a ground attack out of Kuwait to eliminate Iraqi forces, isolate Saddam's government in Baghdad (as well as Ba'ath Party support in Tikrit, if needed), remove the régime from control of the country, and transition to security operations once combat had terminated. The ground attack was supported by a maritime coalition of U.S., British, Australian, and other naval assets that provided air support and safeguarded vessels entering the theater of operations from the sea. Air assets consisted of 1,800 aircraft from various nations. SOF conducted two supporting operations. In northern Iraq, SOF and conventional forces operated with Iraqi Kurds to attack Iraqi units, then moved south to attack Tikrit. SOF, along with air support, conducted operations in western Iraq to prevent the use of missiles against Jordan, Turkey, or Israel. Additional SOF assets provided reconnaissance and direct-action strike missions, as needed.[99]

The primary concern, however, was the urban war in Baghdad, the center of gravity for Saddam's political power and home to 5 million people. Forty other Iraqi cities also had to be reckoned with. Planners tried to design operations to prevent extensive damage, for the mission aimed to liberate the populace, not conquer it. Thus, planning focused on how to disrupt but not destroy the country through "systems-based planning." FM 3-0 did not describe that technique, which was informal practice. It involved analyzing how a city functioned and identifying what must be destroyed (without causing extensive damage) to compel the leaders of a modern urban area to capitulate. The planners viewed Saddam's power as emanating from his ability to control Baghdad through economics, human factors, political mechanisms, infrastructure, internal security and intelligence organizations, and the military. Affecting each of those areas required discovering what characteristics formed each one, identifying what was vulnerable and targetable, what was essential, and then figuring out what type of weapon (electronic, physical, or psychological) to use against it. In the case of human factors, for example, making use of ethnic, clan, and tribal friction, religious beliefs, culture, and education all played a role in a psychological approach to targeting. Infrastructure included understanding Iraqi telecommunications, the media, transportation methods, power generation and distribution, water purification and distribution, food production and distribution, and medical support. Some of those areas were to be destroyed or controlled while others remained untouched so as to avoid needless suffering. The intent was to attack real and symbolic urban control mechanisms with "shock and awe" and bring about submission without extensive house-to-house fighting.[100]

On 16 March 2003, when President George W. Bush made his decision to begin combat operations, army doctrine again came to the fore. A shaping operation, or "decapitation attack," commenced on 19 March, using a combination of F-117 aircraft and Tomahawk missiles in an effort to kill Saddam Hussein and other top members of the country's leadership. Hussein escaped and, on 20 March, the ground invasion began.

Operationally, the liberation of Iraq was a near-simultaneous attack from multiple directions, as FM 3-0 described. Central Intelligence Agency elements and SOF had infiltrated Iraq prior to the border crossing by the main attacking force to provide intelligence assets on the ground. Air attacks struck Baghdad on 21 March. Bombing continued to Tikrit, Mosul, and Kirkuk. In a noncontiguous linear operation on 26 March, 1,000 army paratroopers of the 173rd Airborne Brigade parachuted into a Kurdish-controlled area to open the northern front and secure a local airfield. Ground combat forces of the V Corps and 1 MEF attacked from Kuwait while SOF forces entered Iraq from the west.

The combat operation lasted until 14 April. Over the course of three weeks, V Corps and 1 MEF combat divisions and other forces, along with SOF and coalition allies, secured numerous Iraqi cities and crossing sites over the Euphrates River, conducted thousands of air strikes, secured oil wells, stabilized the Kurdish zone, protected neighboring countries from missile attacks, and brought down Saddam Hussein's government after thirty years of dictatorship. However, Iraqi troops did not surrender in mass as expected. Instead, most regular army troops just disappeared, while the Republican Guards offered a mixed resistance. A greater threat was posed by the unconventional Fedayeen, a tenacious and fanatical paramilitary organization. While they did not disrupt the advance, they inflicted higher than expected casualties. The operation succeeded in part because the Army had properly planned, prepared, and executed Phase III (Decisive Operations) of the campaign according to full-spectrum operations doctrine. However, the nature of the Iraqi conflict and doctrine's utility in providing direction for conducting the war soon changed.

Full-Spectrum Operations and a Different War

CENTCOM and Third Army leaders had never fully developed Phase IV (Transition), the mission following Hussein's removal. This situation contravened doctrine that had underscored *sequels*, future operations that follow the current phase. FM 3-0 advised that "commanders consider sequels early and revisit them throughout an operation. Without such planning, current operations leave forces poorly positioned for future opportunities, and leaders are unprepared to retain the initiative." In the Iraq war, current operations trumped future operations planning.[101]

Between June 2003 and January 2005, the conventional war became a complex insurgency. The immature Phase IV plan had not considered this situation. It envisioned instead the final destruction of Saddam's forces, the search for weapons of mass destruction, hunting terrorists, and providing limited support for the Office of Reconstruction and Humanitarian Assistance (ORHA), all in the context of rapid handover to an Iraqi government. Army leaders expected to leave Iraq by September 2003, once ORHA, a new joint task force, international organizations, and a new national government assumed control. By summer, the planned early exit was no longer realistic.[102]

Even if Defense Department officials had considered the possibility of an insurgency, the 2001 army keystone doctrine offered little in how to counter it given Iraq's unique circumstance. Within FM 3-0, counterinsurgency (COIN) was part of foreign internal defense (FID), which was defined as "participation by civilian and military agencies of one government in programs taken by another government to free and protect its society from subversion, lawlessness, and insurgency." The Army assisted "host governments [to] deal with two principal groups: the insurgents and the people. Army forces help host governments protect the people from insurgent violence and separate them from insurgent control [and aid] the host government police, paramilitary, and military forces perform counterinsurgency." Although FM 3-0 had stressed protecting the people from insurgents, in the Iraq case neither the host government nor its various forces were capable of countering an insurgency. The manual had not anticipated such a circumstance.[103]

From August 2003 until April 2004, most Department of Defense (DoD) civilians and some coalition leaders denied that an insurgency existed. Among those who recognized the unfolding violence as an insurgency, few believed that it was sustainable. In July 2003, Lieutenant General Ricardo S. Sanchez, commanding general of Combined/Joint Task Force 7 (CJTF-7), had issued a directive to "conduct offensive operations to defeat remaining noncompliant forces and neutralize destabilizing influences in the AO [area of operations] in order to create a secure environment in direct support of the Coalition Provisional Authority.

Concurrently conduct stability operations to support the establishment of government and economic development in order to set the conditions for a transfer of operations to designated follow-on military or civilian authorities." Sanchez drew upon full-spectrum operations doctrine (offensive operations and stability operations in particular) to target recalcitrant forces as the object of military action. Well into 2005, most army forces conducted operations as "enemy-centric/search-and-destroy/kill-capture."[104]

The coalition lacked an appropriate campaign plan in part because Secretary of Defense Donald Rumsfeld and Sanchez had envisioned an early exit. In 2004, most commanders achieved little, if any, success in executing full-spectrum operations. As one veteran noted, "Each division was operating so differently, right next to each other—absolutely hard-ass here, and hearts-and-minds here." In Fallujah and Ramadi, Major General Charles Swannack, commanding general of the 82d Airborne Division, took a kinetic approach to seek out and kill the enemy. The war worsened in his area of operations; by August 2004 Fallujah was under insurgent control. Others, however, modified doctrinal guidelines and resorted to informal practice. One such commander was Major General David H. Petraeus, the commanding general of the 101st Airborne Division (Air Assault) in Mosul. Petraeus understood the situation early on. While hunting insurgents, Petraeus also engaged and protected the population. He sought to erode the insurgent base of support, a technique that special operations forces had advocated for years. Success soon followed.[105]

Where FM 3-0 was deficient in COIN guidance suitable for Iraq, a number of obsolete and current army field manuals existed that broached the subject, if taken in context. The FM 27-5, *United States Army and Navy Manual of Military Government and Civil Affairs* (1943), guided army operations until the early 1960s. The *Field Service Regulations* FM 100-5 *Operations* (1962) had described guerrilla, evasion and escape, and subversion activities but not COIN. FM 31-22 *Counterinsurgency Forces* (1963) offered guidance on a wide range of unconventional operations. The Army's FM 100-5 *Operations of Army Forces in the Field* (1968) had devoted one chapter to unconventional operations, in light of the service's involvement in Vietnam. But with U.S. military operations in Vietnam ending in 1973, the growing emphasis on defending Europe caused the FM 100-5 series to be written in the 1970s and 1980s as guidelines for defeating Soviet forces. In the 1990s, FM 100-25 *Doctrine for Army Special Operations Forces* (1991) and FM 100-23 *Peace Operations* (1993) described unconventional operations. Yet, for field commanders consumed by ongoing missions, finding and reading manuals competed with tackling daily realties. As army leaders struggled to cope with the Iraqi insurgency, a movement was under way to produce a relevant COIN field manual.[106]

Toward FM 3-24 *Counterinsurgency*

Beginning in the spring of 2004 and into 2005, army leaders realized the seriousness of the insurgency. Multinational Force–Iraq (MNF-I)/Multinational Corps–Iraq (MNC-I) replaced CJTF-7 (primarily U.S. V Corps troops) with the III Corps from Fort Hood, Texas. Combat operations continued to focus on seeking out and killing insurgents, with mixed results.[107]

Throughout the twentieth century and into the next, certain conventional units used keystone doctrine to train for COIN missions, but doctrine was often meager or woefully obsolete. Other priorities, personnel turnover, and skill erosion over time complicated matters. From the 1950s until the end of the Cold War in 1991 and beyond, army forces conducted training exercises in support of allies. After 1973, in putting Vietnam behind them, most senior army leaders, more conventional than not in mind-set, became Cold War warriors fixated on Europe. Doctrine reflected that situation. Just as U.S. Army keystone doctrine of the nineteenth century imbedded French precepts within the service from 1815 until the 1890s, the 1949–1986 editions of FM 100-5 infused generations of army leaders with a Cold War mentality. Although the 1993 manual attempted to alter that mind-set with its brief discourse on OOTW and the 2001 FM 3-0 gave equal weight to offense, defense, stability, and support operations, no doctrine could expunge an entrenched Cold War mentality in the short term.[108]

Change, however, came nonetheless. In June 2004, soon after returning from Iraq with the 101st, Petraeus was promoted to lieutenant general. With a third star came the responsibility for creating Iraqi security forces under the Multinational Security Transition Command–Iraq (MSTC-I). General George Casey had also assumed command of MNF-I from Sanchez. Both Petraeus and Casey waged a war, trained indigenous forces, and assisted in holding elections by January 2005, an event that signaled cessation of the terrorism, insurgency, and sectarian violence that pervaded the country.[109]

After fifteen months, Petraeus returned to the United States to assume command of the Combined Arms Center at Fort Leavenworth, Kansas.

With the army's intellectual center at his disposal, Petraeus made COIN his top priority, joining with Marine Corps Lieutenant General James Mattis, who had commanded the 1st Marine Division in the attack on Baghdad in 2003 and operated in Al Anbar province in 2004. Both generals shared a vision of reforming their services to understand and wage COIN warfare.[110]

Before Petraeus's arrival, Leavenworth's doctrine division had produced an updated COIN manual, FM (Interim) 3-07-22, on 1 October 2004. Although revision slowly continued, Petraeus invigorated the process by turning to his West Point classmate (1974), Conrad Crane, a retired army lieutenant colonel with a doctorate in history from Stanford University. With Crane in charge of producing a manual, the work effort accelerated.[111]

With Petraeus's support, Crane drew on a contingent of academics, as well as army and Marine Corps veterans of the Afghanistan and Iraq conflicts. He also took advantage of the Information Operations Conference, held at Leavenworth in December 2005, to assemble twenty authors. The writing team outlined the manual's contents while identifying fundamental concepts and determining chapter authors and deadlines. After two months of effort, a collection of academics, civil rights advocates, journalists, and practitioners attended a Fort Leavenworth conference in February 2006 to scrutinize the draft manual. Attendee input altered and improved the manual's content as revisions continued into the summer. Central Intelligence Agency operatives, military personnel, politicians, State Department employees, and the press scrutinized every word. After analysis and much anticipation by an estimated 600,000 "editors," the manual was released on 15 December 2006. FM 3-24, the new numerical designation, was subsequently published by the University of Chicago Press and subjected to a *New York Times* review. *Newsweek* featured Crane in its 25 December 2006 edition and in the "Who's Next 2007" section.[112]

FM 3-24 Overview

FM 3-24 was a team effort between army and Marine Corps writers with the works of Dan Baum, David Galula, and Sir Robert Thompson providing the theoretical basis. Practioners and interested parties contributed additional thoughts. Produced for battalion-level command and staffs and above, the manual's salient point was that the population mattered in COIN and soldiers must learn to adapt accordingly. The preface discussed

General David H. Petraeus, Conrad Crane, and Colonel John Martin, Baghdad, Iraq, 2007 (Courtesy of Conrad Crane)

how COIN had been historically neglected in broader (keystone) army doctrine. Army thinking had failed to address the post–Cold War realities of "technological advances, globalization, and the spread of extremist ideologies, some of them claiming the authority of religious faith." Infused with historical examples and vignettes, each chapter was a tutorial on various aspects of insurgency and counterinsurgency warfare. Chapter 1 described insurgencies and counterinsurgencies, as well as Jominian-like principles and imperatives necessary for the conduct of successful COIN operations. The intellectual thrust was that COIN operations are counterintuitive, for conventional military forces must shun accepted practice and adapt to ever-changing circumstances. Chapter 2 discussed COIN nonmilitary organizations and principles for integrating military and civilian activities, as well as how they combine to achieve success. Chapter 3 addressed intelligence operations from the perspective of COIN requiring a supportive population to identify the location and identity of insurgents, not simply firepower alone. Chapters 4 and 5 described how to design COIN operations, introducing new doctrinal ideas not included in the 2001 version of FM 3-0. In reflecting Marine Corps thinking, the manual identified an assortment of enemies, sources of insurgency, and how operational art assists in the identification of threats. "Logic lines of operation" were introduced; these included combat operations, building host nation security forces, essential services, good governance, economic development, and information operations. The logic lines of operation were to be treated holistically rather than as separate means to an end. Chapter 6 further stressed that viable host nation security forces ultimately win or lose COIN campaigns and that third-party nations can only assist. Leadership and ethical concerns were discussed in Chapter 7, with emphasis on the uniqueness of a war without boundaries with a chameleonlike enemy that blends into the local population. Chapter 8 addressed how to sustain COIN operations. Several appendices provided further details on planning, preparing, executing, and assessing a COIN operation, as well as additional intelligence information guidance, legal concerns, and the role of airpower.[113]

Reaction to FM 3-24

In 1962, army keystone doctrine reflected President John F. Kennedy's contention that insurgencies would dominate future war. Both Crane and Marine Corps Lieutenant Colonel Lance McDaniel advanced that thesis by arguing that a COIN campaign was more complex than a traditional, conventional one. Crane stated, "It takes a lot more analysis before you jump into it, because if you do the wrong thing, it could have major implications." In truth, all army operations are complex, require considerable analysis, and have major implications, as the Army discovered during Cold War missions in Korea, the Dominican Republic, Vietnam, Grenada, Panama, the Gulf War, Somalia, Rwanda, Haiti, Bosnia, Kosovo, and elsewhere. COIN does not require more analysis than conventional warfare. It simply necessitates an intellectual approach appropriate to its context and the nature of the conflict. FM 3-24 provided such an approach by giving practitioners "the tools to do [their] analysis and the guidelines to apply it with the understanding that every situation is going to be unique." As with the 1779 and 1813 *Regulations* and the 1862 *Infantry Tactics*, FM 3-24 was written to regulate the chaos of an ongoing problem, in this case COIN in Afghanistan and Iraq. As Crane put it, "This is not a doctrine that is being jammed down people's throats, this is doctrine that they are demanding." Where field commanders in the American Revolution, War of 1812, and Civil War required and demanded appropriate doctrine, the field also requested a COIN manual in 2005. That this occurred at all is indicative of doctrine's significant role within the Army.[114]

As with many doctrinal publications throughout the Army's history, FM 3-24 generated controversy. Some saw the manual as precisely what was needed, while others claimed it was too dogmatic. Within the Army's journal *Military Review*, lieutenant colonels Dale Kuehl and James R. Crider praised the manual for its appropriateness in addressing large-scale insurgencies. Kuehl, with personal experience in northwest Baghdad, noted that from what he read of the manual (published after his deployment), Galula's essay on communist and colonial insurgencies "is relevant to our current fight in Iraq." Kuehl concluded, "From our experience, it appears that Galula was correct in asserting that gaining the support of the population is essential for the counterinsurgent." Crider, also an Iraq veteran, observed that with "the implementation of a new strategy in Iraq based on the tenets of FM 3-24, *Counterinsurgency*, our military has proven that it can effectively conduct counterinsurgency operations on a large scale." Others were less convinced. Bing West, a former assistant secretary of defense, acknowledged that Petraeus had taken proper steps in Mosul in protecting the population. In Bing's opinion, however, the manual went too far in its implicit assumption that "U.S. commanders had the authority or power to persuade the host nation's leaders to carry out benevolent Western tenets. But we are not colonialists with power to accomplish those tasks. Instead, we gave back sovereignty in both Iraq and Afghanistan. Our soldiers cannot build those nations. With limited leverage, they can only advise." He added, "in our military writings, we have overemphasized theories about nation building and understated the practical effect of aggressive tactics on the ground." Within the *Small Wars Journal*, military and security consultant Dr. Hamid Hussain argued that "we are walking into a dangerous arena if we simply overlook the past experience by arguing that Vietnam was lost due to bad tactics and if we had a good cook book of a counterinsurgency manual we could have won it. And now that we have mastered a cook book [FM 3-24] we can simply march into any country whenever we want." He further noted that "asking a platoon and company commander to be master of language and culture, negotiator, diplomat, economic advisor as well as a first rate soldier is quite a tall order." Such skills require experience, talent, and time to attain, not a few months of scrutinizing manuals and undergoing exercises, as Hussain pointed out. As of this writing, controversy continues, and the impact of FM 3-24 on COIN operations in Iraq, as well as Afghanistan, remains to be fully assessed. The manual, however, contributed to the next edition of army keystone doctrine.[115]

DoD Directive 3000.05

Before FM 3-24 came into being, the DoD published Directive 3000.05 on 28 November 2005. Its purpose was threefold: "to provide guidance on stability operations that will evolve over time as joint operating concepts, mission sets, and lessons learned develop; to establish DoD policy and assign responsibility within DoD for planning, training, and preparing to conduct and support stability operations pursuant to Secretary of Defense authority; and to supersede any conflicting portions of existing DoD issuances." Stability operations were defined as "military and civilian activities conducted across the spectrum from peace to conflict to establish order in States and regions," significantly modifying the 2001 FM 3-0 definition.[116]

Where the Army's 2001 version of FM 3-0 *Operations* had intended to make stability operations the peer of offensive, defensive, and support operations, the DoD directive took matters one step further. The policy made it clear that "stability operations are a core U.S. military mission that the Department of Defense shall be prepared to conduct and support. They shall be given priority comparable to combat operations and be explicitly addressed and integrated across all DoD activities including doctrine, organizations, training, education, exercises, materiel, leadership, personnel, facilities, and planning." Further, DoD established that "stability operations are conducted to help establish order that advances U.S. interests and values. The immediate goal often is to provide the local populace with security, restore essential services, and meet humanitarian needs. The long-term goal is to help develop indigenous capacity for securing essential services, a viable market economy, rule of law, democratic institutions, and a robust civil society." Moreover, "U.S. military forces shall be prepared to perform all tasks necessary to establish or maintain order when civilians cannot do so."

The Department of Defense directive cut to the chase. From 1779 until 2001, army leaders had carefully avoided including stability operations within keystone doctrine. When the Army approved the 2001 version of FM 3-0 *Operations*, the service had admitted that stability operations were legitimate missions. Still, experience in such matters was lacking, and basic combat skills dominated training. Now, DoD had directed all services, not just the Army, to prepare to assist in "rebuilding indigenous institutions including various types of security forces, correctional facilities, and judicial systems necessary to secure and stabilize the environment; revive or build the private sector, including encouraging citizen-driven, bottom-up economic activity and constructing necessary infrastructure; and to develop representative governmental institutions." The Army had undertaken similar tasks during reconstruction of the South after the Civil War, the Philippines in 1899–1902, and in post–World War II Germany and Japan. Yet no policy had gone so far as DoD 3000.05 in systemizing stability operations among the services. Given the directive, as well as ongoing COIN operations in Iraq and Afghanistan, the Army's keystone doctrine was in need of revision.[117]

Field Manual 3-0 *Operations* 2008

When the Army published FM 3-0 *Operations* in 2001, it did so with the understanding that it would be replaced within seven years. Several factors ensured that a replacement was realized: planned obsolescence, the ongoing conflicts in Afghanistan, the Philippines, and Iraq (the latter with its surge in 2007), DoD Directive 3000.05, and the publication of FM 3-24.

Under the direction of General William S. Wallace (commanding general of TRADOC), Lieutenant General William B. Caldwell IV (Combined Arms Center commander), and a writing team led by Lieutenant Colonel Steve Leonard, the Army revamped and replaced the 2001 edition of FM 3-0 *Operations* on 28 February 2008. As with the 1923 edition of FM 100-5, the 2008 version drew heavily upon recent combat experience and input from various sources.

The 2008 manual emerged from a series of army issue papers that began in 2005. The papers addressed "unified action, the design of warfighting functions, the continuum of operations, and the Army's operational concept." Over two hundred organizations, media groups, and individuals reviewed the material, including "the Army Staff; Army commands; Army service component commands; Army corps and division headquarters; training divisions; and TRADOC commands and centers, proponents, and staff, as well as Air Force, Marine, and Navy doctrine centers." Themes, concepts, and the field manual's organization soon took form.[118]

With a blueprint in hand, the Combined Arms Directorate at Leavenworth produced three drafts, each one discussed at "councils of colonels." Doctrine writers sought to generate consensus from over 4,000 comments. Once agreement was reached, the final step, a TRADOC commander-hosted doctrine and concepts conference, finalized the manual's contents.[119]

The 2008 manual was announced as both evolutionary and revolutionary in design. Insofar as it embellished the concept of full-spectrum operations from the 2001 version, the new manual was evolutionary. From the service's perspective, however, as the manual placed more emphasis on stability operations and COIN, it was deemed to be revolutionary.

Significantly, the 2008 manual regulated much that had been informal practice. In the 2001 manual, stability operations were considered to be "other joint missions that occurred within an army context." In the 2008 manual, stability operations and civil-support operations received additional emphasis in deference to combat experience and DoD Directive 3000.05. As the 2008 manual indicated, army forces now considered offensive operations and defensive operations to occur simultaneously with stability operations, the focus being on nonlethal means to accomplish missions as much as lethal ones.[120]

The uniqueness of the 2008 version of FM 3-0 is that it argued that winning battles and engagements (tactical actions) were not decisive alone. Army forces must also shape the civil situation in concert with other government agencies, international organizations, civil authorities, and multinational forces. Unlike past keystone doctrine where emphasis was placed on force destruction through fire and maneuver as a means to impose one's will on the enemy, the Army must now operate in and among civilian populations to prevent misunderstanding and build trust. In truth, the Army had done this throughout its history: Indian Wars of various eras, Mexico 1846–1848, the Spanish-American War 1898, the Philippines and China in the early 1900s, and numerous other operations saw American soldiers operating among civilian populations. The Army's principal doctrine had historically paid little attention to acknowledging that fact, for such interaction was often regarded as informal practice. With the 2008 edition, what had been informal practice was now regulated.[121]

For its claims to novelty, the 2008 version has strong historic ties to past manuals. In 1867, Emory Upton raised the idea that army leaders and soldiers must exercise maximum individual and small-unit initiative. That idea was adopted in the 1891 *Infantry Tactics*. Battle command and operational art concepts date to the 1982 manual. In the 1820s–1830s, Carl von Clausewitz described uncertainty, chaos, and friction as inherent parts of military operations. His views permeated twentieth-century U.S. Army doctrine. Arthur Wagner discussed similar notions in the 1890s. All of those ideas carried forward into the 2008 manual.

Still, the 2008 doctrine's organization demonstrates that the service sought a new direction in addressing current and future problems. The foreword announced that "America is at war where global terrorism and extremist ideologies are realities." For the Army, America faced an era of "persistent conflict" (not unlike the Indian Wars period 1865–1890) with "protracted confrontation among states, nonstate, and individual actors increasingly willing to use violence to achieve their political ends," a situation that reflected the aftermath of the Cold War, globalization, and increased technology to access once remote populations. The preface provided a chapter and appendix synopsis targeting an audience of middle and senior leadership (officers in the rank of major and above).[122]

The introduction acknowledged that this was the fifteenth edition of the Army's capstone (a change from keystone) operations manual, stretching back to 1779 with ideas and concepts that carried forward through the various manuals and their updates. This manual's purpose was revealed immediately as giving equal weight to stability and civil support as traditional offense and defense operations and immediately included a portion of DoD Directive 3000.05. The ability to conduct nonlethal actions was deemed "vital," although deterrence remained the Army's primary purpose. In carrying forward a long-established service truism dating to 1794, the doctrine implied that properly led and trained conventional troops were fully capable of conducting stability and civil-support operations, missions typically considered to be informal practice before 2001.[123]

Chapter 1 ("The Operational Environment") established the context of land operations in terms of a global environment that required army expeditionary forces as part of a joint, interagency, and multinational undertaking with areas of joint interdependence. The operational environment was projected to be both unstable and characterized by persistent conflict well into the future. Globalization, technology, population growth, urbanization, demand for energy, water, and food, as well as climate change and the proliferation of weapons of mass destruction meant that army forces had to address adversaries who combined "traditional, disruptive, catastrophic, and irregular capabilities with the capacity to change the nature of the conflict to strike where the U.S. was least prepared." Operational variables (political, military, economic, social, information, infrastructure, physical environment, and time), combined with six mission variables (mission, enemy, terrain and weather, troops and support available, time available, and civil considerations), expanded on decision-making processes that developed in the twentieth century. More emphasis was placed on interagency coordination and cooperation with other organizations than previous manuals. Of note was how campaigns were defined, moving beyond more traditional joint service combat operations to include the reestablishment of civil authority after joint operations end, "even when combat is not required." This change reflected the thinking found in FM 3-24 and indicated how much influence that manual had over FM 3-0.[124]

The second chapter ("The Continuum of Operations") echoed the established 2001 concept by identifying a "spectrum of conflict," this time with stable peace at one pole and general war at the other. As violence increased, stable peace moved to unstable peace, then insurgency (reflecting the influence of FM 3-24), then general war. The manual cautioned that violent conflict is not progressive and that an unstable peace may move to general war and back again. Army forces, however, were to operate anywhere along the spectrum to advance U.S. national interests. Army leaders must continuously evaluate the situation and adapt to what is appropriate. "Operational themes," a means to describe "the character of the dominant major operation being conducted within a land force commander's area of operations" included peacetime military engagement, limited intervention, peace operations, and irregular warfare. Where FID had been a separate category in the 2001 manual and COIN part of FID, the 2008 version identified FID, support to insurgency, COIN, combating terrorism, and unconventional warfare as parts of irregular warfare. Each operational theme intellectually corresponded to a portion of the spectrum of conflict, with a diagram describing how each theme fit within the spectrum. Each theme was explained in detail, along with historical examples.[125]

Chapter 3 ("Full Spectrum Operations") was the most important chapter, for it described the Army's operational concept and role in offense, defense, stability, and civil-support operations (a change from the previous 2001 "support operations"). As with the 2001 manual, full-spectrum operations remained "the core" doctrinal concept, but stability operations received more emphasis. "Lethal" and "nonlethal actions" replaced "kinetic" and "nonkinetic," with nonlethal actions receiving considerable attention to include how they provide constructive support to civil authorities. Commanders must "apply force without killing or crippling the enemy." This was crucial, for soldiers operated among civilian populations and civil organizations where influence and persuasion were important means to an end. Well-trained troops held a "moral advantage" that increased fear and doubt among enemy forces. The concept of "mission command" was introduced as the "conduct of military operations through decentralized execution based on mission orders." In what can be considered reminiscent of Emory Upton's 1867 logic, "commanders focus their orders on the purpose of the operation rather than the details of how to perform assigned tasks." This is what Theodore Roosevelt attempted in Cuba in 1898, but the terrain proved too difficult for units to operate independently. Still, the 2008 manual's focus was on establishing freedom of action within the higher commander's intent as a way to restrain micromanagement. Offensive and defensive tasks retained many concepts from previous manuals. Stability operations expanded to five tasks: civil security, civil control, restoration of essential services, support to governance, and support to economic and infrastructural development. Many of these tasks had been previously addressed in Special Forces doctrine under various terminology; SOF units in Haiti 1994 tackled many of these missions. FM 2008 adopted many Special Forces ideas and instilled them in the conventional army, a significant shift in bringing what was once informal practice under "capstone doctrine" regulation.[126]

Chapter 4 focused on combat power. The chapter replaced the "battlefield operating systems" with eight "warfighting functions." Combat power stemmed from these functions. Leadership and information amplified the effects of movement and maneuver, intelligence, fires, sustainment, command and control, and protection. As with many previous manuals, combined arms (infantry, armor, artillery, aviation, and other assets) were considered necessary for combat power to achieve its full potential. Combined arms both complemented and reinforced the capabilities of one another, but the long-held concept of combat arms, combat support, and combat service support vanished. "Tenets of operations" were deleted, for they were now intrinsic warfighting functions and elements of combat power.

Chapter 5 ("Command and Control") contained numerous precepts from previous manuals and especially the 2001 version. "Command" now included leadership. Battle command (with its previously identified inherent characteristics championed by General Frederick M. Franks Jr. in the early 1990s) was redefined as how to visualize and understand the situation, determine a desired end state, and create a broad sequence of events to achieve the end state. The "operational framework" disappeared but carried forward certain AirLand Battle ideas of the 1980s, such as area of influence and area of interest, as well as the new concept of mission variables. Under civil considerations, commanders were introduced to a new acronym, ASCOPE, or area, structures, capabilities, organizations, people, and events, which reinforced the importance of civil-support and stability operations. To provide effective command and control, commanders had to consider "lines of effort." Lines of effort (explained in Chapter 6) linked the execution of tactical tasks to end-state conditions or the actions that must be taken by various entities to achieve the desired outcome. Above all, however, mission command was deemed the preferred method of battle command with commanders allowing their subordinates to demonstrate initiative through mission orders, mutual trust, and understanding.[127]

The concept of operational art had matured enough to warrant its own chapter, and in Chapter 6, operational art was defined as "the application of creative imagination by commanders and staffs—supported by their skill, knowledge, and experience—to design strategies, campaigns, and major operations and organize and employ military forces." Operational art was reconceptualized to include "problem framing" or how to deduce what must be resolved. It blended ends, ways, and means across the three levels of war (strategic, operational, and tactical), as defined in earlier manuals. In the 2008 FM 3-0, operational art was deemed to act as a bridge between the strategic and the tactical levels, a way for commanders to derive the essence of an operation. Operational art was described as how commanders translated their concept of operations into operational design and ultimately into appropriate tactical tasks. Where the chapter embellished previous theoretical ideas concerning operational design, the Army was introduced to lines of effort that complemented lines of operation, the latter concept dating back to Antoine Henri de Jomini in the early nineteenth century. Lines of effort, however, carried over from FM 3-24 ideas in that they became essential when "positional references to an enemy or adversary have little relevance." They were considered to be of great value in situations "involving multinational forces and civilian organizations, where unity of command is elusive, if not impractical." Lines of effort drive the operational design for stability and civil-support operations, "linking military actions with the broader interagency effort across the levels of war." Where operational art had focused primarily on combat operations in previous manuals, FM 3-0 2008 now mandated its use in noncombat situations. Lines of operation and lines of effort were not mutually exclusive, however. Both were deemed essential in conducting full-spectrum operations throughout a campaign or major operation.[128]

Chapter 7 ("Information Superiority") continued to address concepts contained within Chapter 11 of the 2001 FM 3-0. The chapter was updated, however, to include knowledge management as a contributor and provided historical examples of recent operations in Afghanistan and Iraq. The 2008 FM 3-0 established five army information tasks (information engagement, command and control warfare, information protection, operations security, and military deception). Each task reflected globalization and technological advances since the previous manual and recognized that the side best capable of leveraging information has a distinct advantage. Included was how information superiority draws from cultural awareness and relevant social and political factors, yet another reference to FM 3-24. Intelligence, surveillance, and reconnaissance were essential for understanding the operational environment and crucial to the information superiority effort. Information not only must be accurate, timely, usable, complete, precise, reliable, and secure but also relevant.[129]

The last chapter, Chapter 8 ("Strategic and Operational Reach") expanded on what had been Chapter 12 ("Combat Service Support") in the previous manual. Strategic reach was introduced as providing "the capability to operate against complex, adaptive, threats operating anywhere." In a return to the 1949 FM 100-5 line of thought, "Army forces extend joint force's strategic reach by securing and operating bases far from the United States." Strategic reach required joint assets to move army forces, as first mentioned in the 1944 manual. Characteristics of strategic reach included ability to project power globally and at times directly from the United States into an operational area. It also included expeditionary campaigns under austere conditions, such as those Brigadier General John J. Pershing conducted in Mexico (1916–1917). Under contemporary circumstances, the doctrine explained that an expeditionary army is subject to U.S. Transportation Command assets to move the force to the operational area. Expeditionary warfare demanded generating and sustaining combat power not only from U.S. assets but also from host-nation and contracted support. Operational reach, defined in the 2001 manual, was embellished to include more detail on lodgments (a designated area in a hostile territory that makes the continuous landing of troops possible) and forward operating bases (an area used to support tactical operations without establishing full support facilities).[130]

FM 3-0 2008 contained several appendices. Appendix A ("Principles of War and Operations") restated the Army's nine principles of war (objective, offensive, mass, economy of force, maneuver, unity of command, security, surprise, and simplicity) while adding additional principles of joint operations (perseverance, legitimacy, and restraint), the latter having confused the service when listed as OOTW principles in the 1993 version of FM 100-5. Appendix B ("Command and Support Relationships") explained what constituted a chain of command, combatant commands, and joint task forces and service components. Joint command relationships received additional definition, particularly in explaining terms such as combatant command, operational control, joint assignment and attachment, coordinating authority, and direct liaison authorized. Various other relationships were discussed, as well as regulatory agencies. Appendix C ("The Army Modular Force") explained how the Army's brigade-based force concept originated and acts as its basic maneuver element. The appendix further described how a corps headquarters serves as an intermediate land force headquarters and the role of the division as the Army's primary tactical warfighting headquarters. Brigade combat teams form the basic building block of the Army's tactical formations and consist of three versions: heavy, infantry, and Stryker. Brigades are composed of battalion-sized maneuver, fires, reconnaissance, and sustainment units. Modular support brigades consist of five variants: battlefield surveillance brigade, fires brigade, combat aviation brigade, maneuver enhancement brigade, and sustainment brigade. The appendix explained each formation's role and organization and provided a diagram of each. Appendix D ("The Role of Doctrine and Summary of Changes") defined doctrine as "a body of thought on how Army forces intend to operate as an integral part of a joint force," reinforcing the 1986 Goldwater-Nichols Act. Army doctrine was explained as consisting of the Army's collective wisdom (best available thought) regarding past, present, and future operations. Appendix D also included a chapter-by-chapter description of changes from the 2001 manual (some of which are described above) and a table of new terms and modified or rescinded definitions. Source notes, a glossary, references, and an index followed Appendix D.[131]

At the present, it is not possible to determine what effect FM 3-0 2008 has had on the Army. Initial reviews were for the most part positive, describing the manual "as a quality document and a worthy descendant of similar works from the last century, most notably FM 100-5." What is known is that the manual is apparently already under revision. If approved, Chapter 7 ("Information Operations") will be altered to include two army information tasks in support of full-spectrum operations: to conduct inform and influence activities (IIA) and to conduct cyber/electromagnetic activities (EM). Given the history of army doctrine and generations of service leaders seeking to envision and prepare for future conflict, a new FM 3-0 or equivalent should appear five to eight years after the 2008 edition.[132]

CONCLUSION

REGULATING CHAOS

Pity the theory that conflicts with reason!—
Carl von Clausewitz, *On War*

Shall a theory be pronounced absurd because it has
only three-fourths of the whole number of chances of
success in its favor?—Antoine Henri de Jomini,
The Art of War

Analyzing over 225 years of U.S. Army doctrine reveals that the American Army has been far more adaptive and innovative than scholars have acknowledged. Far from belonging to a rigid institution bent upon replicating the past, army leaders created a system that blended education and individual/group recognition of ideas to frame doctrine within both a national and service culture. Although technology and other factors affected doctrine, it was the individual soldier who mattered most. When the Army focused its doctrine on the American soldier, a pattern established by Major General George Washington and von Steuben in 1779, it was far more successful in developing appropriate precepts than when deviating from it.

The relationship between doctrine and the U.S. Army reveals the purpose of the keystone manuals to be a philosophical methodology for winning wars and accomplishing other tasks, as determined by civilian authority. The manuals contain the essence of how the army leadership has envisioned regulating the chaos of armed conflict through military operations. Although compliance was often an issue and imperfection all but assured, doctrine was and is essential for army forces to plan, train, and carry out national-security missions. Generations of officers and enlisted men have consulted and used doctrine in peacetime and war.

But doctrine is much more than the best available thought. The manuals spurred the development of strategy, operations, and tactics through theories about war and nonwar activities. They educated the officer corps.

They imposed control over the Army and sought to do the same to other services and allies. They justified agendas, as well as defended the service's relevancy.

Keystone doctrine was the outcome of national and institutional values and expectations, yet the manuals' precepts were hardly based upon American thought alone. More often than not, authors borrowed foreign ideas and recast them in an American light. Over the span of their history, the publications gleaned ideas taken from British, French, and German sources, as well as Italian and Soviet thought, among others. The first American doctrine established that precedent. From 1607 until 1779, no American warfighting doctrine existed. Militia and Ranger units followed informal practice. If any doctrinal manuals were used to train soldiers, they were primarily British. But the American Revolution demanded the establishment of a conventional Continental Army that, in turn, mandated a cohesive tactical system capable of vanquishing the British juggernaut. Major General George Washington and others knew this, but the tumultuous events of the rebellion hindered efforts to write a manual. Fortunately, at Valley Forge (1777–1778), Washington found an author in von Steuben, a Prussian officer with expertise in European drill procedures. Steuben, under Washington's direct supervision, drew upon British and Prussian methods in crafting a doctrine specifically for the American soldier. There was no publication like it. Precisely what the Army required, it was a conventional tactical approach to war with the American soldier in mind. The manual, however, did not consider unconventional forces, which continued to follow informal practice.

After the Revolution, the U.S. government required a doctrine for defending a fledgling United States from foreign invasion and for fighting unconventional forces, Indians, on the frontier. The 1779 *Regulations* was too cumbersome for the militia to learn quickly. Two victories by Indians over army forces intensified ongoing debate over how to organize and train troops to fight them. But the Army's subsequent decisive victory at Fallen Timbers (1794), one made possible by taking time to train a mixed force to the 1779 doctrine, convinced service leaders that properly led and prepared conventional soldiers can defeat Indians. What the nation required was a conventional doctrine to turn rabble into a fighting force during a crisis. By the early 1800s, the army leadership looked to French doctrine for two reasons: Napoleon's success in Europe and France's effectiveness in turning a mob into an army during the years of the French Revolution. In 1812, the Army replaced its 1779 instructions with a

French-based doctrine. That manual proved so unpopular that it was replaced in 1813, again containing French precepts. Many officers found the 1813 version to be repugnant and returned to the obsolete 1812 version or its predecessor, the 1779 manual. Some imported and used French doctrinal manuals. In 1815, Brigadier General Winfield Scott's board of officers produced another manual that also adopted French methods. Subsequent revisions of that manual did the same; the Army slowly became *de facto* French. The manuals reflected debates over the column or line, which ultimately were settled by recommending one formation or the other under certain battlefield situations. Successful campaigns against Indians from the 1830s to 1850s demonstrated that conventional doctrine worked against unconventional forces, albeit with difficulties and informal practice augmentation. Still, by the 1840s, the Army had pacified most of the eastern United States. The notion that properly led and trained conventional forces could defeat unconventional ones became a sacrosanct belief of the service. With the American victory in Mexico (1846–1848), the Americanized French-based doctrine had proven capable of not just defeating Indians but also a conventional opponent, the Mexican Army.

French ideas continued to dominate doctrine after 1848. As rifle technology was integrated into European armies, the U.S. Army adopted the rifle-musket and again borrowed tactics from France. Secretary of War Jefferson Davis adopted not only a new infantry weapon, but also a new emphasis in doctrine. The 1855 doctrine produced by a board of officers headed by Brevet Lieutenant Colonel William J. Hardee was specifically written to exploit a weapon. Thus, it placed more importance upon technology as a tool for enhancing combat performance than the soldier. During the American Civil War (1861–1865), technology overpowered tactics; thousands of rapid reloading rifled weapons made a mockery of offensive maneuver. A revised 1862 manual, devised by Brigadier General Silas Casey and other officers, again based upon French ideas, proved no better in countering the slaughter. A postwar 1867 manual written by Brevet Major General Emory Upton proved unpopular, as did its 1875 revision. Its methods to counter weapon technology in the defense placed too much emphasis upon lower-rank initiative to suit control-minded officers. With the French defeat in the Franco-Prussian War (1870), however, army intellectuals such as Lieutenant Colonel Arthur Wagner rethought the issue of initiative given the advantages of the decentralized German system over the more controlling French one. With the 1891 and 1895 manuals, German precepts had entered doctrine, influencing tactics in the War with Spain (1898) and beyond.

Operations in Cuba and later in the Philippines (1899–1902) and China (1900) not only propelled the United States to global power but also prompted doctrinal changes. For most of its existence, the War Department had little to say about doctrine or its implementation. In 1905, Secretary of War Elihu Root strengthened the War Department's role. Now, the newly formed general staff wrote the manuals. Yet, the Army continued to lack the means to enforce its contents; the War Department could not hold field commanders accountable. Still, keystone manuals changed from tactical drill to the *Field Service Manual* (*FSR*), a tome that contained tactics, government legislation, and secretary of war directives crafted by general staff officers. The *FSR* also served to educate officers in their wartime responsibilities. The 1905 manual included military government administration, a mission previously considered to be informal practice. This indicated that the War Department now aimed to impose more control over field forces by prescribing not only how to conduct war but also how to handle its aftermath. Rather than simple maneuvers using line or column, new ideas appeared by 1914 such as combined arms. General John J. Pershing's open warfare was the crux of a 1923 edition that described the offense and defense in more detail than ever. The *FSR* also gained a numerical designation and title change in 1939, *FSR* Field Manual (FM) 100-5 *Operations*. By the 1941 version, the Army had included guerrilla war, fighting in deserts, mountains, and other locals, principles of war, leadership, and decision-making. Doctrine also included changes in technology, including airplanes, machine guns, poison gas, radios, and tanks. The 1944 edition included amphibious assaults and airborne operations. By 1945, doctrine was no longer just a tactics manual but a sophisticated approach to war that adjusted with change. Doctrine was also heavily influenced by the maturing army education system with many army educators contributing to the writing process.

The advent of atomic weapons in 1945 sent the army leadership scrambling to justify the service's existence. Doctrine became a means to survive. With the creation of the Department of Defense (1947), the Army joined the Air Force and Navy (to include the Marine Corps), as well as the Coast Guard, in competing for missions and resources. While retaining operational and tactical fundamentals of warfare, doctrine now expressed a more strategic vision and portrayed the service as an essential part of a multiservice team. The 1949 edition, heavily influenced by General Omar Bradley, described the Army as a force that stabilized areas attacked by strategic nuclear weapons delivered by the Air Force. The 1954 doctrine rearticulated the Army's role as the decisive land force. In the 1956 and 1958 updates to that manual, Generals Matthew B. Ridgway, Maxwell D. Taylor, and James M. Gavin stressed that the army division could employ tactical nuclear weapons successfully and also survive on a nuclear battlefield. But pushing technology came at a price, not only budgetary but also by raising questions among the service as to the viability of the Pentomic concept. In reorganizing the Army in the early 1960s, maneuver advocates used doctrine to downplay tactical nuclear weapons and shift support to armored and mechanized forces as more mobile and survivable.

With the Kennedy administration's growing interest in the increased frequency of insurgencies, the 1962 doctrine addressed unconventional warfare and the use of Special Forces to counter such activities. Coalition warfare also came to the fore. However, a conventional approach to warfare still dominated due to European defense priorities, with irregular (unconventional) warfare doctrine being little more than an afterthought than mainstream thinking. America's Vietnam War (1965–1973) was fought under the auspices of ideas formulated to fight the Soviets, although modifications were made on the ground in Southeast Asia. In 1968, the manual's title deleted *FSR* to become *Operations of Army Forces in the Field*. The 1968 doctrine addressed airmobility and the use of helicopters but said little about how to fight in Vietnam. Once the war ended for the United States in 1973, Vietnam was less important than the Arab-Israeli conflict (1973), for its conventionality offered more insight into what might occur in Europe. In the Middle East, Soviet equipment outperformed American weapons. In response, doctrinal innovators such as General William E. DePuy with the 1976 manual (now entitled FM 100-5 *Operations*) and General Donn Starry with the 1982 and 1986 versions devised methods to defeat massed or echeloned Soviet forces primarily in Europe but also globally. The North Atlantic Treaty Organization (NATO) and the German government influenced those manuals. Doctrine now included theories about campaign design, operational art, and other concepts that allowed officers not only to design large-unit multiservice and multinational operations more effectively, but also to receive a War College and Command and General Staff College education through the field manual.

With the end of the Cold War in 1991, the army leadership struggled to find both relevancy and a credible threat. Intellectually, the service had difficulty crafting a doctrine that reconciled what modern war might become: a rapid and simultaneous takedown of a sovereign nation (Panama 1989) or more conventional force-on-force military operations (the Gulf War 1990–1991). Discussions ensued over the role of the United Nations and humanitarian relief operations in northern Iraq (1991). The Army also had to address the 1986 Goldwater-Nichols Act, which meant including more guidance for joint operations in doctrine. An increase in peace operations led to attempts to define differences and similarities between war and nonwar operations. In reconciling the two, Generals Gordon R. Sullivan and Frederick J. Franks framed the 1993 doctrine to separate war and operations other than war (OOTW). But the limited guidelines for conducting OOTW demonstrated that the service understood the former more than the latter. The manual unleashed a flurry of intellectual debates that ultimately found more commonality than differences. Many Cold War army leaders wedded to conventional warfare precepts found that the discussion often confused the officer corps over what was or was not war. Moreover, the discussion increasingly raised awareness of the Army's ability to conduct nonwar missions, something that army leaders were reluctant to see incorporated into keystone doctrine. Instead, the army leadership held to its conviction that a conventional soldier could perform OOTW with proper leadership and training. But that did not hold true in Haiti (1994) where Special Forces soldiers proved to be more effective than conventional ones. Somalia (1992–1993) and Rwanda (1994) also indicated that doctrine must be more inclusive of what the Army was obligated to do by law, not what its leaders desired to do. The result was the 1997 concept of offense, defense, stability, and support (ODSS). With Lieutenant General William J. Steele prodding the Army to accept ODSS, the concept became part of army doctrine in 2001. OOTW became MOOTW to place more emphasis upon the military aspects of nonwar missions. The manual became FM 3-0, *Operations,* reflecting the joint forces publication system. In its 2008 edition, FM 3-0 retained ODSS but was heavily influenced by war on terror operations in Afghanistan, Iraq, and the Philippines, as well as Department of Defense Directive 3000.05 and FM 3-24 *Counterinsurgency.* With its strong emphasis on stability operations and civil support, the 2008 manual regulated far more informal practice than any previous edition. What this means for the Army remains to be seen.

Taken collectively as a body of work, the Army's doctrine has evolved through four distinctive eras. Each expanded upon the previous period but did not totally replace it. Thus, doctrine is far more evolutionary than revolutionary. Tactical drill manuals constituted the first era (1779–1904), a period where guidelines for fire and maneuver were written by single authors or a board of officers. This changed in 1905 when the Root Reforms fixed doctrinal responsibility with the general staff, and a second era (1905–1944) emerged with the *Field Service Regulations* (*FSR*). Doctrine became the product of staff officers, as well as army educators, and was approved by army leaders who elevated the manuals from tactics to operations in the field, both in the United States and overseas, as a single service. In 1944, however, doctrine acknowledged that army troops required the Army Air Forces and the Navy to transport soldiers and equipment to wartime theaters of operation. Thus began the era of multiservice doctrine (1944–1962). In 1962, coalition (now multinational) warfare entered the manuals, unleashing an era that has continued into the twenty-first century. Interagency cooperation has joined this period, for the present.

While strategy, operations, and tactics had become the hallmark of army doctrine by 2008, doctrinal manuals have also been used to impose order upon the service. Army leaders used doctrine written by the War Department and later the Department of the Army to assert more control over Regulars, militia, Reserves, and the National Guard. Doctrine writers sought to devise precepts for every conceivable military operation, thereby reducing the commander's use of unregulated informal practice or, for want of a better word, freedom. While commanders since 1779 had the latitude to use doctrine or not, this prerogative was reduced with the creation of the Training and Doctrine Command (1973). That headquarters not only supervised doctrine development but also enforced its use. By the 1980s, the Army had established doctrinal standards to periodically evaluate commanders, units, and soldiers. With the founding of the National Training Centers and the Battle Command Training Program, teams of observer/controller experts edified doctrine and enforced compliance, albeit not always stringently. Doctrine is, however, the basis of instruction in the army school system, as it has been for generations, thereby ensuring that all officers and enlisted soldiers share a common intellectual framework.

Since 1779, the Army's doctrine has also furthered various agendas. Certain manuals argued for conventional over unconventional forces; Regulars and militias; foreign theoretical ideas; the column, line, or mixed formations; two or three ranks; firepower or maneuver; and élan. They have advanced the beliefs of airpower, armor, counterinsurgency, mechanized advocates, atomic weapons, technologists, and maneuverists. They have celebrated combined arms, open warfare, AirLand Battle, ODSS, and other theoretical constructs such as operational art. They have served to encourage modernization.

The Army's doctrine has also attempted to regulate the behavior of other services and allies. This line of reasoning emerged primarily after World War II and proved to be unattainable. The U.S. Air Force leadership agreed with army doctrine only when it suited their interest, such as having a say in the AirLand Battle doctrine of the 1970s and 1980s and more recent editions. The Navy, Marines, and Coast Guard had little interest in shaping army doctrine, though they made contributions to the 2001 edition. The Marine Corps, however, contributed significantly to the 2008 manual regarding stability operations. Nor has the conventional Army reconciled its doctrinal differences with the SOF community, which is often viewed as a de facto separate service. Controlling allies has also been problematic. While the 1962 doctrine discussed coalitions, the 1968 doctrine stressed the importance of NATO and how the service complied with its standard agreements. In the 1970s and 1980s, doctrine attempted to convince Europeans in particular that the American Army way of warfare was their way. However, it was the Army that recanted its position. The Germans approved of the 1976 manual, but the Army rejected it. The 1982 version catered to Army desires more than NATO and Germany. The 1986 edition placated Allies. Since the 1990s, army doctrine has yet to gain control over the conduct of UN operations.

These setbacks aside, the army leadership has been savvy in using doctrine to remain relevant and justify the future of the service. Doctrine did not prepare the service to fight the last war, as some have argued, for such a conclusion overlooks the breadth of service doctrinal history. Instead, the civilian and military leadership has ruminated on the domestic and international security environment and published doctrine with an eye to the future, not the past. When doctrine has become obsolete, it has been replaced by current and forward-leaning thought to keep the Army viable. At times, the Army has been subject to pundits who questioned the service's continued existence. When attacked, the Army's leadership has used doctrine to tie the service to federal law and national-security policy. This reminded the government and the people of the Army's legal and historic role in defending the nation. In this way, the American Army has been no different than many of its domestic and global counterparts by publishing doctrine not only to prepare a service for future conflict but also to justify its survival.

Since 1779, the Army's keystone doctrine has evolved from being a tactical drill manual to its present form: a collection of strategic, operational, and tactical precepts. Generations of army personnel have used doctrine to intellectually prepare themselves and their units to accomplish the nation's business and to conduct the operations necessary to do so. Army doctrine has its own history, one that must be considered in order to attain a fuller appreciation of the service and its contributions to national defense. Doctrine will continue to be the Army's principal way to regulate chaos in peacetime and war.

NOTES

INTRODUCTION. U.S. ARMY DOCTRINE IN HISTORICAL PERSPECTIVE

1. For a comprehensive list of works on this subject, see the bibliography.

2. For an overview of doctrine, see Bert Chapman, *Military Doctrine: A Reference Handbook* (Santa Barbara, CA: Praeger Security International, 2009). Among scholars and practitioners of doctrine, there is much debate over use of the term "capstone" or "keystone" to describe the dominant manual of a particular era. In masonry work, from which both terms originate, the capstone is the crowning point of brickwork, which implies that it is dependent upon subordinate stonework for support. The keystone, however, is a wedge-shaped piece at the crown; other associated stones depend upon it for support. Intellectually, as far as doctrine is concerned, keystone is the appropriate analogy to indicate the dominant manual upon which all others depend. However, where a particular manual uses the capstone descriptively, that term will be used for accuracy.

3. That said, drill manuals are doctrinal publications, as my definition asserts.

4. Michael Howard, "Military Science in an Age of Peace," *Journal of the United Services Institute for Defence Studies* 119, no. 1 (March 1974): 3–9. Andrew J. Birtle, *U.S. Army Counterinsurgency and Contingency Operations Doctrine, 1860–1941* (Washington, DC: U.S. Army Center of Military History, 1998), 5–6. Birtle's definition of informal practice is used here.

5. Rangers evolved over time from using unconventional to conventional methods of operation. Special-operations forces is a twentieth-century term that groups a number of similar units such as army Rangers, army Special Forces, and navy SEALS. Doctrine occasionally refers to unconventional forces as "irregulars." In such cases, the doctrinal term will be used.

6. Archer Jones, *The Art of War in the Western World* (New York: Oxford University Press, 1987), 221–222. *The Dictionary of Battles: The World's Key Battles from 405 BC to Today,* ed. David Chandler (New York: Henry Holt & Co., 1987), 78–79. Bruce D. Porter, *War and the Rise of the State: The Military Foundations of Modern Politics* (New York: Free Press, 2002), 65. Military literatures available during the sixteenth century include Aelianus, *Tacticus;* Sextus Julius Frontinus, *De re militari, sive Strategematicon* (Bologna: Franciscus [Plato] de Benedictis, 10 July 1495–17 January 1496); Sextus Julius Frontinus, *De re militari. Flavius Vegetius vir illustris, De re militari. Aelianus, De instruendis aciebus. Modesti, Libellus de vocabulis rei militaris* (Bologna: Plato de Benedictis, 1495–96); Polybius, 205?–125? B.C. *De primo bello Punico* [Latin adaptation by Leonardus Aretinus Brunus], *Plutarchi Parallela minora,* [trans. Guarinus Veronensis.] (Brescia: Jacobus Britannicus, October 1498); Gaius Julius

Caesar, 100–44 B.C. *Commentariorum de bello Gallico* (Venice, 1503); and *Polyaenus, Stratagemi dell'arte della guerra . . . Dalla greca nella volgar lingua italiana tradotti da m. Nicolo Mutoni* (Venice: Al segno d'Erasmo, 1551).

7. Why Americans never developed an approved army doctrine during the colonial period has much to do with colonial society and the nature of security issues. For insights, see William T. Allison, Jeffrey Grey, and Janet G. Valentine, *American Military History: A Survey from Colonial Times to the Present* (Upper Saddle River, NJ: Pearson Prentice Hall, 2007), 3. Joyce Appleby, "Value and Society," in *Colonial British America: Essays in the New History of the Early Modern Era*, ed. Jack P. Greene and J. R. Pole (Baltimore: Johns Hopkins University Press, 1984), 301. Don Higginbotham, *The War of American Independence: Military Attitudes, Policies, and Practice, 1763–1789* (Boston: Northeastern University Press, 1983), 7. Virgil Ney, *Evolution of the United States Army Field Manual: Valley Forge to Vietnam* (Fort Belvoir, VA: Combat Operations Research Group, 1966), 3. Military literature was available; between 1643 and 1644 in Plymouth Colony, Myles Standish consulted William Barrisse's *Military Discipline or the Young Artillery-Man* to help organize and discipline colonists. The English colonists certainly did not need to be reminded of the failed Roanoke, Virginia, supply colony, which disappeared sometime between 1587 and 1590. Plymouth Colony established four conventionally organized militia companies within weeks of arrival in North America, and Massachusetts formed three regiments five years after its founding. Maryland raised militia companies by 1638, and others such as Virginia by the 1640s and South Carolina in 1708 with its two infantry regiments followed suit. In 1652, New England militia units numbered about sixty-four men plus commissioned officers such as a captain, a lieutenant, and an ensign who were oftentimes elected. If the population was large enough, some towns had more than one company; Boston alone had ten regiments of various types by 1680, and some colonies had county militias, as well. Militias had set drill days established by their colonial governments, where the members underwent training in the use of firearms and basic marching procedures. They also learned how to move into ranks and to "give fire" as a group. While useful on a parade field, such methods were incongruous when fighting Indians in the woods. John Ferling, "The New England Soldier: A Study in Changing Perceptions," *American Quarterly* 33, no. 1 (Spring 1981): 30. John W. Wright, "Some Notes on the Continental Army," First Installment, *William and Mary College Quarterly Historical Magazine*, 2nd ser., 11, no. 2 (April 1931): 81. Only a company of British regulars briefly served the duke of York in the colonies, while other British troops momentarily served in Virginia in 1677. William L. Shea, "The First American Militia," *Military Affairs* 46, no. 1 (February 1982): 17. Richard P. Gildrie, "Defiance, Diversion, and the Exercise of Arms: The Several Meanings of Colonial Training Days in Colonial Massachusetts," *Military Affairs* 52, no. 2 (April 1988): 53. Boston regiments consisted of eight infantry, one cavalry, and one artillery unit.

8. John Grenier, *The First Way of War: American War Making on the Frontier* (New

York: Cambridge University Press, 2005), 5, 22–29, 39–43, 87–114. Extirpative war is more akin to the European concept of *petite guerre* or "small war." See Thomas Auguste le Grandmaison, *La Petite Guerre, ou traite du service des troupes légères en campagne* (n.p., 1756). Despite the ongoing debate over its validity as war or criminal activity, Europeans formed numerous types of units for fighting *petites guerres*, to include *pandours, croats, hussars, freikorps, jaegers, chasseurs,* light infantry, and partisans. While the British used burning to great effect against the Scots and Irish, they never officially adopted such methods until the Boer War. Brian M. Linn and Russell F. Weigley, "The American Way of War Revisited," *Journal of Military History* 66, no. 2 (April 2002): 501–533. William T. Allison, Jeffrey Grey, and Janet G. Valentine, *American Military History,* also investigate Weigley's thesis in *American Military History.* Jeremy Black, *America as a Military Power: From the American Revolution to the Civil War* (Westport, CT: Praeger Publishers, 2002), 6–7. Shea, "The First American Militia," 18. George Percy, "A Trewe Relacyon," *Tyler's Quarterly Historical and Genealogical Magazine* 3, no. 4 (April 1922): 270–273, quoted in Shea, "The First American Militia," 16. For a description of colonial-style warfare, albeit from a slightly later period, see Michael O. Siochru, "Atrocity, Codes of Conduct and the Irish in the British Civil Wars, 1641–1653," *Past and Present,* no. 195 (May 2007): 55–86. Also see John Smith, "The Proceedings of the English Colonies in Virginia," in *The Complete Works of John Smith,* ed. Philip L. Barbour, 3 vols. (Chapel Hill: University of North Carolina Press, 1986), 1:206.

9. Rangers trace their origin to thirteenth-century England and a term to describe foresters or borderers. To "range" is to patrol. By the seventeenth century, titles of units such as "Border Rangers" appeared. Church provides the first evidence of the term's use in North America. Benjamin Church, *Diary of King Philip's War, 1675–76,* ed. Alan Simpson and Mary Simpson (Chester, CT: Pequot Press, 1975).

10. Thomas Church, *The History of the Great Indian War of 1675 and 1676, Commonly Called Philip's War,* ed. Samuel G. Drake (reprint, Whitefish, MT: Kessinger Publishing, 2007). John J. Tierney Jr., *Chasing Ghosts: Unconventional Warfare in American History* (Washington, DC: Potomac Books, Inc., 2007), 23. For discussions of Rangers, see Grenier, *The First Way of War,* 134. In 1756, the New Hampshire Ranger unit was first called the Ranger Company of the New Hampshire Provincial Regiment. Later, the Independent Company consisted of nine companies. Grenier has claimed that Rogers produced "the first formal doctrine of war making composed in North America," with a written set of orders for "ranging" in 1757. Roger's *Standing Orders* were not doctrine but standing operating procedures that fall under informal practice, for the *Orders* applied to one unit and not all Ranger forces that existed in North America.

11. Allison, Grey, and Valentine, *American Military History,* chap. 2.

12. Sylvia R. Frey, *The British Soldier in America: A Social History of Military Life in the Revolutionary Period* (Austin: University of Texas Press, 1981), 94. Peter E. Russell, "Redcoats in the Wilderness: British Officers and Irregular Warfare in Europe

and America, 1740 to 1760," *William and Mary Quarterly*, 3rd ser., 35, no. 4 (October 1978): 629–652. James Wolfe, quoted in Alan Rogers, *Empire and Liberty: American Resistance to British Authority, 1755–1763* (Berkeley: University of California Press, 1974), 63. Russell Weigley, *History of the United States Army* (New York: Macmillan, 1967), 11. Wright, "Some Notes," 83.

13. Steven T. Ross, *From Flintlock to Rifle: Infantry Tactics, 1740–1866*, 2nd ed. (London: Frank Cass, 1996), 33–35.

14. Higginbotham, *The War of American Independence*, 22. John Shy, *A People Numerous and Armed: Reflections on the Military Struggle for American Independence* (Ann Arbor: University of Michigan Press, 1990), 39–40. Nathaniel Greene quoted in Robert K. Wright Jr., *The Continental Army* (Washington, DC: U.S. Army Center of Military History, 1983), 122.

15. Clifford K. Shipton, "Literary Leaven in Provincial New England," *New England Quarterly* 9, no. 2 (June 1936): 269–301. For a detailed list of manuals in America during this era, see Sandra L. Powers, "Studying the Art of War: Military Books Known to American Officers and Their French Counterparts during the Second Half of the Eighteenth Century," *Journal of Military History* 70, no. 3 (July 2006): 781–814. Although the military publications provided expert information, they also gave conflicting advice concerning how to organize, train, and wage war tactically, as well as explaining in some cases the relationship between war and society. Bland incorporated civil control of the military, arguing "the practice of all nations, ancient and modern, everywhere the People have been blessed with the highest liberty, never to admit of a military independence upon their military superiors." Humphrey Bland, *Treatise of Military Discipline: in which is laid down and explained the duty of the officer and soldier, through the several branches of the service* (London: Printed for R. Baldwin, 1759), 3–4. Conversely, Turpin de Crisse espoused that French soldiers fought for "love of duty, the desire for glory, and zeal for our king and country," thus stressing loyalty to a monarch. Lancelot Turpin de Crisse, *An Essay on the Art of War*, trans. Joseph Otway (London: Printed by A. Hamilton for W. Johnson, 1761), i–ii. Frederick the Great argued that war demanded every soldier to possess an unshakable sense of duty to the state, in this case, Frederick. Frederick the Great, quoted by Jay Luvass, *Frederick the Great on the Art of War* (New York: Free Press, 1966), 72–77. Christopher Duffy, *Frederick the Great: A Military Life* (New York: Routledge, 1993), 290.

16. *A Plan for establishing and Disciplining a National Militia in Great Britain, Ireland, and in all the British Dominions of America, to which is added An Appendix containing Proposals for improving the Maritime Power of Great Britain, a New Edition, with a Preface suited to the Present State of Affairs* (London: Author unidentified, printed for A. Millar, over-against Katherine-street in the Strand; and sold by M. Cooper in Paternoster Row, 1745). Fred K. Vigman, "A 1745 Plan for . . . A National Militia in Great Britain . . . and America," *Military Affairs* 9, no. 4 (Winter 1945): 360. Hugh Jameson, "Equipment for the Militia of the Middle States, 1775–1781," *Journal of the American Military Institute* 3, no. 1 (Spring 1939): 27. Joyce Appleby,

"Value and Society," in *Colonial British America: Essays in the New History of the Early Modern Era*, Jack P. Greene and J. R. Pole, eds. (Baltimore: Johns Hopkins University Press, 1984), 301. Given that Thomas Pickering and George Washington had scrutinized the *Plan*, it is interesting to speculate why a universal American militia doctrine never evolved from it. Arguably, failure had much to do with what Don Higginbotham calls the provincialism of the era with its transplanted English culture centered upon local leadership that met local requirements. Under such conditions, a standardized militia doctrine acceptable to independent-minded colonial governments was problematic. In 1771, Massachusetts published *A Plan of Exercise for the Militia of the Province of Massachusetts* but not all colonies followed suit. Also likely was the American reluctance to adopt tactics that seemed unsuitable for the terrain, the weather, and the enemy. Higginbotham, *The War of American Independence*, 2–3, 7.

17. Wright, *The Continental Army*, 121–152.

18. Weigley, *History of the United States Army*, 23–24. In 1755, Braddock's army was well trained but had difficulty maneuvering by battalions on the narrow wilderness paths. Shy, *A People Numerous and Armed*, 86. Grenier, *The First Way of War*, 128.

19. Vigman, "A 1745 Plan," 360. John Shy, "A New Look at Colonial Militia," *William and Mary Quarterly*, 3rd ser., 20, no. 2 (April 1963): 175–185.

20. Charles Patrick Neimeyer, *America Goes to War: A Social History of the Continental Army* (New York: New York University Press, 1996), xiii. Joseph J. Ellis, *His Excellency, George Washington* (New York: Random House, 2004), 65. Powers, "Studying the Art of War," 789–800. Thomas Webb, *A Military Treatise on the Appointments of the Army Containing Many Useful Hints, Not touched upon by any Author, and proposing some New regulations in the Army, which will be particularly Useful in Carrying out the War in North-America, A Short Treatise with Military Honors* (Philadelphia: W. Dunlop, 1759). Also popular was Thomas Simes, *The Military Guide for Young Officers* (Philadelphia: Reprinted by J. Humphries, R. Bell, and R. Aitken, 1776).

21. George Washington, *The Writings of George Washington, 1745–1799*, ed. John C. Fitzpatrick, 39 vols. (Washington, DC: Government Printing Office, 1931–1944), vol. 3, "To the President of Congress, Cambridge, August 4, 1775."

22. Weigley, *History of the United States Army*, 68–69.

23. E. Wayne Carp, *To Starve the Army at Pleasure: Continental Army Administration and American Political Culture, 1775–1783* (Chapel Hill: University of North Carolina Press, 1984). Kenneth Schaffel, "The American Board of War, 1776–1781," *Military Affairs* 50, no. 4 (October 1986): 185.

24. Washington, *The Writings of George Washington, 1745–1799*, vol. 3, "To the President of Congress, Cambridge, August 4, 1775"; vol. 8, "To Brigadier General Alexander McDougall, Head Quarters, Morris Town, May 23, 1777"; vol. 10, "Remarks on Plan of Field Officers For Remodeling the Army, November 1777."

25. William B. Willcox, "Too Many Cooks: British Planning before Saratoga," *Journal of British Studies* 2, no. 1 (November 1962): 56–90. Powers, "Studying the Art of War," 791.

26. James Mitchell Varnum to Nathaniel Greene, February 12, 1778, *The Spirit of 'Seventy-Six': The Story of the American Revolution as Told by Participants*, ed. Henry Steele Commager and Richard B. Morris, 2 vols. (Indianapolis: Bobbs-Merrill, 1958), 1:650–651. Royster, *A Revolutionary People at War*, 190–192.

CHAPTER ONE. MIMICS AT WAR: TACTICAL DRILL, 1778–1848

1. George Washington, *The Writings of George Washington, 1745–1799* (Washington, DC: Government Printing Office, 1931–1944), vol. 10, "To Baron Steuben, January 9, 1778," and "To Brigadier General George Weedon, February 10, 1778."

2. Timothy Pickering, *An Easy Plan of Discipline for a Militia* (Salem, MA: Printed by Samuel and Ebenezer Hall, 1775), xi. Pickering's comment was uttered prior to Steuben's arrival but serves as an example of prevalent attitudes toward overt regimentation.

3. Charles Patrick Neimeyer, *America Goes to War: A Social History of the Continental Army* (New York: New York University Press, 1996), 160–161. A. Van Doren Honeyman, ed., "The Revolutionary War Record of Samuel Sutphin, Slave," *Somerset County Historical Quarterly*, no. 3 (1914): 186–190. Scholars disagree over who was the instigator, Conway or Washington's former aide, Thomas Mifflin. Joseph J. Ellis, *His Excellency George Washington* (New York: Vintage Books, 2004), 117. Philander Dean Chase, "Baron von Steuben in the War of Independence" (Ph.D. diss., Duke University, 1972), 57.

4. John Shy, *A People Numerous and Armed: Reflections on the Military Struggle for American Independence*, rev. ed. (Ann Arbor: University of Michigan Press, 1990), 155.

5. Neimeyer, *America Goes to War*, 136–140, 159–165. Washington, *The Writings of George Washington, 1745–1799*, vol. 11, "To Brigadiers and Officers Commanding Brigades, March 19, 1778." Chase, "Baron von Steuben," 59.

6. Washington, *The Writings of George Washington, 1745–1799*, vol. 11, "General Orders of George Washington, March 17, 1778." John Sullivan, *The Letters and Papers of Major General John Sullivan*, vols. 13–15, ed. Otis G. Hammond, Collections of the New Hampshire Historical Society, printed in 3 vols. (Concord: New Hampshire Historical Society, 1930–1939), 14:32–33.

7. Washington, *The Writings of George Washington, 1745–1799*, vol. 11, "General Orders March 22, 1778." Samuel C. Vestal, "Frederick William von Steuben," *Coast Artillery Journal* 75, no. 4 (July–August 1932), 289–293.

8. Washington, *The Writings of George Washington, 1745–1799*, vol. 11, "General Orders, March 28, 1778."

9. Robert K. Wright Jr., *The Continental Army* (Washington, DC: U.S. Army Center of Military History, 1983), 132–135. Baron Friedrich Wilhelm Ludolf Gerhard

Augustin Steuben, *Regulations for the Order and Discipline of the Troops of the United States, Part 1* (Albany, NY: Backus & Whiting, 1897), appendix A. Soldiers were provided with individual movements for how to stand at attention, to load and fire weapons, and to march. Divisions and brigades were instructed to use the closed column for speed and control. When necessary, the formations were to change into line for firing or charging. Skirmishers detailed from either the line or light infantry units covered the main force. As the main force deployed into combat formation, the skirmishers withdrew through gaps in the line and re-formed to await orders.

10. Washington, *The Writings of George Washington, 1745–1799*, vol. 11, "To the President of Congress, April 30, 1778." Chase, "Baron von Steuben," 71, 302–314. Russell Weigley, *History of the United States Army* (New York: Macmillan, 1967), 63–64. Conway resigned on 23 April 1778 and was assigned to Major General Alexander McDougall. Conway later went to France and continued his career in the French Army. Steuben ultimately received 16,000 acres of land in the Mohawk Valley of New York in June 1786. Within a few years, he had spent most of his money on the good life. His attempts at land speculation failed. He eventually received an annual pension of $2,500 per year from the Congress for his services. He died on 28 November 1794 from a stroke and was buried in an unmarked grave near his cabin per the instructions in his will. His body was later moved to a more secluded spot on his farm and a monument placed over it with the inscription "Indispensable to the Achievement of American Independence."

11. Theodore P. Savas and J. David Dameron, *A Guide to the Battles of the American Revolution* (El Dorado Hills, CA: Savas Beatie, 2006), 170–178.

12. Washington, *The Writings of George Washington, 1745–1799*, vol. 14, "To Baron Steuben, February 26, 1779," "To Baron Steuben, 11 March 1779," and "General Orders, March 28, 1779."

13. Sandra L. Powers, "Studying the Art of War: Military Books Known to American Officers and Their French Counterparts during the Second Half of the Eighteenth Century," *Journal of Military History* 70, no. 3 (July 2006): 794–795. Washington, *The Writings of George Washington, 1745–1799*, vol. 14, "General Orders, April 12, 1779." While updated for use by militias, only part 1 was published. The first printing was by Styner & Cist of Philadelphia in 1779 and consisted of 154 pages. Virtually every officer received a copy, and seven editions were published by 1785.

14. James Thatcher, *A Military Journal During the American Revolutionary War from 1775–1783* (Boston, MA: Cottons & Barnard, 1827), 312. Baron von Closen, quoted in Don Higginbotham, *The War of American Independence: Military Attitudes, Policies, and Practice, 1763–1789* (Boston: Northeastern University Press, 1983), 413. By 22 December 1780, Steuben also wrote "Baron von Steuben Regulations for the Cavalry or Corps Legionaire [*sic*]," but it was never published.

15. John Grenier, *The First Way of War: American War Making on the Frontier* (New York: Cambridge University Press, 2005), 152. Savas and Dameron, *A Guide to the Battles of the American Revolution*, 182–194, 198–203, 217–222, 249–252, 276–282,

286–292, 329–345. Col. *George Rogers Clark's Sketch of His Campaign in the Illinois in 1778–9 with an Introduction By Hon. Henry Pirtle, of Louisville and an Appendix containing The Public and Private Instructions to Col. Clark and Major Bowman's Journal of the Taking of Post St. Vincent* (reprint; Bedford, MA: Applewood Books, 2001).

16. Ludwig von Closen, *The Revolutionary Journal of Baron Ludwig von Closen, 1780–1783*, ed. and trans. Evelyn M. Acomb (Chapel Hill: University of North Carolina Press, 1958), 242. Michael S. Warner, "General Josiah Harmar's Campaign Reconsidered: How the Americans Lost the Battle of Kekionga," *Indiana Magazine of History* 83, no. 1 (March 1987): 45.

17. Weigley, *History of the United States Army*, 78–83. William T. Allison, Jeffrey Grey, and Janet G. Valentine, *American Military History: A Survey from Colonial Times to the Present* (Upper Saddle River, NJ: Pearson Prentice Hall, 2007), 68.

18. The Army chain of commanding commanders and generals went through a period of title changes. George Washington was general and commander in chief from 15 June 1775 to 23 December 1783. Major General Henry Knox was senior officer from 23 December 1783 to 20 June 1784, followed by U.S. Army Senior Officer Major John Doughty from 20 June 1784 to 12 August 1784, when Harmer replaced him.

19. Weigley, *History of the United States Army*, 83. National Archives and Records Administration (NARA), Record Group 98, Entry 126, Jonathan Heart, "Letter Book of Captain Jonathan Heart, 1st American Regiment, 17 April 1787–26 January 1788."

20. NARA, Record Group 107, Entry 1, "Report from Henry Knox, Secretary of War, 15 June 1789"; "Letter of Arthur St. Clair to George Washington, 14 September 1789." Wars also broke out in the Old Southwest to include the Franklin-Chickamauga War (1788–1794) and the Creek War (1792–1793). Grenier, *The First Way of War*, 172–193.

21. Grenier, *The First Way of War*, 172–193. Howard K. Peckham, "Josiah Harmar and His Indian Expedition," speech given at the Ohio State Archeological and Historical Society, 11 April 1946. Warner, "General Josiah Harmar's Campaign Reconsidered," 45–47.

22. Grenier, *The First Way of War*, 197.

23. Warner, "General Josiah Harmar's Campaign Reconsidered," 47–54. Basil Meek, "General Harmar's Expedition," *Ohio History*, no. 20 (January 1911): 74–108.

24. Warner, "General Josiah Harmar's Campaign Reconsidered," 55.

25. William O. Odom, "Destined for Defeat: An Analysis of the St. Clair Expedition of 1791," *Northwest Ohio Quarterly* 65, no. 2 (Spring 1993): 74–75. St. Clair, a veteran of the American Revolution and first governor of the Northwest Territory, was appointed a major general in 1791.

26. Frazer Ells Wilson, *Arthur St. Clair, Rugged Ruler of the Old Northwest: An Epic of the American Frontier* (Richmond, VA: Garrett & Massie, 1944), 71–76. Odom, "Destined for Defeat," 81–82. James T. Currie, "The First Congressional Investigation:

St. Clair's Military Disaster of 1791," *Parameters* 20, no. 4 (December 1990): 85, 96. It is unclear whether Little Turtle and Blue Jacket were present, although they have been given credit for being there.

27. Currie, "The First Congressional Investigation," 97. Leroy V. Eid, "American Indian Military Leadership: St. Clair's 1791 Defeat," *Journal of Military History* 57, no. 1 (January 1993): 83–85. Grenier, *The First Way of War*, 198–199.

28. Patrick J. Furlong, "The Investigation of General Arthur St. Clair, 1792–1793," *Capitol Studies*, no. 5 (Fall 1977): 85.

29. Furlong, "The Investigation of General Arthur St. Clair," 73. Odom, "Destined for Defeat," 81.

30. Alan D. Gaff, *Bayonets in the Wilderness: Anthony Wayne's Legion in the Old Northwest* (Norman: University of Oklahoma Press, 2004), 10–12.

31. Andrew J. Birtle, "The Origins of the Legion of the United States," *Journal of Military History* 67, no. 4 (October 2003): 1249–1262.

32. Gaff, *Bayonets in the Wilderness*, 22–23. Of note, Washington selected a conventional commander, Wayne, above all possible former Revolution-era candidates. Among the major candidates, all were conventionally trained Regulars or militia to include major generals Benjamin Lincoln (deemed past his prime), Steuben (too ambitious and a foreigner), and William Moultrie (Washington knew little of him). Also considered and bypassed were brevetted major generals, including Lachlan McIntosh, George Weedon, Edward Hand, Charles Scott, and Jedediah Huntington. Of the brigadier generals, Mordecai Gist, William Irvine, Daniel Morgan, Otho Holland Williams, Rufus Putnam, Charles Pinckney, Henry Lee, Thomas Sumter, and Andrew Pickens were all rejected.

33. Gaff, *Bayonets in the Wilderness*, 62–64.

34. Jay C. Ruston, "Anthony Wayne and the Indian Campaign, 1792–1794," *Indiana Military History Journal*, no. 7 (May 1982): 21–25.

35. Gaff, *Bayonets in the Wilderness*, 299, 301.

36. Francis Paul Prucha, *The Sword of the Republic: The United States Army on the Frontier, 1783–1846* (Bloomington: Indiana University Press, 1977), 36–37. Gaff, *Bayonets in the Wilderness*, 301–313. R. C. McGrane, "William Clark's Journal of General Wayne's Campaign," *Mississippi Valley Historical Review* 1, no. 3 (December 1914): 418–444.

37. Brian McAllister Linn, *The Echo of Battle: The Army's Way of War* (Cambridge, MA: Harvard University Press, 2007), 11. Wayne turned over command to Brigadier General James Wilkinson on 15 December 1796.

38. Grenier, *The First Way of War*, 206–207.

39. Allison, Grey, and Valentine, *American Military History*, 85–86.

40. Grenier, *The First Way of War*, 208. *American State Papers Indian Affairs (AS-PIA)*, 2 vols. (Washington: Gales & Seaton, 1832), 1:776, "Harrison to Secretary of War William Eustis," 18 November 1811.

41. Major Harry D. Tunnell IV, *To Compel with Armed Force: A Staff Ride Handbook*

for the Battle of Tippecanoe (Fort Leavenworth, KS: Combat Studies Institute Press, 2000), http://www.cgsc.edu/carl/resources/csi/tunnell/tunnell.asp. *ASPIA*, 1:776, "Harrison to Secretary of War William Eustis," 18 November 1811.

42. Tunnell, *To Compel with Armed Force.* "Daviess" is also spelled "Daveiss."

43. Ibid.

44. Grenier, *The First Way of War*, 87, 92–93. Johann von Ewald, *Abhandlung über den Kleinen Krieg* (Treatise on Partisan Warfare) (Cassel: J. J. Cramer, 1785).

45. Weigley, *History of the United States Army*, 66. Steuben, *Regulations*, 9–10.

46. Linn, *The Echo of Battle*, 11. William B. Skelton, *An American Profession of Arms: The Army Officer Corps, 1784–1861* (Lawrence: University Press of Kansas, 1992), 95–96.

47. Steven T. Ross, *From Flintlock to Rifle: Infantry Tactics, 1740–1866*, 2nd ed. (London: Frank Cass, 1996), 51–52.

48. Ibid., 52–56.

49. Kevin Riley, "Theorist, Instructors, and Practitioners: The Evolution of French Doctrine in the Revolutionary and Napoleonic Wars, 1792–1815," *The Napoleon Series*, http://www.napoleon-series.org/military/organization/c_tactics.html. Ross, *From Flintlock to Rifle*, 52–56.

50. Gunther E. Rothenberg, *The Art of War in the Age of Napoleon* (Bloomington: Indiana University Press, 1980), 101–102. The *levée en masse* was replaced with the Constitution of the Directorate in the Year III of 22 August 1795, which made volunteers the focus of recruiting.

51. Ibid., 115–116. Ross, *From Flintlock to Rifle*, 88–122.

52. Skelton, *An American Profession of Arms*, 121. David S. Heidler and Jeanne T. Heidler, *The War of 1812* (Westport, CT: Greenwood Press, 2002), 1–4. Richard V. Barbuto, *Niagara 1814: America Invades Canada* (Lawrence: University Press of Kansas, 2000), 125.

53. William Duane, *The System of Discipline and manoeuvers of Infantry, Forming the Basis of Modern Tactics: Established for the National Guards and Armies of France Translated for the American Military, from the edition published by authority in 1805* (Philadelphia: B. Graves, 1807).

54. William Duane, *Military Dictionary, or, Explanation of the Several Systems of Discipline of Different Kinds of Troops, Infantry, Artillery, and Cavalry; The Principles of Fortification, and All the Modern Improvements in the Science of Tactics; Comprising the Pocket Gunner, or Little Bombardier; The Military Regulations of the United States; The Weights, Measures, and Monies of all Nations; The technical Terms and Phrases of the Art of War in the French Language, Particularly Adapted to the Use of the Military Institutions of the United States* (Philadelphia: William Duane Printer and Publisher, 1810), vii–xi.

55. Mark Pitcavage, "Ropes of Sand: Territorial Militias, 1801–1812," *Journal of the Early Republic* 13, no. 4 (Winter 1993): 481–500.

56. Irenée Amelot de Lacroix, *Rules and Regulations for the Field Exercise and*

Manoeuvers of the French Infantry, Issued August 1, 1791. And the manoeuvers added, which have been since adopted by the Emperor Napoleon. also, the Manoeuvers of the Field Artillery with Infantry. By Col. Irenée Amelot de Lacroix Late Chief of Brigade in the French service In Three Volumes The third consisting of plates (Boston: T. B. Wait & Co., 1810). Robert Smirke, Review of a Battalion of Infantry, including the Eighteen Manoeuvers, illustrated by a series of engraved diagrams; together with the words of command and an accurate description of each manoeuver, explaining the duty and ascertaining the situation of the officer, through the various movements of the corps: forming an easy introduction to the part of the system of military discipline (New York: M. and W. Ward, 1810). Smirke (1752–1845), who became famous as an English painter and illustrator of such works as The Scriptures, Arabian Knights, Shakespeare, Milton, and Gay, first published his manual in England in 1799. Isaac Maltby, The Elements of War by Isaac Maltby Brigadier General in the Fourth Massachusetts Division (Boston: Thomas B. Wait & Co., 1811).

57. Barbuto, Niagara 1814, 125.

58. Alexander Smyth, Regulations for the Field Exercise, Maneuvers, and Conduct of the Infantry of the United States Drawn Up and Adapted to the Organization of the Militia and Regular Troops (Philadelphia: Anthony Finley, 1812). A file is a column of soldiers lined up one behind the other.

59. Julius W. Pratt, "Fur Trade Strategy and the American Left Flank in the War of 1812," American Historical Review 40 (January 1935): 251. Ranger companies consisted of four commissioned officers, eight noncommissioned officers, and sixty privates. One-year enlistments were typical to fight mounted or on foot depending upon local requirements. They operated primarily in the Old Northwest to keep Indians neutral, make safe the frontier, and patrol from the Missouri River to the Ohio. Rangers were raised in addition to frontier militia and cavalry units.

60. Heidler and Heidler, The War of 1812, 58–60. Skelton, An American Profession of Arms, 121. Donald E. Graves, "Dry Books of Tactics: U.S. Infantry Manuals of the War of 1812 and After," Military Collector and Historian 38, no. 2 (Summer 1986): 51–61, and 38, no. 4 (Winter 1986): 173–77.

61. Charles E. Heller and William A. Stofft, ed., America's First Battles, 1776–1965 (Lawrence: University Press of Kansas, 1986), 48. Allan Peskin, Winfield Scott and the Profession of Arms (Kent, Ohio: Kent State University Press, 2003), 26.

62. Noxon Toomey, "The History of the Infantry Drill Regulations of the United States Army" (St. Louis, [1917]), 2, Special Collections, Northern Regional Library Facility, University of California, Richmond, California. William Duane, Regulations to be Received and Observed for the Discipline of Infantry in the Army of the United States (Philadelphia: Printed for the Author, 1814).

63. Virgil Ney, Evolution of the United States Army Field Manual: Valley Forge to Vietnam (Fort Belvoir, VA: Combat Operations Research Group, 1966), 8. Peskin, Winfield Scott, 40. Donald R. Hickey, The War of 1812, A Forgotten Conflict (Urbana: University of Illinois Press, 1990), 107. William Duane, Regulations. In 1812, Duane

wrote to Thomas Jefferson concerning issues with army doctrine and received the reply "I thank you for the military manuals. This is the sort of book most needed in our country, where even the elements of tactics are unknown. The young have never seen service, the old are past it, and of those among them who are not superannuated themselves, their science is become so." Thomas Jefferson, *The Jefferson Cyclopedia: A Comprehensive Collection of the Views of Thomas Jefferson*, ed. John P. Foley (New York: Funk & Wagnalls, 1900), 52–53.

64. Jeffrey Kimball, "The Battle of Chippawa: Infantry Tactics in the War of 1812," *Military Affairs* 31, no. 4 (Winter 1967–68): 169. "Chippawa" is also spelled "Chippewa," depending upon the source.

65. Ibid., 172–175.

66. John A. Lynn, *Bayonets of the Republic: Motivation and Tactics in the Army of Revolutionary France, 1791–94* (Boulder, CO: Westview Press, 1996), 261–263. Ross, *From Flintlock to Rifle*, 67.

67. Kimball, "The Battle of Chippawa," 185–186.

68. Pratt, "Fur Trade Strategy," 263–264. Heidler and Heidler, *The War of 1812*, 107–108.

69. Heidler and Heidler, *The War of 1812*, 96–97.

70. Ibid., 98. Robert V. Remini, *Andrew Jackson and His Indian Wars* (New York: Viking, 2001), 60–61.

71. Remini, *Andrew Jackson*, 64.

72. Ibid., 66–67.

73. Ibid., 75–79. Heidler and Heidler, *The War of 1812*, 100–101.

74. The *Landwehr*, formed in February 1813, mobilized able-bodied men between the ages of eighteen and forty-five and later modified in March of that year to include men ages seventeen to forty who were not in the regular Army or volunteers. The *Landstrum* of April 1813 included men ages eighteen to sixty who were not in the *Landwehr*. Ross, *From Flintlock to Rifle*, 113–114, 145–146.

75. William A. Ganoe, *The History of the United States Army* (New York: D. Appleton & Co., 1924), 143; Ney, "Evolution of the United States Army Field Manual," 9. Skelton, *An American Profession of Arms*, 120–136.

76. War Department, *Rules and Regulations For the Field Exercise and Manoeuvers of Infantry Compiled and adapted to the Organization of the Army of the United States, Agreeably to A resolve of Congress, Dated December 1814. Published by order of the War Department* (New York: T. and W. Mercein, 1815). *The Infantry Exercise of the United States Army, abridged, for the use of the militia of the United States* (Poughkeepsie, NY: Printed and published by P. Potter, for himself, and for S. Potter & Co., Philadelphia, PA, 1817). Grady McWhitney and Perry D. Jamieson, *Attack and Die: Civil War Military Tactics and the Southern Heritage* (Tuscaloosa: University of Alabama Press, 1984), 31–33.

77. Roger J. Spiller, Joseph G. Dawson III, and T. Harry Williams, eds., *Dictionary*

of *American Military Biography*, 3 vols. (Westport, CT: Greenwood Press, 1984), 974. Ross, *From Flintlock to Rifle*, 180.

78. Peskin, *Winfield Scott*, 66–67.

79. Unconventional forces continued to be raised but without a tactical doctrine for employing them until the twentieth century. In 1832, the Congress authorized a six-company mounted Ranger battalion for the regular Army to be commanded by Major (formerly militia colonel) Henry Dodge. Later, in 1833, the authorization changed to a full regiment of mounted Rangers that eventually became the 1st Dragoons, ten mounted companies totaling 1,832 men who rode into battle but were expected to fight on foot. Weigley, *History of the United States Army*, 98, 159, 161. Linn, *The Echo of Battle*, 12–20. Robert Wooster, *The American Military Frontiers: The United States Army in the West, 1783–1900* (Albuquerque, NM: University of New Mexico Press, 2009), 49.

80. War Department, *By Authority Infantry Tactics; or Rules for the Exercise and Manoeuvers of the Infantry of the U.S. Army* (Washington, DC: Davis & Force, 1825). War Department, *Abstract of Infantry Tactics; including exercises and manoeuvers of light-infantry and riflemen; for use of the militia of the United States. Published by the Department of War; under the Authority of an Act of Congress of the 2d of March, 1829.* (Boston: Hillard, Gray, Little, & Wilkins, 1830).

81. War Department, *By Authority Infantry Tactics; or Rules for the exercise and manoeuvers of the United States Infantry. By Major-General Scott, U.S. Army* (New York: G. Dearborn, 1835). War Department, *By Authority Infantry Tactics; or Rules for the exercise and manoeuvers of the United States Infantry. By Major-General Scott, U.S. Army*, 3 vols. (New York: Harper & Brothers, 1840). War Department, *By Authority Infantry Tactics; or Rules for the exercise and manoeuvers of the United States Infantry. New Edition, 3 vols. By Major-General Scott, U.S. Army.* (New York: Harper & Brothers, 1846).

82. Allison, Grey, and Valentine, *American Military History*, 142. Francis Paul Prucha, "The United States Army as Viewed by British Travelers, 1825–1860," *Military Affairs* 17, no. 3 (Fall 1953): 113–124.

83. In 1822, Lieutenant Colonel Pierce Darrow published *Cavalry Tactics* (Hartford, CT: Oliver D. Cooke), a manual that conformed to the Army's *Infantry Tactics*. Scott later (1826) assisted in developing cavalry and artillery manuals to supplement the Army's infantry doctrine.

84. Dennis Hart Mahan, *An Elementary Treatise on an Advanced Guard, Out-post and Detachment Service of Troops and Handling Them in the Presence of an Enemy, With a Historical Sketch of the Rise and Progress of Tactics, &c., &c.* (New York: Wiley & Long, 1847, 1853, 1863).

85. Henry Wagner Halleck, *Elements of Military Art and Science: or, Course of Instruction in Strategy, Fortification, Tactics of Battles &c. Embracing the distance of staff, infantry, cavalry, artillery and engineers adapted to the use of volunteers and militia* (1846;

reprint, New York: B. Appleton & Co., 1862). Edward Hagerman, *The American Civil War and the Origins of Modern Warfare: Ideas, Organization, and Field Command* (Bloomington: Indiana University Press, 1992), 9–10.

86. Remini, *Andrew Jackson*, 130–131. John K. Mahon, *History of the Second Seminole War, 1835–1842*, rev. ed. (Gainesville: University of Florida Press, 1985), 22–23, 324–325.

87. Weigley, *History of the United States Army*, 182–183.

88. Heller and Stofft, ed., *America's First Battles*, 60–61. K. Jack Bauer, *The Mexican War, 1846–1848* (Lincoln: University of Nebraska Press, 1986), 33–34.

89. Heller and Stofft, eds., *America's First Battles*, 61. James M. McCaffrey, *Army of Manifest Destiny: The American Soldier in the Mexican War, 1846–1848* (New York and London: New York University Press, 1992), 42–43. Samuel Cooper, *A Concise System of Instructions and Regulations for the militia and volunteers of the United States: comprehending the exercises and movements of the infantry, light infantry, and riflemen, cavalry and artillery: together with the manner of doing duty in garrison and in camp, and the forms of parades, reviews, and inspections, as established by authority for the government of the regular Army* (Philadelphia: Robert P. Desilver, 1836).

90. Richard Bruce Winders, *Mr. Polk's Army: The American Experience in the Mexican War* (College Station: Texas A&M University Press, 1997), 68–69, 94. Relatively few militia units served in the war and mostly along the Rio Grande. Most of their valuable service was administrative.

91. McCaffrey, *Army of Manifest Destiny*, 43, 132, 142–146. Ross, *From Flintlock to Rifle*, 180.

92. Milton Jamieson, *Journal and Notes of a Campaign in Mexico: Containing a History of Company C, of the Second regiment of Ohio Volunteers* (Cincinnati, OH: Ben Franklin Printing House, 1849), 17–18. James D. Elderkin, *Biographical Sketches and Anecdotes of a Soldier of Three Wars* (Detroit: Record Printing, 1899), 67.

93. Robert M. Utley, *Frontiersmen in Blue: The United States Army and the Indian, 1848–1865* (Lincoln: University of Nebraska Press, 1967), 2.

CHAPTER TWO. FROM FRENCH DRILL TO TEUTONIC INITIATIVE: *INFANTRY TACTICS AND REGULATIONS*, 1855–1905

1. Matthew Moten, *The Delafield Commission and the American Military Profession* (College Station: Texas A&M University Press, 2000), 71. Robert M. Utley, *Frontiersmen in Blue: The United States Army and the Indian, 1848–1865* (Lincoln: University of Nebraska Press, 1967), 57.

2. Moten, *The Delafield Commission*, 87.

3. Paddy Griffith, *Battle Tactics of the Civil War* (New Haven, CT: Yale University Press, 2001), 100–102. Author unknown, "A Brief History of U.S. Infantry Tactics (1855–1865)," http://www.usregulars.com/drill_history.html.

4. Vigil Ney, *Evolution of the United States Army Field Manual: Valley Forge to Vietnam* (Fort Belvoir, VA: Combat Operations Research Group, 1966), 20. William J.

Hardee, *Rifle and Light Infantry Tactics for the Exercise of Maneuvers of troops when acting as light infantrymen or riflemen*, 2 vols. (Philadelphia: Lippincott, Grambo, & Co., 1855), 1:12–13, 16. Steven T. Ross, *From Flintlock to Rifle: Infantry Tactics, 1740–1866*, 2nd ed. (London: Frank Cass, 1996), 181.

5. National Archives and Records Administration (NARA), Record Group 94, Roll M567, "Letter to Colonel S. Cooper, Adjutant General, Washington DC, 22 March 1855." NARA, Record Group 94, Roll M567, "Letter to Colonel W. J. Hardee from Lippincott, 2 April 1855."

6. Hardee, *Rifle and Light Infantry Tactics*. Geoff Walden and Dom Dal Bello, "Manual of Arms for Infantry: A Re-examination, Part I." http://216.247.222.222/vpp/cgg/maualarms_1.htm. Although the 1855 Springfield was manufactured as a rifle, the term "rifle-musket" was a carryover from the practice of rifling old muskets.

7. Moten, *The Delafield Commission*, 108–110. Brevet Major Silvanus Thayer, future superintendent of West Point, also visited Europe in July 1815 and remained for two years.

8. Ibid., 178–180.

9. Ibid., 210.

10. Hardee, *Rifle and Light Infantry Tactics*, 1:36–65. Volume 2 contains fire and maneuver directions for battalions that would be executable on a Napoleonic battlefield.

11. Utley, *Frontiersmen in Blue*, 205–207.

12. Tony R. Mullis, *Peacekeeping on the Plains: Army Operations in Bleeding Kansas* (Columbia: University of Missouri Press, 2004), 164–165.

13. William A. Ganoe, *The History of the United States Army* (New York: D. Appleton & Co., 1924), 256–257.

14. Hermann Hattaway and Archer Jones, *How the North Won: A Military History of the Civil War* (Champaign: University of Illinois Press, 1991), 11–17. Paddy Griffith, *Battle Tactics of the Civil War* (New Haven, CT: Yale University Press, 2001), 99–105. Vigil Ney, *Evolution of the United States Army Field Manual*, 25. W. J. Wood, *Civil War Generalship: The Art of Command* (New York: Da Capo Press, 2000), 9–19.

15. NARA, Record Group 94, Roll M567, "Special Order #23, dated 16 February 1858." NARA, Record Group 94, Roll M567, "Letter from Samuel Colt to John B. Floyd, dated 10 May 1860."

16. Perry D. Jamieson, *Crossing the Deadly Ground* (Tuscaloosa: University of Alabama Press, 1994), 1–2, 10. General Francois V. A. de Chanal, *The American Army in the War of Secession* (Leavenworth, KS: G. A. Spooner, 1894), 26.

17. Earl J. Hess, *The Union Soldier in Battle: Enduring the Ordeal of Combat* (Lawrence: University Press of Kansas, 1997), 56–57, 113–114.

18. Robert Underwood Johnson and Clarence Clough Buell, *Battles and Leaders of the Civil War: Being for the Part Contributions by Union and Confederate Officers* (New York: Castle Books, 1956), 594–603.

19. D. W. C. Baxter, *Baxter's Volunteer's Manual: Containing Full Instructions for*

the Recruit, in the Schools of the Soldier and Squad, with One Hundred Illustrations of the Different Positions in the Facings and Manual of Arms and the Loadings and Firings: Arranged According to Scott's System of Infantry Tactics (Philadelphia: King & Baird, 1861).

20. Archer Jones, Civil War Command and Strategy: The Process of Victory and Defeat (New York: Free Press, 1992), 26–30, 33–34.

21. NARA, Record Group 94, Roll M567, "Letter from Silas Casey to Colonel S. Cooper, dated 10 May 1860." NARA, Record Group 94, Roll M619, "Letter Received by the Adjutant General, Appointments to Brigadier General, dated 10 September 1861."

22. Silas Casey, Infantry Tactics For the Infantry, Exercise, and Maneuvers of the Soldier, a Company, Line of Skirmishers, Battalion, Brigade, or Corps d'Armée (New York: D. Van Nostrand, 1862), 5. After commanding a division at the Battle of Fair Oaks during the Virginia Peninsula Campaign of 1862, Casey was transferred to Washington. He was given command of a provisional brigade formed to train newly formed regiments prior to their being sent to the front. Casey later published a third volume for "negro troops."

23. Griffith, Battle Tactics of the Civil War, 102–103. Gerald F. Linderman, Embattled Courage: The Experience of Combat in the American Civil War (New York: Free Press, 1987), 136. Hess, The Union Soldier in Battle, 13.

24. Stephen E. Ambrose, Upton and the Army (Baton Rouge: Louisiana State University Press, 1992), 29–34.

25. Linderman, Embattled Courage, 142–147. "Cover" protects the individual from being struck by bullets and projectiles. "Concealment" is protection from observation only.

26. The 1861 War Department-approved Cavalry Tactics provided guidance more akin to executing various maneuvers from platoon through squadron level than raids. See Cavalry Tactics or Regulations For the Instruction, Formations, and Movements of the Cavalry of the Army and Volunteers of the United States (Philadelphia: J. B. Lippincott & Co., 1862).

27. Ambrose, Upton and the Army, 58–59. Robert M. Utley, Frontier Regulars: The United States Army and the Indian, 1866–1891 (reprint, Lincoln: University of Nebraska Press, 1984), 11, 69.

28. Ulysses S. Grant, Personal Memoirs of U. S. Grant (New York: J. J. Little & Co., 1885), 2:545–546. René Chartrand, The Mexican Adventure, 1861–1867 (Oxford: Osprey Publishing, 1994), 19–20. The French Army fielded the breech-loading Chassepot in December 1866 but the weapon was not sent to Mexico.

29. William H. Morris, "To the Editor of the New York Times dated 19 February 1867," New York Times, 2 March 1867, 2.

30. Ambrose, Upton and the Army, 60–61, 63–65. T. R. Brereton, Educating the Army: Arthur L. Wagner and Reform, 1875–1905 (Lincoln: University of Nebraska Press, 2000), 30–31.

31. Ambrose, *Upton and the Army*, 65. Charles J. Ardant du Picq, "Battle Studies," in *Roots of Strategy*, book 2 (Harrisburg, PA: Stackpole Books, 1987). Peter De-Montravel, "The Career of Lieutenant General Nelson A. Miles from the Civil War through the Indian Wars" (Ph.D. diss., Ohio State University, 1986), 130–131. William Ulrich, "The Northern Military Mind in Regard to Reconstruction, 1865–72: The Attitude of Ten Leading Union Generals" (Ph.D. diss., Ohio State University, 1959), 207, 365. War Department, General Order No. 73, 1 August 1867.

32. Ambrose, *Upton and the Army*, 65. Perry D. Jamieson, *Crossing the Deadly Ground: United States Army Tactics, 1865–1899* (Tuscaloosa: University of Alabama Press, 1994), 2–3.

33. Utley, *Frontier Regulars*, 48–49.

34. Ibid., 219–235.

35. Andrew J. Birtle, *U.S. Army Counterinsurgency and Contingency Operations Doctrine, 1860–1941* (Washington, DC: U.S. Army Center of Military History, 1998), 60–63.

36. Thomas A. Sutherland, "Howard's Campaign against the Nez Perce Indians, 1877" (Portland, OR: A. G. Walling, 1878), in Peter Cozzens, *Eyewitnesses to the Indian Wars, 1865–1890*, vol. 2, *The Wars for the Pacific Northwest* (Mechanicsburg, PA: Stackpole Books, 2002), 388–389. Similar actions occurred against the Apaches. See Frank D. Reeve, ed., "Puritan and Apache: A Diary [of Lieutenant H. M. Lazelle]," *New Mexico Historical Review* 23, no. 4 (October 1948): 269–301.

37. Birtle, *U.S. Army Counterinsurgency and Contingency Operations Doctrine*, 64–66. Randolph B. Marcy, *The Prairie Traveler, a handbook for overland expeditions, with maps, illustrations, and the itineraries of the principal routes between the Mississippi and the Pacific* (Philadelphia, 1859; reprint, Williamstown, MD: Corner Publications, 1968). W. P. Clark, *The Indian Sign Language* (reprint, Lincoln: University of Nebraska Press, 1982).

38. John Bigelow quoted in Ambrose, *Upton and the Army*, 406.

39. Brereton, *Educating the U.S. Army*, 31. Jamieson, *Crossing the Deadly Ground*, 71–73.

40. Ibid., 32.

41. Ambrose, *Upton and the Army*, 76–79.

42. Ibid., 83–84. Brereton, *Educating the U.S. Army*, 33.

43. Helmuth Karl Bernhard von Moltke, *Moltke on the Art of War: Selected Writings*, ed. and trans. Daniel J. Hughes and Harry Bell (Novato, CA: Presidio Press, 1993), vii–viii.

44. Utley, *Frontier Regulars*, 44.

45. Hajo Holborn, "The Prusso-German School: Moltke and the Rise of the General Staff," in *Makers of Modern Strategy: From Machiavelli to the Nuclear Age*, ed. Peter Paret (Princeton, NJ: Princeton University Press, 1986), 281–295.

46. Ambrose, *Upton and the Army*, 88–98, 110–134. Brereton, *Educating the U.S. Army*, 33. Brevet Major General Emory Upton, *The Military Policy of the United States*,

ed. Joseph P. Sanger, William D. Beach, and Charles D. Rhodes (Washington, DC: Government Printing Office, 1904), 336.

47. Edward M. Coffman, *The Old Army: A Portrait of the American Army in Peacetime, 1784–1898* (New York: Oxford University Press, 1986), 271–273.

48. John Bigelow, *Mars-la-Tour and Gravelotte* (Washington, DC: Government Printing Office, 1880), 105–106. Michael D. Krause, "Arthur L. Wagner: Contributions from the Past" (unpublished research paper, 5 May 1978, Combined Arms Research Library, Leavenworth, KS).

49. Jerry Cooper, *The Rise of the National Guard: The Evolution of the American Militia, 1865–1920* (Lincoln: Bison Books, 2002), xv, 11–21.

50. Cooper, *The Rise of the National Guard*, 90–91. L. W. V. Kennon, "Some Points on Tactics," *Army and Navy Register*, no. 9 (August 25, 1888): 540–541.

51. Carl von Clausewitz, *On War*, ed. and trans. Michael Howard and Peter Paret (Princeton, NJ: Princeton University Press, 1984), 85–87.

52. Brereton, *Educating the U.S. Army*, 34. Philip H. Sheridan, *The Personal Memoirs of P. H. Sheridan* (New York: Charles L. Webster & Co., 1888), 2:451–452. Roger Spiller, "Doctrine, Military," in *The Oxford Companion to American Military History*, ed. John Whiteclay Chambers II (New York: Oxford University Press, 1999), 231–234.

53. Brereton, *Educating the U.S. Army*, 34–35. Board members included lieutenant colonels John C. Bates (infantry) and George B. Sanford (cavalry), majors Henry C. Hasbrouck (artillery) and John C. Gilmore (assistant adjutant general), captains John T. Haskell (infantry), Edward S. Godfrey (cavalry), and James M. Lancaster (artillery), and two recorders, initially First Lieutenant George Andrews (infantry) and later First Lieutenant J. T. French Jr. (artillery).

54. Ibid., 35–36.

55. Ibid., 36–37.

56. Ibid., 38–39.

57. Ibid., 39–40.

58. Ibid., 40. Arthur L. Wagner, *The Service of Security and Information* (Kansas City: Hudson-Kimberly Publishing Co., 1893). Arthur L. Wagner, *Organization and Tactics*, 6th ed. (Kansas City: Hudson-Kimberly Publishing Co., 1905), 1.

59. Wagner, *Organization and Tactics*, 43.

60. Clausewitz, *On War*, 100–122. Wagner, *Organization and Tactics*, 48.

61. Jamieson, *Crossing the Deadly Ground*, 108.

62. Ibid., 77–78, 154. Ardant du Picq, "Battle Studies," in *Roots of Strategy*, book 2, 135–68. General de Negrier, "Some Lessons of the Russo-Japanese War," *Journal of the Royal United Service Institution* 50, no. 339 (July–December 1906): 910–919. War Department, *Regulations of the Army of the United States, 1889* (Washington, DC: Government Printing Office, 1889). War Department, *Troops in Campaign: Regulations for the Army of the United States* (Washington, DC: Government Printing Office, 1892).

63. Graham A. Cosmas, *An Army for Empire, The United States Army in the Spanish American War* (College Station: Texas A&M University Press, 1994), 152–154. Theodore Roosevelt, *The Rough Riders*, additional text by Richard Bak (Dallas: Taylor Publishing, 1997), 119.

64. Cosmas, *An Army for Empire*, 128–129. Stan Cohen, *Images of the Spanish American War: April–August 1898* (Missoula, MT: Pictorial Histories Publishing Co., 1997), 138. Camp locations included Camp Alger, VA; Camp Black, NY; Camp Cuba Libre, FL; Clarks Point, MA; Columbus Barracks, OH; Dutch Island, RI; Hilton Head, SC; Camp Hamilton, KY; Jackson Barracks, LA; Jefferson Barracks, MO; Key West, FL; Long Island Head, MA; Camp McKenzie, GA; Miami, FL; Camp Meade, PA; Plum Island, NY; Sheridan Point, VA; Sullivans Island, SC; San Diego Barracks, CA; Camp Shipp, AL; Camp Thomas, GA; Tampa, FL; Tybee Island, GA; Washington Barracks, DC; Camp Wheeler, AL; Willets Point, NY; Winthrop, MA; and Camp Wikoff, NY.

65. *Harper's Pictorial History of the War with Spain with an Introduction by Maj-Gen Nelson A. Miles, Commanding United States Army, in Two Volumes* (New York: Harper & Brothers, 1899), 1:174–184. Jamieson, *Crossing the Deadly Ground*, 148–149.

66. Stephen D. Coats, *Gathering at the Golden Gate: Mobilizing for War in the Philippines, 1898* (Fort Leavenworth, KS: Combat Studies Institute Press, 2006), 44, 96–105. In addition to Regulars, volunteer units arrived from California, Iowa, Kansas, Minnesota, Montana, New York, Tennessee, Utah, and Wyoming.

67. Jamieson, *Crossing the Deadly Ground*, 136–137.

68. G. J. A. O'Toole, *The Spanish War: An American Epic* (1984; reprint, New York: W. W. Norton, 1986), 272–273. Richard Harding Davis, *Notes of a War Correspondent* (New York: Charles Scribner's Sons, 1914), 88.

69. Richard Harding Davis, *The Cuban and Porto Rican Campaigns* (New York: privately published, 1898), 218–220. Charles Johnson Post, *The Little War of Private Post* (Boston: Little, Brown, 1960), 182–186.

70. Post, *The Little War of Private Post*, 182–186. O'Toole, *The Spanish War*, 319, 391. Cosmas, *An Army for Empire*, 216–218, 227.

71. Cosmas, *An Army for Empire*, 234–237.

72. Brian McAllister Linn, *The Philippine War: 1899–1902* (Lawrence: University Press of Kansas, 2000), 42–44.

73. Birtle, *U.S. Army Counterinsurgency and Contingency Operations Doctrine*, 111–112. Robert D. Ramsey III, *Savage Wars of Peace: Case Studies of Pacification in the Philippines, 1900–1902* (Fort Leavenworth, KS: Combat Studies Institute Press, 2008), 53.

74. Quoted in Birtle, *U.S. Army Counterinsurgency and Contingency Operations Doctrine*, 112.

75. Ibid., 114–116.

76. Ibid., 116–117.

77. Ramsey, *Savage Wars of Peace*, 23.

78. Birtle, *U.S. Army Counterinsurgency and Contingency Operations Doctrine*, 128. General Order 100, enacted on 24 April 1863, was entitled "Instructions for the Government of Armies of the United States in the Field." Known as the "Lieber Code," it was written by Francis Lieber L.L.D. and revised by an army board of officers. General Order 100 was approved by the president of the United States and regulated the Army's behavior when deployed. It discusses martial law, the nature of war, military necessity, prisoners, and other topics. Ramsey, *Savage Wars of Peace*, 135–157.

79. "Proclamation, Office of U.S. Military Governor in the Philippine Islands and Headquarters Division of the Philippines, Manila, P.I., December 10, 1900," in Ramsey, *Savage Wars of Peace*, 159–162.

80. Birtle, *U.S. Army Counterinsurgency and Contingency Operations Doctrine*, 130–131. Linn, *The Philippine War*, 214–215.

81. Birtle, *U.S. Army Counterinsurgency and Contingency Operations Doctrine*, 130–132. Linn, *The Philippine War*, 195–196.

82. Birtle, *U.S. Army Counterinsurgency and Contingency Operations Doctrine*, 130–148. Diana Preston, *The Boxer Rebellion: The Dramatic Story of China's War on Foreigners That Shook the World in the Summer of 1900* (New York: Berkley Books, 2000), 22–23.

83. Max Boot, *The Savage Wars of Peace: Small Wars and the Rise of American Power* (New York: Basic Books, 2002), 87.

84. Ibid., 87–88.

85. Ibid., 93.

86. Birtle, *U.S. Army Counterinsurgency and Contingency Operations Doctrine*, 148–150.

87. Thomas H. Ruger, *Extended Order Drill: Infantry Drill Regulations* (Washington, DC: Government Printing Office, 1898). War Department, *Infantry Drill Regulations, United States Army. Revised 1904* (Washington, DC: Government Printing Office, 1904). The chief of staff position was created by Secretary of War Elihu Root and implemented under the General Staff Act of 1903. In addition, forty-five officers were to serve on the general staff under the War Department.

88. Russell Weigley, *History of the United States Army* (New York: Macmillan, 1967), 314.

CHAPTER THREE. DOCTRINE FOR ARMY OPERATIONS: FROM *FIELD SERVICE REGULATIONS* TO FIELD MANUAL 100-5, 1905–1945

1. Graham A. Cosmas, *An Army for Empire: The United States Army in the Spanish American War* (College Station: Texas A&M University Press, 1994), 320. Philip C. Jessup, *Elihu Root* (New York: Dodd, Mead, 1938; reprint, New York: Archon Books, 1964), 215 (page citations are to the reprint edition). While reminiscing in March 1915, Root remarked that his selection had been made specifically for his legal expertise in order to direct American military government over the former Spanish islands.

2. Cosmas, *An Army for Empire*, 320–326. Russell F. Weigley, *History of the United States Army* (New York: Macmillan, 1967), 314. Elihu Root, "Address by the Secretary of War at the Marquette Club," 4. Edward M. Coffman, *The Regulars: The American Army, 1898–1941* (Cambridge, MA: Harvard University Press, 2007), 191.

3. Weigley, *History of the United States Army*, 325. At Root's urging, Congress had placed service oversight responsibility not only upon the secretary of war but also the commanding general and the army general staff. Collectively, as the Army's overseers, they required a means to convey official policy concerning the administration, organization, operations, and sustainment of the service to the service, as a whole, in peacetime and war. This obligation applied not only to the regular Army, for the Dick Act of 1903 had changed the structure of the American militia system. Although Root's reformist ideas generated resistance among civilians and officers alike, they nonetheless came to fruition with the War Department's general orders of 27 November 1901. The same orders tackled a second Root priority, education. On 14 February 1903, the General Staff Act abolished the office of the commanding general while defining the composition and duties of general staff officers. It provided for a military chief of staff within the War Department who acted under the orders of the president. When representing the chief executive, the chief of staff also answered to the secretary of war. A newly founded general staff corps consisted of forty-four assigned officers relieved of all other duties. War College instruction commenced in 1904; a permanent building opened at Washington Barracks on 9 November 1908. In Root's *Report of the Secretary of War 1903*, the army staff's efforts to revise service regulations and produce the 1904 infantry manual were highlighted. "Extract of the Report of the Secretary of War for 1903," *The Military and Colonial Policy of the United States*, 430. Timothy K. Nenninger, *The Leavenworth Schools and the Old Army: Education, Professionalism, and the Officer Corps of the United States Army, 1881–1918* (Westport, CT: Greenwood Press, 1978).

4. T. R. Brereton, *Educating the U.S. Army: Arthur L. Wagner and Reform, 1875–1905* (Lincoln: University of Nebraska Press, 2000), 111. Under Root's direction, the army general staff now gave considerable attention to the roles and missions of the Army in wartime. To articulate their view of future war to the service, the general staff wrote the *FSR*. As the creators of army keystone doctrine, the Army staff replaced boards of officers designed for such purposes. Among the many contributors to the *FSR* was Captain Joseph T. Dickman, a subordinate of Arthur Wagner at Fort Leavenworth. After much composition and editing, what emerged was a tactical army doctrine that also prescribed the administrative and organizational conduct of an army in the field. With its focus on the intricacies of command and staff functions, as well as combat and its many nuances, the intended audience was clearly the army officer corps. Individual and unit drill movements were removed and relegated to new manuals designed for those purposes. Branch-specific tactics such as infantry, artillery, and cavalry methods were eventually developed by the training schools at various forts and, at least theoretically, complied with the ideas advanced

in the *FSR*. However, the general staff lacked the means to evaluate every service manual. It would take years before compliance was actualized.

5. The twelve articles were Article I *Organization*, Article II *Orders*, Article III *The service of information*, Article IV *The service of security*, Article V *Marches*, Article VI *Combat*, Article VII *Ammunition supply*, Article VIII *Subsistence*, Article IX *Transportation*, Article X *Shelter*, Article XI *Medical and sanitary service*, and Article XII *Instructions for government of armies of the United States in time of war*. U.S. Army, General Staff, *Field Service Regulations, 1905* (Washington, DC: Government Printing Office), 1905. Richard W. Stewart, gen. ed., *American Military History: The United States Army and the Forging of a Nation, 1775–1917* (Washington, DC: U.S. Army Center of Military History, 2005), 1:23–25.

6. U.S. Army, *Field Service Regulations, 1905*, 11–39, Articles III, IV, IX, XII. Also see Baron Antoine Henri de Jomini, *The Art of War*, Greenhill ed. (Mechanicsburg, PA: Stackpole Books, 1996). When Winfield Scott assumed control of Mexico City in 1847, he was left to his own devices, for army doctrine did not provide any guidance. Other American commanders faced similar situations during Reconstruction (1865–1877), as well as in Cuba (1898), the Philippines (1898–1902), and Peking (1900).

7. U.S Army, *Field Service Regulations, 1905*, 101. According to Jomini, the decisive point is the most important place on a battlefield, a position where topography combines with strategy and is dependent upon force arrangement. When enemy lines are extended and the defense thinner, the center of the enemy line is the decisive point. When enemy forces are concentrated, then the flank is the decisive point. Jomini, *The Art of War*, Greenhill ed., 187.

8. U.S. Army, *Field Service Regulations, 1905*, 102. Light artillery was capable of distant effects at more than 4,500 yards, serious effects at 4,500 to 3,500 yards, effective at 3,500 to 2,000 yards, and decisive at less than 2,000 yards. Heavy artillery effectiveness by category includes distant at more than 6,000 yards, serious at 6,000 to 4,000 yards, effective at 4,000 to 2,500 yards, and decisive at less than 2,500 yards.

9. Ibid., 102–103.

10. Ibid., 103–104.

11. Ibid., 107–109.

12. Ibid., 109–116.

13. Ibid., 116–117.

14. Stewart, *American Military History,* 25, 375.

15. War Department, *Reports of Military Observers Attached to the Armies in Manchuria during the Russo-Japanese War* (Washington, DC: Government Printing Office, 1907), 137, 209.

16. Ibid., 97–99, 101–104. Charles Ardant du Picq, "Battle Studies," in *Roots of Strategy*, book 2 (Mechanicsburg, PA: Stackpole Books, 1987), 135–136. General de

Negrier, "Some Lessons of the Russo-Japanese War," *Journal of the Royal United Service Institution* 50, no. 339 (July–December 1906): 910–919.

17. War Department, *Reports of Military Observers*, 215.

18. War Department, Office of the Chief of Staff, *Field Service Regulations, U.S. Army, 1910* (Washington, DC: Government Printing Office, 1910), 162–165.

19. Ibid., 20–23, 67, 73, 111, 141, 149, 169, 177, 219. Edward M. Coffman, *The Regulars: The American Army, 1898–1941* (Cambridge, MA: Belknap Press of Harvard University Press, 2004), 161–162.

20. War Department, Office of the Chief of Staff, *Field Service Regulations, 1913, with Corrections to May 21, 1913* (Washington, DC: Government Printing Office, 1913), 1–11. The new article organization now became Article I *Organization*, Article II *The service of information*, Article III *Orders*, Article IV *The service of security*, Article V *Marches and convoys*, Article VI *Shelter*, Article VII *The service of supply*, Article VIII *Transportation*, Article IX *Combat*, Article X *The sanitary service*, and Article XI *The laws of war—Instructions for the government of the armies of the United States in time of war*. The manual also included eight appendices, A–H, covering such subjects as flags and pennants of field armies, field message blanks, field map and sketch symbols, field orders, road space and camp dimensions, weights, measures and practicality of slopes, forms of infantry trenches, and additional instructions for the administration and supply of troops in war. Clearly, the army staff had evolved enough to become more involved in minutia than previous editions, one indicator that the service was using doctrine to increase the regulation of field service behavior rather than leaving such matters up to the informal practice of field commanders.

21. War Department, Office of the Chief of Staff, *Field Service Regulations, 1913*, 55.

22. Coffman, *The Regulars*, 163–164. Stephen Budiansky, *Air Power: The Men, Machines, and Ideas That Revolutionized War, from Kitty Hawk to Gulf War II* (New York: Viking, 2004), 31. For a discussion of "air-mindedness" in the early to mid-twentieth century and the growing importance of aviation as an expression of national power and culture, see Scott W. Palmer, *Dictatorship of the Air: Aviation Culture and the Fate of Modern Russia* (New York: Cambridge University Press, 2009), 2–3. The first head of the Aeronautical Division of the Signal Corps was Captain Charles DeForest Chandler. By December 1907, Brigadier General James Allen was in the process of accepting bids for a heavier-than-air flying machine, although pricing and other matters had already been discussed with the Wright brothers. Early flight advocates included army officers Benjamin D. Foulois, Frank P. Lahm, and Thomas E. Selfridge. Lieutenant Selfridge was killed in a crash on September 1908 while a passenger aboard a Wright Flyer piloted by Orville Wright. Orville ended up in the hospital for six weeks. Lieutenant Foulois, also flying as a passenger, completed Wright Flyer acceptance testing for the Army on 30 July 1909. The initial price paid was $25,000 plus a $5,000 bonus and the Army was the first service to adopt the aeroplane.

23. War Department, Office of the Chief of Staff, *Field Service Regulations, 1913*, 54–55.

24. Ibid., 55, 78, 110.

25. Weigley, *History of the United States Army*, 334–335. Combat Studies Institute, "The History of Transformation," *Military Review* 80, no. 3 (May–June 2000), 17–29.

26. War Department, Office of the Chief of Staff, *Field Service Regulations, United States Army (Including Questions), 1914, Text Corrections to August 18, 1917 (Changes No.7)* (Menasha, WI: Collegiate Press, George Banta Publishing Co., 1918), 10.

27. War Department, Office of the Chief of Staff, *Field Service Regulations, United States Army 1914*, 78–79.

28. Ibid., 80–81.

29. Ibid., 82–88.

30. Ibid., 17–18, 89–102.

31. Ibid., 68.

32. Ibid., 2.

33. Ibid., 72.

34. James W. Hurst, *Pancho Villa and Black Jack Pershing: The Punitive Expedition in Mexico* (Westport, CT: Praeger, 2008), xvii–xx. Stewart, *American Military History*, 377.

35. Coffman, *The Regulars*, 195.

36. Hurst, *Pancho Villa and Black Jack Pershing*, xxi, 21–33, 42.

37. Ibid., 44. Jerry Cooper, *The Rise of the National Guard: The Evolution of the American Militia, 1865–1920* (Lincoln: Bison Books, 2002), 156.

38. Hurst, *Pancho Villa and Black Jack Pershing*, 45–46, 75–76.

39. Ibid., 71–73. Villa was assassinated by gunmen on 20 July 1923, in Parral, Mexico.

40. Ibid., 45–46, 90–95. Andrew J. Birtle, *U.S. Army Counterinsurgency and Contingency Operations Doctrine, 1860–1941* (Washington, DC: U.S. Army Center of Military History, 1998), 205.

41. Birtle, *U.S. Army Counterinsurgency and Contingency Operations Doctrine*, 207–208.

42. War Department, Office of the Chief of Staff, *Field Service Regulations United States Army 1914, Corrected to July 31, 1918 (Changes Nos. 1 to 11)* (Washington, DC: Government Printing Office, 1918), 1–3. The Army entered the Great War under the doctrinal regulations established in the 1914 manual. This fact stands in contrast to two sources that mistakenly report the existence of a 1917 doctrine. In *An Uncertain Trumpet*, Kenneth Finlayson asserts, "Based upon the *Field Service Regulations 1917*, American tactical doctrine placed the rifleman at the center of the battle." The *FSR* referred to by Finlayson was not a new doctrinal manual but merely the 1914 edition with changes No. 7 dated to 18 August 1917. A 1917 doctrinal manual is not listed

in Finlayson's bibliography. See Kenneth Finlayson, *An Uncertain Trumpet: The Evolution of U.S. Army Infantry Doctrine, 1919–1941* (Westport, CT: Greenwood Press, 2001). The existence of a new 1917 doctrine is also claimed by William O. Odom in *After the Trenches: The Transformation of U.S. Army Doctrine, 1918–1939* (College Station: Texas A&M University Press, 1999). Although Odom refers to a 1917 *FSR* on page 6 of his book, he, too, does not list the manual in his bibliography. Instead, he references the 1914 edition with editorial changes No. 1–7, published in 1917. This version did not constitute a new *FSR*. Rather, it was a slightly edited version of the 1914 doctrine (as the manual's title page denotes), containing no significant change in intellectual content.

43. Finlayson, *An Uncertain Trumpet*, 34. The square division also dropped the cavalry regiment, added a machine-gun battalion and trench mortar battalion, as well as military police and supply trains.

44. John J. Pershing, *My Experiences in the World War* (New York: Frederick A. Stokes Co., 1931), 1:8–12.

45. Ibid., 152. Finlayson, *An Uncertain Trumpet*, 41.

46. War Department, Office of the Chief of Staff, *Field Service Regulations United States Army 1914*, 70, 75. Pershing wrote in his memoirs that he wanted to avoid the effect of French Army teaching and thus took over and directed training from an American perspective. As he noted, "For the purpose of impressing our doctrine upon officers, a training program was issued which laid great stress on open warfare methods and offensive action." Pershing, *My Experiences*, 153.

47. For the role of Upton and Wagner on doctrinal development, see the previous chapter. War Department, Office of the Chief of Staff, *Field Service Regulations United States Army 1914*, 67, 75, 78, 82–83, 86–87.

48. Coffman, *The Regulars*, 199, 205–206. Cooper, *The Rise of the National Guard*, 153–156. In addition to creating the ROTC, the National Defense Act of 1916 was a compromise to allow the National Guard to retain its position as the primary augmentation force to the regular Army. Guardsmen took two oaths, one to the state and the other to the Constitution. This allowed them to be legally deployed overseas, if needed. Officers at peak included 3 percent Regulars, 6 percent National Guard, 8 percent from the ranks, and 48 percent from OTC.

49. Edward M. Coffman, *The War to End All Wars: The American Military Experience in World War I* (Madison: University of Wisconsin Press, 1986), 55–57.

50. Pershing, *My Experiences*, 153. Pershing later voiced his pleasure when the French admitted that open-warfare training and rifle proficiency was the right course of action overall.

51. Allan R. Millett, "Cantigny," in *America's First Battles, 1776–1965*, ed. Charles E. Heller and William A. Stofft (Lawrence: University Press of Kansas, 1986), 180. Coffman, *The War to End All Wars*, 156–157.

52. Finlayson, *An Uncertain Trumpet*, 42. Timothy K. Nenninger, "Tactical

Dysfunction in the AEF, 1917–1918," *Military* Affairs 51, no. 4 (October 1987): 177–181. Arthur L. Schlesinger Jr., ed. *The Almanac of American History* (New York: Barnes & Noble Books, 1993), 436.

53. Pershing, *My Experiences*, 159–162. Alfred F. Hurley, *Billy Mitchell: Crusader for Air Power*, new ed. (Bloomington: Indiana University Press, 1975), 28–31.

54. Coffman, *The Regulars*, 219–220.

55. Birtle, *U.S. Army Counterinsurgency and Contingency Operations Doctrine*, 208–210. Coffman, *The Regulars*, 219. Also see George F. Kennan, *Soviet-American Relations: 1917–1920: The Decision to Intervene*, 2 vols. (Princeton, NJ: Princeton University Press, 1958); and Betty Miller Unterberger, *America's Siberian Expedition, 1918–1920* (Durham, NC: Duke University Press, 1956). A journalist's account can be found in E. M. Halliday's *The Ignorant Armies* (New York: Harper Brothers, 1960). Halliday was a war correspondent in Northern Russia with the "Polar Bears," a unit made up of Michigan and Wisconsin soldiers. Also see Norman Saul, *War and Revolution: The United States and Russia, 1914–1921* (Lawrence: University Press of Kansas, 2001).

56. Birtle, *U.S. Army Counterinsurgency and Contingency Operations Doctrine*, 223. For a view from the Russian perspective, see Nicholas V. Riasanovsky, *A History of Russia*, 4th ed. (New York: Oxford University Press, 1984), 483.

57. Virgil Ney, *Evolution of the United States Army Field Manual: Valley Forge to Vietnam* (Fort Belvoir, VA: Combat Operations Research Group, 1966), 72. Odom, *After the Trenches*, 23.

58. Finlayson, *An Uncertain Trumpet*, 52.

59. William Gardner Bell, *Commanding Generals and Chiefs of Staff, 1775–1995, Portraits and Biographical Sketches of the United States Army's Senior Officer* (Washington, DC: U.S. Army Center of Military History, 1997), 32, 110. War Department, *Annual Reports of the Chief of Staff, 1920–1941* (Washington, DC: Government Printing Office, 1920–1941), 1922 ed., 119. C. Joseph Bernardo and Eugene H. Bacon, *American Military Policy: Its Development since 1775* (Harrisburg, PA: Stackpole Co., 1957), 384–389. The National Defense Act of 1920 provided for a smaller, regular army and the rapid mobilization of a citizen army in times of emergency. The act established the Army of the United States, consisted of the regular Army (17,000 officers and 280,000 soldiers), the National Guard (435,000), and the Organized Reserves, a regional skeleton force that could expand during wartime. The assistant secretary of war was charged with mobilization planning and procurement of supplies. The chief of staff presided over the War Department general staff and, under direction of the president or secretary of war, recruited, organized, supplied, equipped, mobilized, trained, employed, and demobilized the Army of the United States. The country was divided into corps areas based upon population, each containing at least one division of citizen soldiers. Congress also created the Chemical Warfare Service, Air Service, Finance Department, and the Office of the Chief of Chaplains. The act established chiefs of Infantry, Field Artillery, Coast Artillery, Cavalry, Air Service, and

Chemical Warfare Service to promote technical and tactical development of each branch of the Army. The act also provided a force to fight American contingencies as they arose, and also to raise a citizen army, if warranted.

60. Hugh A. Drum, "Annual Report of the Director of the School of the Line, 1919–1920," in *Annual Report of the Commandant of the General Service Schools, 1920* (Fort Leavenworth, KS: General Service Schools Press, 1920), 19–20.

61. Odom, *After the Trenches*, 33.

62. Ibid., 30–31.

63. Ibid., 32–33.

64. Ibid., 33–36.

65. Stewart, *American Military History*, 2:56.

66. War Department, *Annual Reports of the Chief of Staff, 1920–1941*, 1922 edition, 115.

67. Antoine Henri de Jomini, *The Art of War* (reprint, Mechanicsburg, PA: Stackpole Books, 1992), 70. For common threads that affect soldiers in war, see John Keegan, *The Face of Battle: A Study of Agincourt, Waterloo & the Somme* (New York: Vintage Press, 1977).

68. Department of the Army, Historical Division, *United States Army in the World War, 1917–1919*, 19 vols. (Washington, DC: Government Printing Office, 1948), 2:296.

69. Chief of Staff of the Army, *Field Service Regulations, United States Army, 1923* (Washington, DC: Government Printing Office, 1924), 5, 11, 115–116.

70. Ibid., 77. Jomini, *The Art of War*, 337. Carl von Clausewitz, *On War*, ed. and trans. Michael Howard and Peter Paret (Princeton, NJ: Princeton University Press, 1984), 198.

71. Chief of Staff of the Army, *Field Service Regulations, United States Army, 1923*, 77–82.

72. Ibid., 82–83.

73. Ibid., 89–90, 96–97.

74. Ibid., 85–86, 94, 96, 99.

75. Cooper, *The Rise of the National Guard*, 174. Finlayson, *An Uncertain Trumpet*, 69–70. Stewart, *American Military History*, 1:377. The National Defense Act of 4 June 1920 established the Army of the United States as the regular Army, the National Guard, and the Organized Reserves, the latter having an Officers Reserve Corps and an Enlisted Reserve Corps. Each of the three components was to be regulated in peacetime so they could prepare for a national emergency. After generations of struggle over a permanent standing and expansible army, the federal government conceded that standing peacetime forces were inadequate to face national emergencies alone and that wartime expansion took too much time. For large-scale mobilizations, a new citizen-soldier force was necessary. Training the National Guard and Organized Reserves in peacetime fell to the regular Army, which authorized 17,726 officers. From a doctrinal perspective, three new branches were established by law,

the Air Service and the Chemical Warfare Service, which reflected Great War experience, and the Tank Corps under infantry branch control, as a support weapon. The War Department was tasked to plan for mobilization and prepare the service for future war that also included planning and supervision of industrial procurement. In 1921, General John J. Pershing, as chief of staff, reorganized the War Department general staff along the lines of his AEF General Headquarters staff in France. Five divisions emerged: G–1, Personnel; G–2, Intelligence; G–3, Training and Operations; G–4, Supply; and a specific War Plans Division for strategic planning and war preparations. In war, the general staff directed combat operations. In 1926, the Air Corps became an equal combat arm. Education was changed to reflect the complexity of modern warfare. The United States Military Academy and the Reserve Officers Training Corps offered most of the precommission schooling for new officers. Thirty-one special service schools provided branch training for officers and enlisted men in the regular Army, National Guard, and Organized Reserves. General service schooling included, from 1922 to 1947, the Command and General Staff School (College) at Fort Leavenworth, Kansas, which educated officers for divisional command and general staff positions. In Washington, D.C., the Army War College and the Army Industrial College (1924) prepared senior officers for higher command and staff positions and war plans development.

76. Odom, *After the Trenches*, 170–171.

77. Ibid., 172–195.

78. Paddy Griffith, *Forward into Battle: Fighting Tactics from Waterloo to the Near Future* (Novato, CA: Presidio Press, 1992), 104–105.

79. Larry H. Addington, *The Patterns of War since the Eighteenth Century*, 2nd ed. (Bloomington: Indiana University Press, 1994), 180.

80. Hurley, *Billy Mitchell*, 84–85. Odom, *After the Trenches*, 88–89, 110.

81. War Department, "Proceedings of a Committee of Officers to Review the Manual for Commanders of Large Units," 25 March 1929, Combined Arms Research Library. General Summerall was the chief of staff of the Army at the time. Major General Frank Parker was the War Department's operations officer or "G3" under the army staff system. Colonel Samuel C. Vestal was Parker's primary action officer for developing the new doctrine.

82. Odom, *After the Trenches*, 121. Chapter 2 discusses doctrine during the War of 1812.

83. Ibid., French doctrine is articulated in Robert A. Doughty, *The Seeds of Disaster: The Development of French Army Doctrine, 1919–1939* (Hamden, CT: Archon Books, 1985). Also see Elizabeth Kier, *Imagining War: French and British Military Doctrine between the Wars* (Princeton, NJ: Princeton University Press, 1997). Colonel Edmund L. Gruber was chief of the Army's Training Branch on the War Department general staff. He later became the principal author of the 1941 doctrine revision as the commandant, U.S. Army Command and General Staff School at Fort Leavenworth.

84. War Department, *Tentative Field Service Regulations* FM 100-5 *Operations* (Washington, DC: Government Printing Office, 1939). Odom, *After the Trenches*, 129.

85. War Department, *Tentative Field Service Regulations 1939*, FM 100-5 *Operations*, cover letter. The manual consisted of thirteen chapters and an index. Chapter 1 ("Organization") discussed the Army of the United States, in deference to the 1920 National Defense Act, while Chapter 2 ("Arms and Services") defined the role of the infantry, cavalry, field artillery, coast artillery corps, air corps, corps of engineers, signal corps, and the chemical warfare service. Chapter 3 ("Conduct of War") delineated the principles of war and leadership, while Chapter 4 ("Command and Staff") discussed the nuances of those functions. Chapter 5 ("Intelligence and Counterinformation") was a tutorial on the gathering of information on the enemy while preventing enemy attempts to do so. The major thrust of Chapter 6 ("Reconnaissance") described the assets required for collecting information on the enemy. Chapter 7 ("Troop Leading") was a new section explaining how to prepare troops for battle and to conduct an estimate of the situation to reach a decision. Chapters 8 ("Security") provided ways to safeguard the force against various threats, and Chapter 9 ("Marches") was devoted to the movement of troops. Chapter 10 ("Shelter") informed the reader about bivouacs, physical protection, sanitation, and other related subjects. Chapter 11 ("The Offensive") described various ways to conduct an attack, while Chapter 12 ("The Defensive") did the same for defending an area from an attacker. Chapters 13 ("Special Operations") covered night combat, river crossings, combat in the woods and towns, combat in mountainous areas, and guerrilla warfare. In covering a variety of situations that included troop leading and guerrillas, the 1939 FM 100-5 increased the service's regulation of wartime activities and thus decreased informal practice. Clearly, in producing three warfighting manuals, the general staff was taking more control over the Army in the field than any previous *FSR* to date.

86. Ibid., 27. Clausewitz, *On War*, 87–88.

87. George Edward Thibault, ed., *The Art and Practice of Military Strategy* (Washington, DC: National Defense University, 1984), 1–7.

88. War Department, *Tentative Field Service Regulations*, FM 100-5 *Operations*, 29–30. Peter D. Haynes, "American Culture, Military Services' Cultures, and Military Strategy" (master's thesis, Naval Post Graduate School, Monterey, California, 1998), 39.

89. War Department, *Tentative Field Service Regulations*, FM 100-5 *Operations*, 128, 135, 137, 139–142, 165.

90. Ibid., 6, 8.

91. Ibid., 64–65.

92. Ibid., 133.

93. Ibid., 228–231.

94. Odom, *After the Trenches*, 130–131.

95. Finlayson, *An Uncertain Trumpet*, 148–149.

96. James S. Corum, *The Roots of Blitzkrieg: Hans von Seeckt and German Military Reform* (Lawrence: University Press of Kansas, 1992), 71.

97. Christopher R. Gabel, *The U.S. Army GHQ Maneuvers of 1941* (Washington, DC: U.S. Army Center of Military History, 1991), 23. Jonathan M. House, "Toward Combined Arms Warfare: A Survey of 20th-Century Tactics, Doctrine, and Organization" (Fort Leavenworth, KS: Combat Studies Institute, 1984), 77.

98. War Department, *Field Service Regulations* Field Manual 100-5 *Operations*, 1941 (Washington, DC: Government Printing Office, 1941), II.

99. Ibid., 1–3.

100. Ibid., 98–99.

101. Ibid., 98–103.

102. Ibid., 12–14. The Army Air Corps became the Army Air Forces per Army Regulation 95-5, 20 June 1941.

103. Ibid., 137–142.

104. During World War II, the U.S. industrial base manufactured 1,415,593 1903 Springfield rifles (in many variations), 4,040,000 M1 Garand Rifles, 347,524 M2 machine guns, and over 40,000 flame throwers. For a complete listing of small-arms equipment, see Bruce N. Canfield, *U.S. Infantry Weapons of World War II* (Lincoln, RI: Andrew Mowbray Publishers, 1996), 278–292. U.S. industry also produced over 88,000 tanks, 2.4 million motor vehicles, and over 300,000 airplanes. See James A. Huston, *The Sinews of War: Army Logistics, 1775–1953* (Washington, DC: Office of the Chief of Military History, 1966), 474–486.

105. War Department, *Field Service Regulations* Field Manual 100-5 *Operations*, 1941, 237–238.

106. Ibid., 213–233. Douglas Porch, *The Path to Victory: The Mediterranean Theater in World War II* (New York: Farrar, Straus & Giroux, 2004), 550–553.

107. Archer Jones, *The Art of War in the Western World* (New York: Oxford University Press, 1987), 561–562.

108. Jones, *The Art of War*, 581–582.

109. War Department, *Field Service Regulations* Field Manual 100-5 *Operations*, 1941, 235–237. Harry Gailey, *MacArthur Strikes Back: Decision at Buna: New Guinea, 1942–1943* (Novato, CA: Presidio, 2000), 102–116.

110. There is discussion over whether or not senior commanders actually shifted their thinking from "on-the-job experience" or that changes in doctrine actually played some role in achieving victory. See House, "Toward Combined Arms Warfare," 128–129; and Peter Mansoor, *The G.I. Offensive in Europe: The Triumph of American Infantry Divisions, 1941–1945* (Lawrence: University Press of Kansas, 1999).

111. War Department, *Field Service Regulations* Field Manual 100-5 *Operations*, 1944 (Washington, DC: Government Printing Office, 1944), 20.

112. Ibid., 22–26.

113. Ibid., 287–289.

114. Ibid., 290–296.

115. Ibid., 313–318. Jonathan M. House, *Combined Arms Warfare in the Twentieth Century* (Lawrence, KS: University Press of Kansas, 2001), 137–138.

CHAPTER FOUR. COLD WAR DOCTRINE: FROM ARMY OPERATIONS TO MULTINATIONAL AND MULTISERVICE OPERATIONS, 1945–1991

1. Russell F. Weigley, *History of the United States Army* (New York: Macmillan, 1967), 475. Churchill quoted in R. Ernest Dupuy, *The Compact History of the United States Army* (New York: Hawthorn, 1956), 265. Marvin A. Kriedberg and Merton G. Henry, *History of Military Mobilization in the United States Army, 1775–1945* (Washington, DC: Department of the Army, 1955), 553–575. Adrian R. Lewis, *The American Culture of War: The History of U.S. Military Force from World War II to Operation Iraqi Freedom* (New York: Routledge, 2007), 39.

2. Jonathan R. Adelman, *Prelude to the Cold War: The Tsarist, Soviet, and U.S. Armies in the Two World Wars* (Boulder, CO: Lynne Reimer Publishers, 1988), 211. Larry H. Addington, *The Patterns of War since the Eighteenth Century*, 2nd ed. (Bloomington: Indiana University Press, 1984), 266. Also see Walter LaFeber, *America, Russia, and the Cold War, 1945–1971* (New York: John Wiley & Sons, 1972), 13–14. John Lewis Gaddis, *We Now Know: Rethinking Cold War History* (Oxford: Oxford University Press, 1997), 26. John Costello, *The Pacific War* (New York: Rawson, Wade Publishers, 1981), 588–593.

3. Robert A. Doughty, *The Evolution of U.S. Army Tactical Doctrine, 1946–76* (Fort Leavenworth, KS: Combat Studies Institute, U.S. Army Command and General Staff College, 1979), 2. David MacIssac, "The Air Force and Strategic Thought, 1945–51," International Security Studies Program Working Paper no. 8 (Wilson Center, Washington, DC, June 1979). Samuel F. Wells Jr., "The Origins of Massive Retaliation," *Political Science Quarterly* 96, no. 1 (Spring 1981): 31–52. Russell F. Weigley, *Eisenhower's Lieutenants: The Campaigns of France and Germany, 1944–1945* (Bloomington: Indiana University Press, 1981), 166. Ingo Trauschweizer, *The Cold War U.S. Army: Building Deterrence for Limited War* (Lawrence: University Press of Kansas, 2008), 18–19. Adrian R. Lewis, *The American Culture of War*, 53.

4. William Gardner Bell, *Commanding Generals and Chiefs of Staff, 1775–1995: Portraits & Biographical Sketches of the United States Army's Senior Staff* (Washington, DC: U.S. Army Center of Military History, 1997), 33.

5. General Joseph W. Stilwell, *Report on War Department Equipment Board*, 19 January 1946, 10, Combined Arms Research Library. Gordon W. Prange, *At Dawn We Slept: The Untold Story of Pearl Harbor* (New York: McGraw-Hill, 1981). Christopher R. Gabel, *Seek, Strike and Destroy: U.S. Army Tank Destroyer Doctrine in World War II* (Fort Leavenworth, KS: Combat Studies Institute, U.S. Army Command and General Staff College, 1985), 63–65.

6. Joint Chiefs of Staff, *Organizational Development of the Joint Chiefs of Staff, 1942–1989* (Washington, DC: Historical Division, Joint Secretariat, Joint Chiefs of Staff,

1989), 16. Saul Landau, *The Dangerous Doctrine: National Security and U.S. Foreign Policy* (Boulder, CO: Westview Press, 1988), 47. The act also established the Central Intelligence Agency, the National Security Council, and other related agencies.

7. Andrew J. Birtle, *U.S. Army Counterinsurgency and Contingency Operations Doctrine, 1942–1976* (Washington, DC: U.S. Army Center of Military History, 2006), 42–66. War Department, *Field Service Regulations* Field Manual 100-5 *Operations*, 1944 (Washington, DC: Government Printing Office, 1944), 284–286.

8. Sam C. Sarkesian, *U.S. National Security, Policymakers, Processes, and Politics* (Boulder, CO: Lynne Rienner Publishers, 1989). General Omar N. Bradley, "Creating a Sound Military Force," *Military Review* 29, no. 2 (May 1949): 3–6.

9. Department of the Army, *Field Service Regulations* Field Manual 100-5 *Operations*, 1949 (Washington, DC: Government Printing Office, 1949), vi. Chapter 2 added a section on the transportation corps, while Chapter 5 was renamed "Combat Intelligence, Reconnaissance, and Counterintelligence."

10. Ibid., 2.

11. Ibid., 21–22. Antoine Henri de Jomini, *The Art of War*, Greenhill ed. (Mechanicsburg, PA: Stackpole Books, 1996), 70–71, 337–345. Jomini defined the fundamental principle of war as being to throw the mass of the army onto the decisive point of the battlefield or enemy line.

12. Department of the Army, *Field Service Regulations* Field Manual 100-5 *Operations*, 1949, 6–7, 80, 120. The 1st Guided Missile Battalion was activated at Fort Bliss, Texas, on 11 October 1945.

13. Birtle, *U.S. Army Counterinsurgency and Contingency Operations Doctrine*, 21.

14. Department of the Army, *Field Service Regulations* Field Manual 100-5 *Operations*, 1949, 264–274. Change 1 was published on 25 July 1952 with no significant shift in the intellectual thrust of the manual.

15. For a concise examination of NATO's formation and politics, see John F. Reichart and Steven R. Sturm, eds., *American Defense Policy*, 5th ed. (Baltimore: Johns Hopkins University Press, 1982).

16. Harry S. Truman, *Memoirs by Harry S. Truman*, 2 vols. (Garden City, NY: Doubleday & Co., 1956), 2:355. Michael Carver, "Conventional Warfare in the Nuclear Age," in *Makers of Modern Strategy: From Machiavelli to Modern Times*, ed. Peter Paret (Princeton, NJ: Princeton University Press, 1986), 780–781. David McCullough, *Truman* (New York: Simon & Schuster, 1992), 792.

17. General Matthew B. Ridgway, *The Korean War* (New York: Doubleday & Co., 1967), 11. The Army leadership was not alone in its thinking. In 1950, an Air Force officer remarked, "When the Fifth Air Force gets to work on them, there will not be one North Korean left in North Korea." Quoted in Norman Polmar, *Strategic Weapons: An Introduction*, rev. ed. (New York: Crane Russack, National Strategic Information Center, 1982), 2.

18. Roy E. Appleman, *South to the Naktong, North to the Yalu* (Washington, DC: U.S. Army Center of Military History, 1992), 2.

19. Ibid., 65–76, 542–606.

20. Walter G. Hermes, *Truce Tent and Fighting Front* (Washington, DC: U.S. Army Center of Military History, 1992), 502–503, 508–509.

21. Ibid., 510–511.

22. Robert A. Doughty and Ira D. Gruber, *American Military History and the Evolution of Western Warfare* (Lexington, MA: D. C. Heath, 1996), 584. For overviews see Bernard Brodie, *Strategy in the Missile Age* (Princeton, NJ: Princeton University Press, 1959); Fred M. Kaplan, *The Wizards of Armageddon* (New York: Simon & Schuster, 1983); and Henry Kissinger, *Nuclear Weapons and Foreign Policy* (New York: Harper, 1957).

23. Doughty and Gruber, *American Military History*, 585–586. A. J. Bacevich, *The Pentomic Era: The U.S. Army between Korea and Vietnam* (Washington, DC: National Defense University Press, 1986), 94–95. James M. Gavin, *War and Peace in the Space Age* (New York: Harper, 1958), 129–35, 147–148. "Project Attack: Hypothetical Use of A-Bombs on Massed Armor Illustrated by 'Operation Totalize,' Phase I, 7–8 August 1944" (Baltimore: Operations Research Office, Johns Hopkins University), 3. In 1950, the Army studied nuclear weapons and their potential for tactical warfare. One study, "Tactical Employment of the Atomic Bomb," envisioned using nuclear devices in open battle and urban terrain. Major General James M. Gavin, the former commanding general of the 82d Airborne Division in World War II, was a member of the Weapons System Evaluation Group. He later wrote an article entitled "The Tactical Use of the Atomic Bomb," which was published in *Combat Forces Journal* in November 1950. See James M. Gavin, "The Tactical Use of the Atomic Bomb," *Combat Forces Journal* 1, no. 4 (November 1950): 9–11. Collectively, the studies examined how tactical nuclear artillery projectiles affected World War II–type targets and concluded that nuclear weapons would prove effective against Soviet and Chinese human-wave assaults and densely packed groupings of armored vehicles.

24. Bacevich, *The Pentomic Era*, 12–13. Robert Gilpin, *War and Change in World Politics* (reprint, New York: Cambridge University Press, 1985), 215. Gordon A. Craig and Alexander L. George, *Force and Statecraft: Diplomatic Problems of Our Time* (New York: Oxford University Press, 1983), 114–131.

25. John L. Frisbee, ed., *Makers of the United States Air Force* (Washington, DC: Office of Air Force History, 1987), 268–269. Major John H. Cushman, "What Is the Army's Story," *Army Combat Forces Journal* 5, no. 3 (October 1955): 49. Walter J. Boyne, *Beyond the Wild Blue: A History of the U.S. Air Force* (New York: St. Martin's Griffon, 1998), 95–97.

26. House, "Toward Combined Arms Warfare," 154. Doughty, *The Evolution of U.S. Army Tactical Doctrine*, 14.

27. Department of the Army, *Field Service Regulations* Field Manual 100-5 *Operations*, 1954 (Washington, DC: Government Printing Office, 1954), 4.

28. Ibid., 6–7. Carl von Clausewitz, *On War*, ed. and trans. Michael Howard and Peter Paret (Princeton, NJ: Princeton University Press, 1984), 80–81, 579–581.

29. Department of the Army, *Field Service Regulations* Field Manual 100-5 *Operations*, 1954, 40, 74–75, 96.

30. Ibid., 104–109.

31. Ibid., 89–94.

32. Ibid., 96.

33. Ibid., 115–117.

34. Ibid., 117.

35. Ibid., 124–125.

36. Bacevich, *The Pentomic Era*, 105. Roger J. Spiller, Joseph G. Dawson III, and T. Harry Williams, *Dictionary of American Military Biography*, 3 vols. (Westport, CT: Greenwood Press, 1984), 1:369–372. Gavin resigned in protest over Eisenhower's foreign policy and later wrote several works concerning nuclear weapons and strategy.

37. James M. Gavin, "The Tactical Use of the Atomic Bomb," *Combat Forces Journal* 1, no. 4 (November 1950): 9–11. Lewis, *The American Culture of War*, 152–163. Trauschweizer, *The Cold War U.S. Army*, 48–49.

38. Bacevich, *The Pentomic Era*, 107–108. Walter E. Kretchik, "Pentomic Division," in *Historical Dictionary of the U.S. Army*, ed. Jerold E. Brown (Westport, CT: Greenwood Press, 2001), 362.

39. The Department of the Army, *Field Service Regulations* Field Manual 100-5 *Operations, Change 2, 1956 and Change 3, 1958* changed pages 74 and 75 to further link atomic fires to maneuver. Change 1 had been published in December 1954, adding an index. Theodore C. Mataxis and Seymore L. Goldberg, *Nuclear Tactics* (Harrisburg, PA: Military Service Publishing Co., 1958), 164, 211.

40. The Armored School, *Final Report of Test—Armored Task Force Preparation—Exercise Desert Rock VI*, 1 August 1955. Defense Technical Information Center (DTIC), Springfield, VA.

41. Bacevich, *The Pentomic Era*, 100.

42. Samuel Lyman Atwood Marshall was a military writer, journalist, and army officer who pioneered combat history techniques in World War II. See Spiller, Dawson III, and Williams, *Dictionary of American Military Biography*, 737–742. S. L. A. Marshall, "Arms in Wonderland," *Army* 7, no. 6 (June 1957): 19. Bacevich, *The Pentomic Era*, 131–132.

43. Doughty, *The Evolution of U.S. Army Tactical Doctrine*, 18. The Command and General Staff College course material exposed officers to methods for calculating a nuclear blast radius, fallout considerations, lethality and casualty projections, and other topics concerning nuclear fires.

44. Arthur S. Collins, *Senior Officers Oral History Project*, Senior Officer Debriefing Program (Carlisle Barracks, PA: Military History Institute, 1982), 237–244.

45. Birtle, *U.S. Army Counterinsurgency and Contingency Operations Doctrine*, 183–191.

46. Francis X. Bradley, "The Fallacy of Dual Capability," *Army* 10, no. 3 (October

1959): 18–19. Arthur S. Collins, "The Other Side of the Atom," *Army* 10, no. 4 (November 1959): 18–19. William E. DePuy, "The Case for a Dual Capability," *Army* 10, no. 6 (January 1960): 32–38.

47. Major General Lionel C. McGarr, "Final Report to the Staff and Faculty, 2 June 1960," compiled within "Addresses by Major General Lionel C. McGarr, Commandant, United States Army Command and General Staff College," 2 vols., vol. 2 (3 May 1958 to 10 August 1960), 328, Combined Arms Research Library. Hamilton Howze, *Senior Officers Oral History Project,* Senior Officer Debriefing Program (Carlisle Barracks, PA: Military History Institute, 1972), 6–10. George H. Decker, *Senior Officers Oral History Project,* Senior Officer Debriefing Program (Carlisle Barracks, PA: Military History Institute, 1976), 60–69. Henry E. Kelly, "Verbal Defense," *Military Review* 35, no. 7 (October 1955): 45–46, 51. William E. Roberts, "Keeping Pace with the Future—Training Officers to Fight on Atomic Battlefields," *Military Review* 37, no. 7 (October 1957): 22–29. Lieutenant Colonel Linwood A. Carleton and Lieutenant Colonel Frank A. Farnsworth, "A Philosophy for Tactics," *Military Review* 40, no. 4 (July 1960): 12, 22.

48. Bacevich, *The Pentomic Era,* 141, 148–149.

49. Trauschweizer, *The Cold War U.S. Army,* 109–113.

50. Robert S. McNamara, *In Retrospect: The Tragedy and Lessons of Vietnam* (New York: Vintage Books, 1996), 20. Doughty, *The Evolution of U.S. Army Tactical Doctrine,* 21. John F. Kennedy quoted in Doughty and Gruber, *American Military History,* 593. Trauschweizer, *The Cold War U.S. Army,* 81.

51. Quoted in William H. Zierdt Jr., "The Structure of the New Army Divisions," *Army* 11, no. 7 (July 1961): 49. Harold K. Johnson, "U.S. Army Command and General Staff College ROAD Briefing, 27 May 1961," Combined Arms Research Library, Leavenworth, KS.

52. Zierdt, "The Structure of the New Army Divisions," 52, 62. James M. Snyder, "ROAD Division Command Staff Relationships," *Military Review* 43, no. 1 (January 1963): 57–62. Armored divisions contained 15,000 soldiers, six armor battalions, and five mechanized infantry battalions. Mechanized divisions had the same number of troops, with seven mechanized infantry battalions and three armor battalions. The infantry division also had 15,000 soldiers, with eight infantry battalions and two armor battalions. The airborne division contained 14,000 soldiers in nine airborne infantry battalions with one assault gun battalion.

53. Department of the Army, *Field Service Regulations* Field Manual 100-5 *Operations, February 1962* (Washington, DC: Government Printing Office, 1962), 3–4.

54. Ibid., 4–6.

55. Ibid., 12–13. Clausewitz, *On War,* 75. The use of Clausewitz is yet another indicator of how the army education system influences doctrine development, as Elihu Root envisioned in 1905.

56. Department of the Army, *Field Service Regulations* Field Manual 100-5 *Operations, February 1962,* 15.

57. Ibid., 24–25.

58. Ibid., 34–40. Close-combat units were infantry, armor, cavalry, and artillery. Combat support included signal, engineers, army aviation, transportation, chemical, military police, electronic warfare, psychological operations, tactical cover and deception, intelligence, air defense. Combat service support units were not specifically mentioned but were included "appropriate to requirement."

59. Ibid., 59–76. Doughty, *The Evolution of U.S. Army Tactical Doctrine*, 24–25. World War II infantry divisions contained 13,207 soldiers. A comparable ROAD division had 13,512 personnel. In World War II, an infantry division defended a front between 4 to 9 miles wide. The ROAD division defended a 12-mile front. The ROAD division was more mechanized than any previous division and had improved weapons systems with longer ranges. However, moving large numbers of soldiers from the defense to the offense remained a difficult undertaking in terrain such as Central Europe.

60. Ibid., 101–109.

61. Daniel Marston and Carter Malkasian, eds., *Counterinsurgency in Modern Warfare* (London: Osprey Publishing Ltd., 2008), 37–46.

62. Department of the Army, *Field Service Regulations* Field Manual 100-5 *Operations, February 1962*, 127–128.

63. Ibid., 129.

64. Ibid., 135–136.

65. Ibid., 138, 152.

66. Lloyd Norman and John B. Spore, "Big Push in Guerrilla Warfare," *Army*, 12, no. 9 (March 1962): 28.

67. Norman and Spore, "Big Push in Guerrilla Warfare," 34. Department of the Army, *Field Service Regulations* Field Manual 100-5 *Operations, February 1962*, 10.

68. Birtle, *U.S. Army Counterinsurgency and Contingency Operations Doctrine*, 199–201.

69. Thomas K. Adams, *U.S. Special Operations Forces in Action: The Challenge of Unconventional Warfare* (Portland, OR: Frank Cass, 1998). Norman and Spore, "Big Push in Guerrilla Warfare," 35. Birtle, *U.S. Army Counterinsurgency and Contingency Operations Doctrine*, 197. Stephen L. Bowman, "The Evolution of United States Army Doctrine for Counterinsurgency Warfare: From World War II to the Commitment of Combat Units in Vietnam" (Ph.D. diss., Duke University, 1985), 184.

70. Trauschweizer, *The Cold War U.S. Army*, 175. The base ROAD infantry division contained just over 13,500 troops but when reinforced could number as high as 36,000 or more by including various attached units.

71. Department of the Army, *Field Service Regulations* Field Manual 100-5 *Operations, February 1962*, 13.

72. Birtle, *U.S. Army Counterinsurgency and Contingency Operations Doctrine*, 202–210.

73. George C. Herring, "Ia Drang Valley," in *America's First Battles, 1776–1965*,

ed. Charles E. Heller and William A. Stofft (Lawrence: University Press of Kansas, 1986), 317–321. Harold G. Moore and Joseph L. Galloway, *We Were Soldiers Once . . . and Young. Ia Drang: The Battle That Changed the War in Vietnam* (New York: Random House, 1992), 306–307. George L. MacGarrigle, *Taking the Offensive: October 1966 to October 1967* (Washington, DC: U.S. Army Center of Military History, 1998), 11. For Air Cavalry development, see Lawrence H. Johnson III, *Winged Sabers: The Air Cavalry in Vietnam, 1965–1973* (Harrisburg, PA: Stackpole Books, 1990). The marines fought the first major ground action, Operation Starlight, 18–21 August 1965. The Ia Drang Valley campaign, part of Operation Silver Bayonet, began on 23 October and lasted until 20 November 1965. Moore's losses were 79 killed and 121 wounded out of 450 men. The Vietnamese losses were estimated at 600 killed with more than 1,200 wounded.

74. Trauschweizer, *The Cold War U.S. Army*, 176–177.

75. George C. Herring, *America's Longest War: The United States and Vietnam, 1950–1975*, 2nd ed. (New York: Alfred A. Knopf, 1986), 150. The nuances between the 1962 FM 100-5 and search-and-destroy operations are subtle. Under the 1962 doctrine, an army force moved to make contact with the enemy or conducted a meeting engagement by accident. Regardless, once enemy contact occurred, units then fired and maneuvered. In Vietnam, the enemy locations were not always known, so units searched for them. In essence, the soldiers were conducting either a movement to contact or a meeting engagement, and General William C. Westmoreland simply coined another term for that doctrinal method. Andrew F. Krepinevich Jr., *The Army and Vietnam* (Baltimore: Johns Hopkins University Press, 1986), 182–183.

76. For North Vietnam's approach, see Douglas Pike, *PAVN: People's Army of Vietnam* (Novato, CA: Presidio Press, 1986). Birtle, *U.S. Army Counterinsurgency and Contingency Operations Doctrine*, 370.

77. Birtle, *U.S. Army Counterinsurgency and Contingency Operations Doctrine*, 376–377.

78. Lieutenant Colonel John A. Nagl, "Counterinsurgency in Vietnam, American Organizational Culture and Learning," in Marston and Malkasian, *Counterinsurgency in Modern Warfare*, 131–148. Abrams was promoted to permanent major general in August 1965 and served as both acting and then vice chief of staff of the Army from August 1964 to April 1967.

79. Herring, *America's Longest War*, 153.

80. Trauschweizer, *The Cold War U.S. Army*, 179–180.

81. Headquarters, Department of the Army, Field Manual 100-5 *Operations of Army Forces in the Field* (Washington, DC: Government Printing Office, September 1968), 13–3.

82. Ibid., 5–6, 5–7 to 5–9.

83. Ibid., 6–4, 6–29.

84. Ibid., 8–1, 11–1.

85. Ibid., 12–1. Interview with Lieutenant Colonel (Ret) John B. Hunt by author, Leavenworth, Kansas, February 2001.

86. *Public Papers of the Presidents of the United States* (Washington, DC: Office of the Federal Register, National Archives and Records Service, 1974), Richard M. Nixon, "Address to the Nation on Vietnam," 14 May 1969, "A Redefinition of the United States Role in the World," Speech 25 February 1971, U.S. Foreign Policy-1971 (Washington, DC: Department of State, Richard M. Nixon, 1972). Paul H. Herbert, *Deciding What Has to Be Done: General William E. Depuy and the 1976 Edition of* FM 100-5 *Operations* (Fort Leavenworth, KS: Combat Studies Institute, U.S. Army Command and General Staff College, 1988), 5. James F. Dunnigan and Raymond M. Macedonia, *Getting It Right: American Military Reforms after Vietnam to the Gulf War and Beyond* (New York: William Morrow, 1993), 96–108. After Vietnam, the Army underwent extensive reductions in force known as RIF. In the early 1960s, the Army contained 960,000 troops, a number that swelled to 1,570,000 by 1968. By 1973, the Army stood at just over 800,000 troops and finally settled to 750,000 by the mid-1970s. The RIF plagued the army leadership in many ways, but arguably the most serious consequence was too few soldiers for global commitments. The Army, like all services, continually turns over personnel because of retirements, discharges, resignations, and accidents, or about 30 percent turnover a year, all the while assimilating new members. The volunteer force, established after Vietnam, required active recruiting to fill the ranks during an era where the military remained unpopular. Training requirements increased to handle the continual flow of new recruits. The Army also required modernization by fielding new equipment to face the heavily armored threats in Europe and the Middle East. Personnel and equipment demands stretched the Army to the limit in attempting to manage its ongoing missions.

87. Herbert, *Deciding What Has to Be Done*, 6.

88. Anne W. Chapman, Carol J. Lilly, John L. Romjue, and Susan Canedy, *Prepare the Army for War: A Historical Overview of the Army Training and Doctrine Command, 1973–1998* (Fort Monroe, VA: Military History Office, U.S. Army Training and Doctrine Command, 1998), 9. Herbert, *Deciding What Has to Be Done*, 6, 11–16.

89. Herbert, *Deciding What Has to Be Done*, 76. General Abrams had very little to do with the new doctrine as he was dying from cancer. His illness prevented him from providing much guidance at all regarding the concepts in the manual and helped to provide DePuy with the autonomy to imprint his ideas on the Army.

90. "Briefing by LTG DePuy, 7 June 1973 [at Fort Polk, Louisiana]," in Richard M. Swain, comp., *Selected Papers of General William E. DePuy* (Fort Leavenworth, KS: Combat Studies Institute, U.S. Army Command and General Staff College, 1994), 60–61. Herbert, *Deciding What Has to Be Done*, 16.

91. Swain, *Selected Papers of General William E. DePuy*, 61–62. Clausewitz, *On War*, 119.

92. Charles Patrick Neimeyer, *America Goes to War: A Social History of the Continental Army* (New York: New York University Press, 1996), 8–26. Lieutenant Colonel J. W. Nicholson, "General William E. DePuy Guidance, Memorandum for Record," Combined Arms Research Library, Fort Leavenworth, Kansas.

93. George W. Gawrych, *The Albatross of Decisive Victory: War and Policy between Egypt and Israel in the 1967 and 1973 Arab-Israeli Wars* (Westport, CT: Greenwood Press, 2000), xiii.

94. Swain, *Selected Papers of General William E. DePuy*, 75–77.

95. Doughty, *The Evolution of U.S. Army Tactical Doctrine*, 41. McNamara, *In Retrospect*, 14–25. Herbert, *Deciding What Has to Be Done*, 101. Lewis Sorley, ed., *Press On! Selected Works of General Donn A. Starry*, 2 vols. (Fort Leavenworth, KS: Combat Studies Institute, U.S. Army Combined Arms Center, 2009), 1:281–284.

96. Sorley, *Press On!* 1:281, 335. John L. Romjue, *From Active Defense to AirLand Battle: The Development of Army Doctrine, 1973–1982* (Fort Monroe, VA: U.S. Army Training and Doctrine Command, 1984), 5. Department of the Army, Field Manual 100-5 *Operations*, 1 July 1976, with Change 1, 29 April 1977, Chapter 8. The chapter notes that modern battle contains both air and land components and that neither the Army nor the Air Force can win the war alone.

97. Department of the Army, Field Manual 100-5 *Operations*, 1 July 1976, with Change 1, 29 April 1977, 4–3, 5–2, 5–12 to 5–13. Trauschweizer, *The Cold War U.S. Army*, 207.

98. Department of the Army, Field Manual 100-5 *Operations*, 1 July 1976, with Change 1, 29 April 1977 (Washington, DC: Government Printing Office, 1976, 1977), i. Ibid., 3–5, 5–3.

99. Ibid., 1–1, 1–2.

100. Ibid., 2–1 to 2–32. Army manuals had, by 1976, adopted the metric system to be more compliant with NATO and other international agreements and forces.

101. Ibid., 3–4.

102. Ibid., 4–1 to 4–12.

103. Ibid., 5–1 to 5–9.

104. Ibid., 5–10 to 5–14.

105. Ibid., 8–1 to 8–7.

106. Chapman, Lilly, Romjue, and Canedy, *Prepare the Army for War*, 57, 71–72.

107. Philip A. Karber, "Dynamic Doctrine for Dynamic Defense," *Armed Forces Journal* 114, no. 2 (October 1976): 28–29. Archer Jones, "The New FM 100-5: A View from the Ivory Tower," *Military Review* 58, no. 2 (May 1984): 27–36. Colin S. Gray, "Force Planning, Political Guidance and the Decision to Fight," *Military Review* 53, no. 4 (April 1978): 27–36. Herbert, *Deciding What Has to Be Done*, 96–97. Conversation with General (Ret) Donn Starry by author, U.S. Army Command and General Staff College, May 2000.

108. Trauschweizer, *The Cold War U.S. Army*, 212–214. An operational maneuver

group was a large combined-arms force with up to 20,000 vehicles and 125,000 personnel specifically trained and equipped to fight as an independent force. It was about the size of a U.S. Army corps.

109. Herbert, *Deciding What Has to Be Done*, 95.

110. John L. Romjue, "AirLand Battle: The Historical Background," *Military Review* 66, no. 2 (March 1986): 52–55. Cavazos was in agreement with Ardant du Picq, *Battle Studies* in *Roots of Strategy*, book 2 (Harrisburg, PA: Stackpole Books, 1987), 135–168. Du Picq argued that a "moral influence" existed in war, one that led to the cohesion or the disintegration of units in battle. Also see John Keegan's *The Face of Battle: A Study of Agincourt, Waterloo & the Somme* (New York: Vintage Books, 1977). His purpose, a history of the psychology of war, becomes evident on page 78, where he suggested how and why soldiers faced their fears and were killed or wounded for their efforts. John L. Romjue, *From Active Defense to AirLand Battle*, 53–54.

111. Roger J. Spiller, "War History and the History Wars: Establishing the Combat Studies Institute," *Public Historian* 10, no. 4 (Fall 1988): 70. "Doctrinal Implications of CSI," Working Group Papers dated January 1979, Combined Arms Research Library. Conversation with General (Ret.) Donn Starry by author, April 2000, Fort Leavenworth, Kansas. Charles R. Schrader, "Such History as Every Young Gentleman Should Be Presumed to Know: Officer Education and Military History— An Opinion," date unknown. Photocopy provided to author in 2000 by Roger J. Spiller, then George C. Marshall Chair of Military History, U.S. Army Command and General Staff College. In truth, Starry's efforts gave him much more than he asked for. His intent was to gather up a few historians to assist in producing doctrine. Instead, opportunistic officers at CGSC, such as Major Charles R. Schrader and others, viewed his query as a mandate. In 1978, CGSC history instruction was part of the Department of Unified and Combined Operations, or DUCO. Building upon Starry's question, Schrader played a significant role in convincing the college leadership to create a history department, one that used history to inform doctrine writers and educate officers. CSI emerged in 1979, a mixture of civilian and army historians prepared to "conduct original, interpretive research on historical topics pertinent to the current doctrinal concerns of the United States Army." Such work was performed "in accordance with priorities established by the Commander, United States Army Training and Doctrine Command in a variety of useful formats." Two other missions included teaching military history and serving as the coordinator and integrator of military history instruction throughout the army school system. In his interview, Starry stated that he never intended for anything like CSI to be established. Once Starry left command, CSI suffered from a lack of patronage and interest. CSI later separated from its mission of military history instruction to focus upon historical studies and staff rides, as a TRADOC asset. The Department of Military History, or DMH, now conducts military history instruction.

112. Romjue, *From Active Defense to AirLand Battle*, 57. It is unclear why Starry

chose the term "AirLand Battle," since it came from the 1976 manual. One might speculate, however, that he did so to appease DePuy and not make it appear that the Army had totally rejected the 1976 doctrine.

113. Trauschweizer, *The Cold War U.S. Army*, 223. One Soviet theorist was Mikhail Nikolayevich Tukhachevsky (1893–1937).

114. Department of the Army, Field Manual 100-5 *Operations*, 1982 (Washington, DC: Government Printing Office, 1982), 2–3.

115. Ibid., *Operations*, 1982, 7–9. The table was a risk analysis for soldiers operating in a nuclear or chemical environment. Using history to push agendas is not new. See Peter Novick, *That Noble Dream: The "Objectivity Question" and the American Historical Profession* (Cambridge: Cambridge University Press, 1988), 207–224. The manual contained numerous ideas and quotes from Clausewitz, Jomini, Liddell Hart, Sun Tzu, and Russian military thinkers to illustrate theoretical concepts that underpinned the doctrine. See page 2–1 for one example.

116. Ibid., preface.

117. Trauschweizer, *The Cold War U.S. Army*, 222–223. The doctrine noted that the strategic level was beyond the scope of the manual. Interview with Lieutenant Colonel (Ret.) John B. Hunt by author, February 2001. Hunt acknowledged that the new doctrine took time to take hold because the Army had become fixated on the task-oriented 1976 doctrine. Operational art was the product of Soviet thinking of the 1930s. For an analysis of operational warfare development, see Robert M. Citino, *Blitzkrieg to Desert Storm: The Evolution of Operational Warfare* (Lawrence: University Press of Kansas, 2004).

118. Department of the Army, Field Manual 100-5 *Operations*, 1982, 7–1 to 7–3.

119. Ibid., 9–1.

120. Ibid., 9–2.

121. Ibid., 10–1 to 10–2, 11–9.

122. Ibid., 10–4.

123. Ibid., 11–1 to 11–11.

124. Frank Kitson described LIC in *Low-Intensity Operations: Subversion, Insurgency, Peacekeeping* (London: Stackpole Books, 1971). LIC was Kitson's label for a wide range of operations that were not war.

125. Thomas K. Adams, *U.S. Special Operations Forces in Action: The Challenge of Unconventional Warfare* (Portland, OR: Frank Cass, 1998), 174. Daniel P. Bolger, *Americans at War, 1975–1986: An Era of Violent Peace* (Novato, CA: Presidio Press, 1988), 261, 308, 317–321, 326–327, 332. Mark Adkins, *Urgent Fury: The Battle for Grenada* (Lexington, MA: D. C. Heath & Co., 1989), 259–263, 339. Grenada is also called the "Spice Island," its nickname due to the exportation of nutmeg.

126. John L. Romjue, *From Active Defense to AirLand Battle: The Development of Army Doctrine, 1973–1982* (Fort Monroe, VA: U.S. Army Training and Doctrine Command, 1984), 34. Major General Robert L. Schweitzer, Congressional Testimony, U.S. Congress, House, 1983, 31 (House Record 2287). Adams, *U.S. Special*

Operations Forces in Action, 192–194. David J. Baratto, "Special Forces in the 1980s: A Strategic Reorientation," *Military Review* 63, no. 3 (March 1983): 2–14.

127. White Paper 1983, "The Security of the Federal Republic of Germany," Federal Ministry of Defense, 1983, 160, quoted in Dennis S. Driggers, "The United States Army's Long March from Saigon to Baghdad: The Development of War-Fighting Doctrine in the Post-Vietnam Era" (Ph.D. diss., Syracuse University, 1995), 106.

128. Sorley, *Press On!* 1:412–423.

129. William R. Richardson, "FM 100-5: The AirLand Battle in 1986," *Military Review* 66, no. 2 (March 1986): 1:4–11.

130. Britt Lynn Edwards, "Reforming the Army: The Formulation and Implementation of AirLand Battle 2000" (Ph.D. diss., University of California Santa Barbara, 1985), 240–242. Chapman, Lilly, Romjue, and Canedy, *Prepare the Army for War*, 113–114.

131. L. D. Holder, "Doctrinal Development, 1975–85," *Military Review* 55, no. 5 (May 1985): 50–52.

132. Department of the Army, Field Manual 100-5 *Operations*, 1986 (Washington, DC: Government Printing Office, 1986), preface. Military Agency For Standardization, Allied Tactical Publication 35 (A), *Land Forces Tactical Doctrine*, Change 5 (Mons, Belgium: Supreme Headquarters Allied Powers Europe, 5 December 1990). Change 5 contains the fundamental principles in effect in the early 1980s, to include calling for a common approach to land warfare among the Alliance.

133. Department of the Army, Field Manual 100-5 *Operations*, 1986, 49.

134. Ibid., 91–94, 129–131.

135. Ibid., 10, 179–182. Operational art was defined as "the employment of military forces to attain strategic goals in a theater of war or theater of operations through the design, organization, and conduct of campaigns and major operations." For the emergence of operational art, see James J. Schneider, "Vulcan's Anvil: The American Civil War and the Emergence of Operational Art," Theoretical Paper no. 4 (Fort Leavenworth, KS: School of Advanced Military Studies, 16 June 1991). Clausewitz defined the center of gravity as "the hub of all power and movement on which everything depends." Clausewitz identified three centers of gravity: the enemy army, for it contained the destructive force that could bring about the friendly force's defeat; the enemy's capital, as it was the seat of government and center of culture; and the army of an enemy's ally if more powerful than the enemy's army. Clausewitz, *On War*, 595–596. Jomini described lines of operation, the routes by which an army moves and sustains itself, in *The Art of War*, 114–115. Clausewitz discussed culmination, the point where an attacker's strength no longer exceeds that of the defender, in *On War*, 528, 566–573.

136. Department of the Army, Field Manual 100-5 *Operations*, 1986, 4–5.

137. For more information on the Iran rescue attempt, see Paul B. Ryan, *The Iranian Rescue Mission* (Annapolis, MD: U.S. Naval Institute Press, 1985); and Bolger,

Americans at War, 99–168. For Lebanon, 1982–1984, see Eric Hammel, *The Root: The Marines in Lebanon, August 1982–February 1984* (San Diego, CA: Harcourt Brace Jovanovich, 1985); and Daniel P. Bolger, *Savage Peace: Americans at War in the 1990s* (Novato, CA: Presidio Press, 1995), 167–219. U.S. Congress, Senate Committee on Armed Services, *Defense Organization and the Need for Change*, Staff Report: Congress, 1st Session. (Washington, DC: Government Printing Office, 1985), 49. Congressional reactions to military operations during the 1970s and 1980s began as early as 1983 when Senators Henry Jackson and John Tower called for a committee to study the organization and decision-making procedures within the Department of Defense. In 1985, the Locher Report, a staff study headed by James R. Locher, announced that the Department of Defense required reorganization. In June 1985, the Reagan administration established a Blue Ribbon Commission known as the Packard Commission (after Chairman David Packard) to examine defense management in an attempt to deflect congressional reform of the Department of Defense. In February 1986, the Packard Commission reported similar findings as the Locher Report, although its recommendations were less radical in scope. By spring 1986, Senator Barry Goldwater used his influence to move Congress toward accepting the Packard reforms. Senator Sam Nunn and Representative Bill Nichols joined Goldwater in building bipartisan support for reform. Goldwater-Nichols was enacted as Public Law 99–433 on 1 October 1986. The law directed the Department of Defense to organize into seven entities with provisions for establishing additional offices if required by the president. The act also controlled military activities by implementing various national-security procedures. Goldwater-Nichols made the combatant commanders (CINCs) directly accountable to the president and secretary of defense instead of through the Joint Chiefs of Staff (JCS) as done previously. To reduce service parochialism, combatant commanders gained in authority over their respective commands, to include more control over assigned service forces within their areas of responsibility such as training, resourcing, utilization, and courts martial authority.

138. The separation of SOF was not absolute, as with the Army-Air Force split in 1947. Army SOF assets remain in the Army but are controlled and funded by USSOCOM. Army SOF has developed their own doctrinal procedures that allegedly mesh with FM 100-5 but not in all cases, as was seen both in Operation Desert Storm and later in Haiti.

139. Chapman, Lilly, Romjue, and Canedy, *Prepare the Army for War*, 82–83.

140. Many Americans have little understanding of what military families undergo during invasions. In 1990, Lieutenant Colonel John J. Moore was an army doctor stationed in Panama during Just Cause. Moore and his Panamanian wife had just gone to bed when he heard intense gunfire outside of the compound. He went out on the balcony to watch as a stream of tracers "came out of the sky and hit the ground about half a mile from my house." He and his wife watched the ongoing

invasion from their bedroom window until someone pounded on his door announcing that he was an American. Moore found several army Rangers asking him to treat a member of the Panamanian Defense Forces that they had just shot. Moore ran downstairs but found that the soldier had lost the back of his head. The Rangers thanked him anyway and Moore ran off to the military hospital to offer his help. Interview with Lieutenant Colonel John J. Moore by author, February 1991, Fort Leavenworth, Kansas. Casualties included 23 Americans killed and 324 wounded or injured. Of the Panamanians, 655 died and 2,000 were injured. See Kenneth J. Jones, *The Enemy Within* (El Dorado, Panama: Focus Publications, 1990), 7.

141. Thomas Donnelly, Margaret Roth, and Caleb Baker, *Operation Just Cause: The Storming of Panama* (New York: Lexington Books, 1991), 108. Malcolm McConnell, *Just Cause: The Real Story of America's High-Tech Invasion of Panama* (New York: St. Martin's Press, 1991), 271. As for Noriega, he fled into the night upon discovering that he was a wanted man, clad only in his red underwear. He left behind both his military uniform and a local prostitute. He was later apprehended while wearing a T-shirt, baggy shorts, and a baseball cap.

142. For a detailed accounting of the invasion, see Donnelly, Roth, and Baker, *Operation Just Cause.*

143. Robert L. Click, "Combat Jump," *Army* 42, no. 8 (August 1992): 24–30. Maxwell R. Thurman and William Hartzog, "Simultaneity: The Panama Case," *Army* 43, no. 11 (November 1993): 16–24. Steven N. Collins, "*Just Cause* up Close: A Light Infantryman's View of LIC," *Parameters* 22, no. 2 (Summer 1992): 55–65. Rules of engagement are the written restrictions that the military places upon its soldiers to curtail excessive force. ROE, as they are known, are subject to change depending upon the circumstances but usually retain that "nothing in these rules prevents you from exercising force for self-defense" or similar words.

144. "Training Center Rotation Reports, 1989," Center for Army Lessons Learned (CALL). Department of the Army, Field Manual 100-5 *Operations*, 1986, 1.

145. Adams, *U.S. Special Operations Forces in Action*, 234. Chapman, Lilly, Romjue, and Canedy, *Prepare the Army for War*, 83. The so-called "big five" weapons systems were the Abrams tank, the Bradley infantry fighting vehicle, the Apache and Blackhawk helicopters, and the Multiple Launch Rocket System (MLRS).

146. Norman Friedman, *Desert Victory: The War for Kuwait*, updated ed. (Annapolis, MD: Naval Institute Press, 1992), 129. Department of the Army, Field Manual 100-5 *Operations*, 1986, 19.

147. Adams, *U.S. Special Operations Forces in Action*, 244. Michael R. Gordon and Bernard E. Trainor, *The General's War* (New York: Little, Brown, 1995), 243. Douglas C. Waller, *The Commandos: The Inside Story of America's Secret Soldiers* (New York: Simon & Schuster, 1994), 251. Department of the Army, Field Manual 100-5 *Operations*, 1986, 17. USSOCOM personnel were familiar with AirLand Battle doctrine but had developed their own procedures to match their missions and capabilities. Such doctrine can be found in U.S. Army Field Manual 100-25, *Doctrine*

for Special Operations Forces (Washington, DC: Government Printing Office, 1991). There, the SOF community states its beliefs: "Special operations [SO] are different from conventional military operations. They have unique characteristics. They can occur in operational environments unsuited for conventional military operations, across the operational continuum. The distinctive roles and missions of SOF dictate that they apply principles of war differently in design and conduct of SO." See page 2–1.

148. Richard M. Swain, *"Lucky War": Third Army in Desert Storm* (Fort Leavenworth, KS: U.S. Army Command and General Staff College Press, 1994), 333.

149. John L. Romjue, *American Army Doctrine for the Post Cold War* (Fort Monroe, VA: Military History Office, U.S. Army Training and Doctrine Command, 1996), 22. The 1986 Goldwater-Nichols Act revised how the Executive Branch formulated national-security strategy. The White House publishes a periodic national security strategy, generally every two years, to inform the people as to the purpose and objectives of U.S. foreign policy and the ways and means for accomplishing them. George Bush, "The Possibility of a New World Order," April 13, 1991, in *Vital Speeches of the Day* 57, no. 15 (May 15, 1991). Dennis S. Ippolito, *Blunting the Sword: Budget Policy and the Future of Defense* (Washington, DC: Institute for National Strategic Studies, National Defense University Press, 1994), 58–70. The White House, *National Security Strategy*, March 1990 (Washington, DC: Government Printing Office, March 1990), 27.

150. For background on international relations theory and strategy, see Margot Light and A. J. R. Groom, eds., *International Relations: A Handbook of Current Theory* (Boulder, CO: Lynne Rienner Publishers, 1985), 141–155. Romjue, *American Army Doctrine for the Post Cold War*, 25.

151. Chapman, Lilly, Romjue, and Canedy, *Prepare the Army for War*, 62–63. Romjue, *American Army Doctrine for the Post Cold War*, 30.

152. Message, Cdr TRADOC to CDR USACAC, 032000Z August 1990, subject: FM 100-5 Revision Study, RC CAC/LVN, SG CGSC, SAMS 002/016. Message, Cdr TRADOC to Cdr CAC, 271312Z, subject: FM 100-5 Revision, Combined Arms Research Library. SAMS was created in 1983 as a second-year course following CGSC for specially selected officers to study the art and science of war. Robert H. Scales, *Certain Victory: The US Army in the Gulf War* (Washington, DC: Office of the Chief of Staff, United States Army, 1993), 27–28. Romjue, *American Army Doctrine for the Post Cold War*, 45. See Frank N. Schubert and Theresa L. Kraus, eds., *The Whirlwind War: The United States Army in Operations Desert Shield and Desert Storm* (Washington, DC: U.S. Army Center of Military History, 1995).

153. Lieutenant Colonel Robert E. Harmon, Ramon A. Malave, Charles A. Miller III, and William K. Nadolski, "Counterdrug Assistance: The Number One Priority," *Military Review* 73, no. 3 (March 1993): 26–35. Adams, *U.S. Special Operations Forces in Action*, 255. Interview with Lieutenant Colonel Alan C. Lowe by author, Fort Leavenworth, KS, May 1999. The interagency consists of the various government

departments and agencies that work together within the National Security Council and are traceable back to the 1947 National Defense Reorganization Act. Various groups within the NSC work together to formulate foreign policy and take actions in support of that policy. For a review of the interagency, see William W. Mendel and David G. Bradford, *Interagency Cooperation: A Regional Model for Overseas Operations* (Washington, DC: Institute for National Strategic Studies, National Defense University, 1995).

154. Major General Stephen Silvasy Jr., "AirLand Battle Future, The Tactical Battlefield," *Military Review* 71, no. 2 (February 1991): 3–4. Reserve component means both the army reserves and the Army National Guard. Lieutenant General (Ret.) Frederic J. Brown, "AirLand Battle Future, The Other Side of the Coin," *Military Review* 71, no. 2 (February 1991): 14–24.

155. Gordon W. Rudd, "Operation Provide Comfort: Humanitarian Intervention in Northern Iraq, 1991." (Ph.D. diss., Duke University, 1993), 100–104. John T. Fishel, *Liberation, Occupation, and Rescue: War Termination and Desert Storm* (Carlisle Barracks, PA: Strategic Studies Institute, 31 August 1992), 51. David S. Elmo, "Food Distribution during Operation PROVIDE COMFORT," *Special Warfare* 5, no. 1 (March 1992): 8–9. Adams, *U.S. Special Operations Forces in Action*, 247–254. Eventually, the relief effort waned as Iraqi attacks and Kurdish factional fighting erupted in late 1996. About 2,000 Kurdish relief workers were evacuated to Guam between November 1996 and January 1997. See *Washington Post*, "U.S. Flies Iraqis to Guam" (7 December 1996), 1. Lieutenant Colonel John P. Abizaid, "Lessons for Peacekeepers," *Military Review* 73, no. 3 (March 1993), 11–19.

156. Rudd, "Operation Provide Comfort," 381.

CHAPTER FIVE. DOCTRINE FOR A POST–COLD WAR WORLD:
MULTINATIONAL AND MULTISERVICE OPERATIONS, 1991–2008

1. General Gordon R. Sullivan, *America's Army into the Twenty First Century*, National Security Paper no. 14 (Cambridge, MA: Institute for Foreign Policy Analysis, 1993), 4–5.

2. Alvin Toffler and Heidi Toffler, *War and Anti-War, Survival at the Dawn of the 21st Century* (Boston: Little, Brown, 1993). The Tofflers argued that civilizations were moving from industrialization to computerization and information and that governments and armies were no better prepared to fight in the new age of war than a medieval swordsman would be today. John L. Romjue, *American Army Doctrine for the Post Cold War* (Fort Monroe, VA: Military History Office, U.S. Army Training and Doctrine Command, 1996), 35.

3. Romjue, *American Army Doctrine for the Post Cold War*, 40–43.

4. James McDonough, "Building the New FM 100-5, Process and Product," *Military Review* 71, no. 10 (October 1991): 5. Steven Metz, "U.S. Strategy and the Changing LIC Threat," *Military Review* 71, no. 6 (June 1991): 22–29.

5. Romjue, *American Army Doctrine for the Post Cold War*, 48–49. Carl W. Stiner,

"The Strategic Employment of Special Operations Forces," *Military Review* 71, no. 6 (June 1991): 3–13.

6. Gordon R. Sullivan, "Doctrine: A Guide to the Future," *Military Review* 72, no. 2 (February 1992): 2–9.

7. U.S. Code Annotated, Title 10 Armed Forces, Sections 3001 to 6480 (Washington, DC: West Group Press, 1998), 82–83. Lawrence A. Yates, "Military and Support Operations: Analogies, Patterns and Recurring Themes," *Military Review* 77, no. 4 (July–August 1997), 51–61. Federal law contained within Title 10 U.S. Code, Section 3062 (a), directs the Army to preserve national peace and security while supporting national policies and objectives. The Army must also defend against and overcome any nation responsible for aggressive acts that imperil the people. The service thus must legally prepare for any contingency, foreign or domestic, that threatens national security. Section 3062 (b) stipulates that the Army must organize, train, and equip itself primarily for land combat. Sections 3062 (a) and (b) compel the Army to prepare for any land-based mission during peacetime or war to meet the nation's security needs.

8. Romjue, *American Army Doctrine for the Post Cold War*, 60. The five dynamics were early entry and lethality, depth and simultaneous attack, battle space, command and control and tempo, and combat service support.

9. Ibid., 82–86. Warrior syndrome is not unique to American senior leaders. For insights into generalship, see Michael Carver, ed., *The War Lords: Military Commanders of the Twentieth Century* (Boston: Little, Brown, 1976). Edward E. Thurman, "Shaping an Army for Peace, Crisis and War: The Continuum of Military Operations," *Military Review* 72, no. 4 (April 1992): 27–35.

10. Romjue, *American Army Doctrine for the Post Cold War*, 88. General Gordon R. Sullivan, "Hurricane Andrew: An After-Action Report," *Army* 43, no. 1 (January 1993): 16–22. Dennis Steele, "Digging Out from Despair," *Army* 42, no. 11 (November 1992): 22–25. Frank R. Finch, "Piercing Winds Hit Hawaii," *Military Engineer* 85, no. 554 (January–February 1993): 4–6. Marjorie Barrell, "Cleanup On Kauai," *Soldiers* 47, no. 12 (December 1992): 6–8.

11. Department of the Army, Field Manual 100-5 *Operations*, 1986, 161.

12. General Frederick M. Franks Jr., "Full-Dimensional Operations: A Doctrine for an Era of Change," *Military Review* 73, no. 12 (December 1993): 8. Romjue, *American Army Doctrine for the Post Cold War*, 96. School of Advanced Military Studies, briefing slides for 3–4 November 1992 conference presented to conference attendees, Combined Arms Research Library. During Panama 1989, army forces struck the Panamanian Defense Forces from the ground and the air simultaneously and gave rise to an attack method that Sullivan later wanted to promote. In 1990, West Germany was about the size of Oregon. In Desert Storm, forces fought a war that would have covered six American eastern states.

13. Department of the Army, Field Manual 100-5 *Operations*, 1993 (Washington, DC: Department of the Army, 1993), iv–v.

14. The service later called the manual "AirLand Operations," although the proper term was full-dimension operations. Department of the Army, Field Manual 100-5 *Operations*, 1993, v, 2–0 to 2–1. The 1993 manual is an anomaly because several chapters begin with page number 1–1 or 3–1 while others are marked 2–0 or 6–0.

15. Ibid., iv–v, Chapters 4–6.

16. Ibid., 1–1 to 1–5.

17. Ibid., 2–0 to 2–24.

18. Ibid., 3–1 to 3–12.

19. Ibid., 5–1 to 5–5.

20. Ibid., 6–1 to 6–23.

21. Ibid., 9–0 to 9–6.

22. Ibid., 10–1 to 10–6.

23. Ibid., 12–1 to 12–13.

24. Ibid., 13–0 to 13–2.

25. Ibid., 13–0 to 13–8.

26. Ibid., 14–1 to 14–5. Roger J. Spiller, "The Tenth Imperative," *Military Review* 69, no. 4 (April 1989): 2–13.

27. Franks, "Full-Dimensional Operations," 5–10. The five warning lights were threats and unknown dangers, national military strategy, history and lessons learned, changing nature of warfare, and technology.

28. Department of the Army, Field Manual 100-5 *Operations*, 1986 (Washington, DC: Government Printing Office, 1993), 164–168. Conversation with Lieutenant Colonel Mike Rampy by author, School of Advanced Military Science (SAMS), Fort Leavenworth, Kansas, October 1991. In January 1991, army forces assisted the Marine Corps and the Air Force plan a noncombatant evacuation (NEO) of Americans, Soviets, and selected third-party foreign nationals in Mogadishu, Somalia. Beginning in August 1992, army Special Forces units provided food-convoy security for an ongoing UN humanitarian assistance operation. Placing emphasis upon UN operations was a sensitive issue in 1991, due to issues over placing American troops under UN command.

29. Department of the Army, *U.S. Army Operations in Support of UNOSOM II, 4 May 1993–31 March 1994*, I-1–2 through I-1–5, CALL.

30. Department of the Army, Field Manual 100-5 *Operations*, 1986 (Washington, DC: Headquarters, Department of the Army, 1986), 161–162, CALL. *U.S. Army Operations in Support of UNOSOM II 4 May 1993–31 March 1994, Lesson Learned Report* (Fort Leavenworth, KS: U.S. Army Combined Arms Center, n.d.), I-2–6 to I-2–7, CALL. The report recommended that a division headquarters such as the 10th Mountain Division not be used as the basis of a JTF staff because "it does not have the joint service staff experience or the staff structure for such duty." Lieutenant Colonel George Steuber had previous joint and multinational experience from working with the UN Transition Authority in Cambodia (UNTAC). Steuber noted that, instead of learning joint procedures, many soldiers were so confused that they

kept doing whatever they were trained to do, whether it was applicable or not. Interview with Lieutenant Colonel George Steuber by author, Fort Leavenworth, Kansas, April 1998.

31. Thomas K. Adams, *U.S. Special Operations Forces in Action: The Challenge of Unconventional Warfare* (Portland, OR: Frank Cass, 1998), 265–266. Department of the Army, *U.S. Army Operations in Support of UNOSOM II*, II-12–2. George Steuber interview. Interview with Major Drew Meyerowich by author, Fort Leavenworth, Kansas, March 1998. When a captain, Meyerowich commanded a rifle company from the 10th Mountain Division (Light) in Somalia and led his troops in battle in Mogadishu on 3 October 1993.

32. Major General S. L. Arnold, "Somalia: An Operation Other than War," *Military Review* 73, no. 12 (December 1993): 26–35. John T. Fishel, ed. *"The Savage Wars of Peace": Toward a New Paradigm of Peace Operations* (Boulder, CO: Westview Press, 1998), 172. Department of the Army, *Operation Restore Hope Lessons Learned Report, 3 December 1992–4 May 1993*, 13, 19.

33. John Garabedian, "Terrorism Kills over 500,000 in Rwanda: Violence and Combating Terrorism Update" (Fort Leavenworth, KS: U.S. Army Command and General Staff College, September 1994), 1. Bradley Graham, "Pentagon Officials Worry Aid Missions Will Sap Military Strength," *Washington Post*, 29 July 1994. On 4 November 1988, President Ronald Reagan signed into law the Prevention of Genocide Act, which was drafted in reaction to Saddam Hussein's use of chemical weapons. The law, U.S. Code Title 18 Section 1091, contains specific legal requirements concerning genocide.

34. Department of the Army, Field Manual 100-5 *Operations*, 1993, 13–5.

35. Joint Task Force Support Hope, Operations Plan 94–004, U.S. European Command, Stuttgart, Germany, 6 August 1994, 7–8, CALL. Stephen O. Wallace, "Joint Task Force Support Hope: The Role of the CMOC in Humanitarian Operations," *Special Warfare* 9, no. 1 (January 1996): 36–41.

36. Interview with William G. O'Neill by author, Port-au-Prince, Haiti, October 1996. O'Neill, a respected veteran of many UN operations and an advocate for human rights, worked with the UN in Rwanda where he observed the Army on a frequent basis. O'Neill believed that the military should become involved in human-rights issues because they had much more capability than civilian agencies. In defense of army forces in Rwanda, law governed support for civilian agencies during humanitarian operations. See Department of Defense, *Department of Defense Directive 5100.46* (Washington, DC: Government Printing Office, 4 December 1975).

37. Department of the Army, Field Manual 100-5 *Operations*, 1993, 13–5.

38. John E. Lange, "Civilian-Military Cooperation and Humanitarian Assistance: Lessons from Rwanda," *Parameters* 28, no. 2 (Summer 1998): 106–122. O'Neill interview. Interview with Major Daniel A. Pinnell by Robert F. Baumann, Combat Studies Institute, U.S. Army Command and General Staff College, 20 April 1999. Pinnell described many of his army comrades as warriors who chafed at the thought

of becoming dispensers of diapers, milk, and food. Others, however, such as the medical community, enjoyed caring for the sick and gaining valuable experience.

39. "Perspective on Rwanda Support: Commander of the 21st Theater Army Area Support Command assesses his command's support for Operation Support Hope," *Army Logistician* 27, no. 3 (May–June 1995): 4–6. USCINCEUR Message, Subject: CINCEUR Assessment of Humanitarian Operations in Rwanda, 30 September 1994, 18–19, Combined Arms Research Library.

40. Sean Naylor, "The Invasion That Never Was," *Army Times*, 26 February 1996, 13. Cedras had led the junta that deposed Aristide in September 1991. John R. Ballard, *Upholding Democracy: The United States Military Campaign in Haiti, 1994–1997* (Westport, CT: Praeger, 1998), 131. Aristide was the first democratically elected president in Haitian history. Walter E. Kretchik, Robert F. Baumann, and John T. Fishel, *Invasion, Intervention, "Intervasion": A Concise History of the U.S. Army in Operation Uphold Democracy* (Fort Leavenworth, KS: U.S. Army Command and General Staff College Press, 1998), ix. Interview with Lieutenant Colonel Edward Donnelly by author, Headquarters U.S. Atlantic Command, Norfolk, Virginia, December 1995. Interview with Lieutenant Colonel Phil Idiart by author, Headquarters U.S. Atlantic Command, Norfolk, Virginia, December 1995. Interview with Major Joseph Doyle by author, Headquarters U.S. Atlantic Command, Norfolk, Virginia, December 1995.

41. Department of the Army, Field Manual 100-5 *Operations*, 1993, 3–1. Interview with Lieutenant Colonel Gordon C. Bonham by author, School of Advanced Military Studies (SAMS), Fort Leavenworth, Kansas, January 1999. Bonham was the chief of plans, XVIIIth Airborne Corps, Fort Bragg, North Carolina, who wrote the plan for invading Haiti. Richard Rinaldo interview by author [telephone], Fort Monroe, Virginia, 11 March 1999.

42. Kretchik, Baumann, and John T. Fishel, *Invasion, Intervention, "Intervasion,"* 49–65.

43. Interview with Lieutenant General Henry H. Shelton by Steve Dietrich, Port-au-Prince, Haiti, 24 October 1994. Cynthia Hayden, ed., *JTF-180 Uphold Democracy: Oral History Interviews* (Fort Bragg, NC: XVIIIth Airborne Corps History Office, 1996), 62. In the 10th Mountain Division's Haiti after-action report, Meade noted that to his soldiers Haiti wasn't OOTW, it was war. He also noted that there was a high correlation between OOTW and warfighting tasks. Soldiers "conducted combat operations on a daily basis." Headquarters 10th Mountain Division (Light), *Operations in Haiti: Planning/Preparation/Execution, August 1994 thru January 1995* (Fort Drum, NY: Headquarters 10th Mountain Division [Light], 1995), 24–2 to 24–3, 24–9. Kretchik, Baumann, and Fishel, *Invasion, Intervention, "Intervasion,"* 79. Interview with Major Kristen Vlahos-Schafer by author, Fort Leavenworth, Kansas, March 1997. Vlahos-Schafer reported that Meade issued orders to the force to avoid patrolling the cities at night for fear that "something might happen." Interview with Major Len Gaddis by author, Fort Leavenworth, Kansas, March 1997. Gaddis, a civil affairs officer, remarked that Meade's policies made it difficult for him to establish proper

liaison with Haitian officials. For Meade's views, see Headquarters 10th Mountain Division (Light), *Operations in Haiti*, Chapter 22, page 30. Interview with Major General (Ret.) David C. Meade by author, Stafford, Virginia, 29 April 2000. Meade stated that a government official told him to keep the invasion "casualty free," thus he had exerted proper caution. Department of the Army, Field Manual 100-5 *Operations*, 1993, 13–0. Interview with Major Berthony Ladouceur by author, Fort Leavenworth, Kansas, May 1997. Interview with Colonel Thomas Miller by Major Christopher Clark, Commander 44th Military History Detachment, Port-au-Prince, Haiti, date unknown. Miller was Meade's operations officer, who stated that peacekeeping and peace enforcement in OOTW means nothing to a rifle squad leader because it is still "a war on the streets."

44. Department of the Army, Field Manual 100-25, *Doctrine for Army Special Operations Forces*, 1991 (Washington, DC: Government Printing Office, 1991), vii, 2–1, 2–8. Department of the Army, Field Manual 100-5 *Operations*, 1993, 2–20 to 2–21. Ed Phillips, "Army SOF: The Right Tool for OOTW," *Special Warfare* 10, no. 3 (Summer 1997): 2–13.

45. Interview with Major John Valledor by author, Fort Leavenworth, Kansas, December 1998. For film footage showing the disparities between the 10th Mountain Division and Special Forces in Haiti, see Walter E. Kretchik, producer, and Robert F. Baumann, *Rhythm of the Street*, VHS, 69 min. (Fort Leavenworth, KS: Combat Studies Institute, 1998). Interview with Colonel Mark C. Boyatt by author, Fort Leavenworth, Kansas, November 1997. Bob Shacochis, "Our Two Armies in Haiti," *New York Times*, 8 January 1995. Steven E. Cook, "Field Manual 100-25: Updating Army SOF Doctrine," *Special Warfare* 9, no. 3 (August 1996): 36–37.

46. U.S. Army Peacekeeping Institute, *Success in Peacekeeping: United Nations Mission in Haiti: The Military Perspective* (Carlisle Barracks, PA: U.S. Army Peacekeeping Institute, 1996), 20.

47. Walter E. Kretchik, "Multinational Staff Effectiveness in UN Peace Operations: The Case of the U.S. Army and UNMIH, 1994–1995," *Armed Forces and Society* 29, no. 3 (Spring 2003): 393–413. Kevin C. Benson and Christopher Thrash, "Declaring Victory: Planning Exit Strategies for Peace Operations," *Parameters* 26, no. 3 (Autumn 1996): 69–80.

48. Robert L. Ord III and Ed Mornston, "Light Forces in the Force-Projection Army," *Military Review* 74, no. 1 (January 1994): 22–33. Lieutenant Colonel (Ret.) John B. Hunt, "OOTW: A Concept in Flux," *Military Review* 76, no. 5 (September–October 1996): 3–9. Hunt represented the camp that saw OOTW as dominating an adversary's behavior, an idea no different than Carl von Clausewitz's concept that wars were fought to impose one's will upon the enemy. Methods, resources, and circumstances varied, but the outcome was to change behavior through direct or indirect force. Carl von Clausewitz, *On War*, ed. and trans. Michael Howard and Peter Paret (Princeton, NJ: Princeton University Press, 1984), 90.

49. Anne W. Chapman, Carol J. Lilly, John L. Romjue, and Susan Canedy, *Prepare*

the Army for War: A Historical Overview of the Army Training and Doctrine Command, 1973–1998 (Fort Monroe, VA: Military History Office, U.S. Army Training and Doctrine Command, 1998), 65. In the 1990s, the Army experimented with information as a weapon by either denying or providing information to adversaries through computers and other means. See Martin C. Libicki, *What Is Information Warfare?* (Washington, DC: Institute for National Strategic Studies, National Defense University, 1995). For technology and OOTW, see Center for Advanced Concepts and Technology, *Operations Other than War: The Technological Dimension*, 2nd printing (Washington, DC: National Defense University, 1997). Force (later Army) XXI was Sullivan's campaign to "reconceptualize and redesign the force at all echelons, from the foxhole to the industrial base, to meet the needs of a volatile and ever changing world." In sum, it was Sullivan's way of modernizing the service. In January 1995, Sullivan initiated a series of laboratory and field experiments to examine new technology as a "lever" for the soldier. The experiments led to digital means to pass sensor and other information between computers at various headquarters nearly simultaneously. Other technology involved communications devices and training aides. The intent was to make the Army better through technological assistance. See Togo D. West Jr. and Gordon R. Sullivan, *Force XXI: Meeting the 21st Century Challenge* (Washington, DC: Government Printing Office, January 1995). Thomas M. Carlin and Mike Sanders, "Soldier of the Future: Assessment and Selection of Force XXI," *Special Warfare* 9, no. 2 (May 1996): 16–21. Gordon R. Sullivan, "A Vision for the Future," *Military Review* 75, no. 3 (May–June 1995): 5–14.

50. Letter, Hartzog to Holder, 27 October 1995, Combined Arms Research Library. Rinaldo interview. Rinaldo described Hartzog as a practical individual and highly intelligent but not interested in scholarly work or anything that did not address army readiness.

51. The White House, *A National Security Strategy of Enlargement and Engagement, February 1996* (Washington, DC: Government Printing Office, 1996), i, 13.

52. For military aspects of the Dayton Accords, see "General Framework Agreement for Peace in Bosnia and Herzegovina," http://www.state/gov/www/current/bosnia/dayframe.html. Department of the Army, Field Manual 100-5 *Operations*, 1993, 3–1.

53. Department of the Army, Field Manual 100-5 *Operations*, 1993, 5–4 to 5–5. Major Bruce E. Akard, "Strategic Deployment: An Analysis of How U.S. Army Europe Deployed VIIth Corps to Desert Storm and the 1st Armored Division to Bosnia" (master's thesis, U.S. Army Command and General Staff College, 1997), 9.

54. Stanley F. Cherrie, "Task Force Eagle," *Military Review* 77, no. 4 (July–August 1997): 63–72.

55. A movement to contact is an offensive operation to develop the situation and to establish or regain contact. At the beginning of the IFOR operation, U.S. forces were not in contact with any of the Former Warring Factions. Under the terms of the Dayton Agreement, U.S. forces moved from Germany to Bosnia, where they

established a zone of separation between the former combatants and disarmed them. See Department of the Army, Field Manual 100-5 *Operations, June 1993*, 7–4. The training area at Tabor Falva, Hungary, became a showcase for American technology. Millions of dollars transformed a decrepit training complex into a modern range complex with maintenance facilities, troop billets, dining halls, and computer technology that bedazzled local Hungarians. Personal notes from visit as Army Component Command Historian, Operation Joint Endeavor, 18 June 1996. Interview with Lieutenant Colonel (Ret.) Glen Boney by author, Leavenworth, Kansas, March 2000. Boney was in charge of the Tabor Falva renovation.

56. Interview with Lieutenant Colonel (Ret.) Michael Burke by author, Fort Leavenworth, Kansas, 29 January 2001. Clausewitz, *On War*, 75.

57. Burke interview. Interview with Lieutenant Colonel (Ret.) Russell Glenn by author [telephone], Santa Monica, California, 1 February 2001. Glenn was a member of the writing team who contributed significantly to the development of ODSS.

58. Glenn interview. Department of the Army, Field Manual 100-5 *Operations, Final Draft*, 5 August 1997 (Washington, DC: Department of the Army, 1997), iii, Chapter 5. "See" meant gaining an understanding of the mission, the enemy situation, the weather conditions, friendly troop status, time available for planning, and civilian considerations such as policy decisions. "Shape" involved maneuvering forces to position them where they had an advantage over the enemy. "Shield" was preventive; it implied protecting the friendly force from enemy interference. "Strike" applied lethal (firepower) and nonlethal (psychological operations) capabilities to achieve objectives. "Move" positioned forces for advantage, then repositioned them as the situation changed. The principles of operations were an edited version of the principles of war to create one list of ideas that merged war and OOTW. The principles included the objective, offensive, maneuver, massed effects, economy of force, simplicity, surprise, unity of effort, security, and morale. For a sampling of the warfighting culture in Bosnia, see Douglas V. Johnson II, ed., *Warriors in Peace Operations* (Carlisle Barracks, PA: Strategic Studies Institute, U.S. Army War College, 1999). Larry E. Bush Jr., "Building Special Forces to Last: Redesigning the Organizational Culture," *Special Warfare* 12, no. 3 (Summer 1999): 16–20.

59. Burke interview. Field Manual 100-5 *Operations, Final Draft*, 5 August 1997, Part V.

60. Burke interview. Field Manual 100-5 *Operations, Final Draft*, 5 August 1997, 17–4 to 17–5.

61. David Fastabend, "The Categorization of Conflict," *Parameters* 27, no. 2 (Summer 1997): 75–85. David Fastabend, "FM 100-5, 1998: Endless Evolution," *Army* 47, no. 5 (May 1997): 45–50.

62. Interview with Lieutenant Colonel John Carmichael by author, Fort Leavenworth, Kansas, September 1997.

63. Burke interview. Burke noted the White House had published a new national strategy, thus Meigs had concerns. See the White House, *A National Security Strategy*

for a New Century, May 1997 (Washington, DC: Government Printing Office, 1997). Meigs's interest in history is evident by his earning a Ph.D. in history from the University of Wisconsin, Madison.

64. The manual did not address the complete Title 10 Section 3062. Section 3062 (a) states that the Army preserves the peace and security of the U.S., its territories, commonwealths, and possessions while supporting national policies and objectives. Section 3062 (b) orders the Army to organize, train, and equip primarily for land warfare. The 1997 draft continued the Army's efforts to focus on section 3062 (b) while limiting discussion of section 3062 (a). See West Group, *United States Code Annotated*, Title 10, Armed Forces, Sections 3001 to 6480 (Washington, DC: Westlaw, 1998), 82–83. *Department of the Army*, Field Manual 100-5 *Operations, Revised Final Draft*, 19 June 1998 (Washington, DC: Department of the Army, 1998), vii.

65. Department of the Army, Field Manual 100-5 *Operations, Revised Final Draft*, 19 June 1998, Chapters 1–6.

66. Ibid., xiii. Burke interview.

67. Ibid., 3–17 to 3–33.

68. Interview with Colonel Michael Parker by author, Fort Leavenworth, Kansas, June 1998. Burke interview. The reviewers included university professors, Department of Defense civilians, general officers, freelance military history writers, and others whom Meigs valued for their opinion.

69. Interview with Lieutenant Colonel Victor Robertson by author, Fort Leavenworth, Kansas, March 1999.

70. Brian R. Sullivan, "Special Operations and LIC in the 21st Century: The Joint Strategic Perspective," *Special Warfare* 9, no. 2 (May 1996): 2–7. Burke interview. Interview with Lieutenant Colonel James Rentz by author, Fort Leavenworth, Kansas, May 2000. Interview with Colonel Steve Rotkoff by author, Fort Leavenworth, Kansas, April 2000. Rotkoff was promoted during his time with the writing team.

71. Benjamin S. Lambeth, *NATO's Air War for Kosovo: A Strategic and Operational Assessment* (Santa Monica, CA: Rand, 2001), v. Wesley K. Clark, *Waging Modern War* (New York: Public Affairs, 2001), xxiv.

72. M. R. Neighbour et al. "Providing operational analysis to a peace support operation: the Kosovo experience," *Journal of the Operational Research Society* 53, no. 5 (2002): 523–543. The mission continues with State National Guard units conducting periodic troop rotations.

73. Department of the Army, *FM 100-5 Content Summary, 1 February 2000* (Fort Monroe, VA: Headquarters, TRADOC, 2000). I joined the team as a historian and doctrine writer in March 2000 and served through September 2000. The "big ideas" included the Army's role in land warfare, unified action, and strategic responsiveness. Steele was not overly concerned about the terms OOTW or MOOTW. He believed that the Army had embraced both terms and that he was not about to "waste time arguing over acronyms." Comment to author during team meeting, March 2000.

74. I, along with Burke, logged the comments, made decisions, and wrote responses for approval by the SAMS director.

75. Lamar Tooke, "Blending Maneuver and Attrition," *Military Review* 90, no. 2 (March–April 2000): 7–13. Force XXI was to be a fully modernized force dependent upon computerization to digitally link combat systems. Both Sullivan and Franks believed that units equipped with digital computers could operate at a higher tempo because they could "see" each other and the enemy and share information faster. On the modern battlefield, speed of action meant doing something quicker than your opponent such as see, strike, and destroy a force.

76. Headquarters, Department of the Army, Field Manual 3-0 *Operations* (Washington, DC: Department of the Army, 2001), vii–viii. See Chapter 7. The Army has other manuals subordinate to FM 100-5 that discuss humanitarian assistance and disaster relief.

77. Comment to author from General (Ret.) Edwin C. Burba, Senior Mentor U.S. Army Battle Command Training Program to the FM 3-0 writing team, Fort Leavenworth, Kansas, June 2000.

78. Headquarters, Department of the Army, Field Manual 3-0 *Operations*, 2001, 1–1 to 1–17.

79. Ibid., 2–1 to 2–24.

80. Ibid., 3–1 to 3–18.

81. Ibid., 4–1.

82. Ibid., 4–1 to 4–32.

83. Ibid., 5–1 to 5–18.

84. Ibid., 6–1 to 6–22.

85. Ibid., 7–1 to 7–28.

86. Ibid., 8–1 to 8–18.

87. Ibid., 9–1 to 9–16.

88. Ibid., 10–1 to 10–14.

89. Ibid., 11–1 to 11–24.

90. Ibid., 12–1 to 12–20.

91. Ibid., 6–15 to 6–17.

92. Ibid., 1–5; 9–11 to 9–12.

93. David P. Fridovich and Fred T. Krawchuk, "The Special Operations Forces Indirect Approach," *Joint Forces Quarterly*, no. 44 (1st Quarter 2007): 24–27.

94. Gregory Fontenot, E. J. Degen, and David Tohn, *On Point: The United States Army in Operation Iraqi Freedom* (Fort Leavenworth, KS: Combat Studies Institute, 2004), 24.

95. Fontenot, Degen, and Tohn, *On Point*, 24–26. Headquarters, Department of the Army, Field Manual 3-0 *Operations*, 2001, 2–6.

96. Fontenot, Degen, and Tohn, *On Point*, 44–45.

97. Ibid., 46.

98. Ibid., 46. Headquarters, Department of the Army, Field Manual 3-0 *Operations*, 2001, 6–3 to 6–5.

99. Fontenot, Degen, and Tohn, *On Point*, 47.

100. Ibid., 49–52.

101. Donald P. Wright, Timothy R. Reese, with the Contemporary Operations Study Team, *On Point II, Transition to the New Campaign: The United States Army in Operation Iraqi Freedom, May 2003–January 2005* (Fort Leavenworth, KS: Combat Studies Institute Press, 2008), 72. Headquarters, Department of the Army, Field Manual 3-0 *Operations* 2001, 6-5.

102. David E. Sanger, "Plans for Postwar Iraq are Re-evaluated as Fast Military Exit Looks Less Likely," *New York Times*, 2 April 2003. Patrick E. Tyler, "Iraq is Planning Protracted War," *New York Times*, 2 April 2003.

103. Headquarters, Department of the Army, Field Manual 3-0 *Operations* 2001, 9-8 to 9-9.

104. Wright, Reese, et al., *On Point II*, 30. Colin Kahl, "The Four Phases of the U.S. COIN Effort in Iraq," e-mail posted online, *Small Wars Journal,* March 18, 2007, http://smallwarsjournal.com/blog/2007/03/the-four-phases-of-the-us-coin/.

105. Wright, Reese, et al., *On Point II*, 31. George Packer, "Letter from Iraq: The Lesson of Tal Afar: Is It Too Late for the Administration to Correct Its Course in Iraq?" *New Yorker* 82, no. 8 (10 April 2006): 48–65.

106. Wright, Reese, et al., *On Point II*, 60–61.

107. Ibid., 173–175.

108. John Nagl, "The Evolution and Importance of Army/Marine Corps Field Manual 3-24, Counterinsurgency," *Small Wars Journal*, June 27, 2007, http://smallwarsjournal.com/blog/2007/06/the-evolution-and-importance-o/.

109. Wright, Reese, et al., *On Point II*, 46.

110. Nagl, "The Evolution and Importance of Army/Marine Corps Field Manual 3-24, Counterinsurgency."

111. Ibid. Crane was the Director of the U.S. Army Military History Institute, Carlisle, Pennsylvania, a position he assumed on 1 February 2003.

112. The official Army and Marine Corps version published 15 December 2006 is Headquarters, Department of the Army, Field Manual 3-24 / Headquarters, U.S. Marine Corps, Marine Corps Warfighting Publication 3-33.5 *Counterinsurgency* (Washington, DC: Government Printing Office, December 2006), http://www.fas .org/irp/doddir/army/fm3-24.pdf. The subsequent 2007 version published by the University of Chicago Press with John Nagl's preface, "The Evolution and Importance of Army/Marine Corps Field Manual 3-24, Counterinsurgency," is *The U.S. Army/Marine Corps Counterinsurgency Field Manual*, U.S. Army Field Manual no. 3-24/Marine Corps Warfighting Publication no. 3-33.5 (Chicago: University of Chicago Press, 2007). Unless otherwise noted, I discuss the 15 December 2006 FM 3-24. Several jihadi websites have reviewed the manual, and copies have been found in Taliban training camps in Pakistan. Dan Ephron, "Conrad Crane: The Military: With Iraq in Flames, a Historian Rethinks the Way We Fight the Enemy," *Newsweek*

148, no. 26 (25 December 2006): 63. Samantha Power, "Our War on Terror," 29 July 2007, Posted at *Small Wars Journal,* July 26, 2007, http://smallwarsjournal.com/blog/2007/07/ny-times-book-review-fm-324-1/.

113. Field Manual 3-24, *Counterinsurgency.* See also Military.com, "New Counterinsurgency Manual" [Comments on FM 3-24], December 18, 2006, http://www.military.com/features/0,15240,120810,00.html; Nagl, "The Evolution and Importance of Army/Marine Corps Field Manual 3-24, Counterinsurgency"; David Galula, *Counterinsurgency Warfare: Theory and Practice* (New York: Praeger, 1964); Dan Baum, "Battle Lessons, What the Generals Don't Know," *New Yorker* 80, no. 43 (17 January 2005): 42–48; and Sir Robert Thompson, *Defeating Communist Insurgency: The Lessons of Malaya and Vietnam* (New York: Praeger, 1966).

114. Military.com, "New Counterinsurgency Manual."

115. Dale Kuehl, "Testing Galula in Ameriyah: The People are the Key," *Military Review* 89, no. 2 (March–April 2009): 72–80; James R. Crider, "A View from inside the Surge," *Military Review* 89, no. 2 (March–April 2009): 81–88; Bing West, "Counterinsurgency Lessons from Iraq," *Military Review* 89, no. 2 (March–April 2009): 2–12; Hamid Hussain, "'Courageous Colonels'—Current History Recap," *Small Wars Journal,* December 23, 2008, http://smallwarsjournal.com/blog/2008/12/courageous-colonels-current-hi/.

116. Department of Defense, Department of Defense Directive 3000.05, 28 November 2005. FM 3-0 2001 defined stability operations as operations that "promote and protect U.S. interests by influencing the threat, political, and information dimensions of the operational environment through a combination of peacetime development, cooperative activities, and coercive actions in response to crisis." Headquarters, Department of the Army, Field Manual 3-0 *Operations,* 2001, 1-15.

117. Department of Defense Directive 3000.05, 28 November 2005; Field Manual 3-0 *Operations,* 2001.

118. William S. Wallace, "FM 3-0 *Operations*: The Army's Blueprint," *Military Review* 88, no. 2 (March–April 2008): 2–7.

119. Ibid., 2.

120. Ibid., 4.

121. Association of the United States Army, "Revolution in Army Doctrine: The 2008 Field Manual 3-0 *Operations,*" February 2008, http://www.ausa.org/publications/torchbearercampaign/torchbearerissuepapers/Documents/TBIP_022508.pdf.

122. Department of the Army, Field Manual 3-0 *Operations,* 2008, v–vi.

123. Ibid., vii–viii. This study establishes a different number of manuals than fifteen based upon how keystone doctrine is defined in the introduction.

124. Ibid., 1-1 to 1-21.

125. Ibid., 2-1 to 2-13.

126. Ibid., 3-1 to 3-22.

127. Ibid., 5-1 to 5-21.

128. Ibid., 6-1 to 6-19.

129. Ibid., 7-1 to 7-12.

130. Ibid., 8-1 to 8-7.

131. Ibid., A-1 to D-6; Source Notes-1; Glossary-1 to Glossary-15; References-1 to References-3; Index-1 to Index-16.

132. Edward B. Atkeson, "FM 3-0: An Assessment," *Army* 58, no. 8 (August 2008): 20–21.

SELECTED BIBLIOGRAPHY

ARCHIVES AND DOCUMENTS

National Archives, Washington, DC.

Microfilm M566, *Letters Received by the Office of the Adjutant General, 1805–1821.*

Microfilm M567, *Letters Received by the Office of the Adjutant General, 1855.*

Record Group 48, *Records of the Office of the Secretary of the Interior.*

Record Group 92, *Office of the Quartermaster General.*

Record Group 94, *Records of the Adjutant General.*

Record Group 98, *Records of the United States Army Commands.*

Record Group 107, *Office of the Secretary of War.*

Record Group 120, *Records of the American Expeditionary Forces.*

Record Group 165, *Records of the War Department General Staff.*

Record Group 177, *Records of the Chief of Arms.*

Record Group 287, *Publications of the United States Government.*

Record Group 319, *Army Staff.*

Record Group 337, *Headquarters Army Ground Forces.*

Record Group 338, *Records of the War Department.*

Record Group 394, *Records of the War Department.*

Record Group 407, *Records of the War Department.*

Battle Command Training Program Archives, Fort Leavenworth, KS.

Center for Army Lessons Learned (CALL), Fort Leavenworth, KS.

Combined Arms Research Library (CARL), Fort Leavenworth, KS.

Library of Congress, Washington, DC.

Special Collections, Northern Regional Library Facility, University of California, Richmond.

Special Collections, Library of the United States Military Academy, West Point, NY.

U.S. Army Heritage and Education Center, Military History Institute, Carlisle Barracks, PA.

U.S. GOVERNMENT, STATE, AND UN PUBLICATIONS

Alberts, David S., and Richard Hayes. *Command Arrangements for Peace Operations.* Washington, DC: Institute for National Strategic Studies, National Defense University, 1995.

American State Papers: Indian Affairs, 2 vols. Washington, DC: Gales & Seaton, 1832.

Appleman, Roy E. *South to the Naktong, North to the Yalu.* Washington, DC: U.S. Army Center of Military History, 1992.

Armed Forces Staff College Pub-1. *The Joint Staff Officers Guide*. Washington, DC: Government Printing Office, 1993.

The Armored School. *Final Report of Test—Armored Task Force Preparation—Exercise Desert Rock VI*, 1 August 1955. Declassified, Defense Technical Information Center (DTIC), Springfield, VA.

Army Historical Series. *American Military History*. Washington, DC: U.S. Army Center of Military History, 1989.

Bacevich, A. J. *The Pentomic Era: The U.S. Army between Korea and Vietnam*. Washington, DC: National Defense University Press, 1986.

Bell, William Gardner. *Commanding Generals and Chiefs of Staff, 1775–1995, Portraits and Biographical Sketches of the United States Army's Senior Officer*. Washington, DC: U.S. Army Center of Military History, 1997.

Bigelow, John. *Mars-la-Tour and Gravelotte*. Washington, DC: Government Printing Office, 1880.

Birtle, Andrew J. *U.S. Army Counterinsurgency and Contingency Operations Doctrine, 1860–1941*. Washington, DC: U.S. Army Center of Military History, 1998.

———. *U.S. Army Counterinsurgency and Contingency Operations Doctrine, 1942–1976*. Washington, DC: U.S. Army Center of Military History, 2006.

de Block, Jean (Ivan). *The Future of War*. Fort Leavenworth, KS: Combat Studies Institute, U.S. Army Command and General Staff College, 1991.

Blumenson, Martin. *Breakout and Pursuit*. Washington, DC: U.S. Army Center of Military History, 1993.

Boutros-Ghali, Boutros. "An Agenda for Peace: Preventive Diplomacy, Peacemaking and Peacekeeping." A Report of the Secretary-General Pursuant to the Statement Adopted by the Summit Meeting of the Security Council. New York: United Nations, 31 January 1992.

Center for Advanced Concepts and Technology. *Operations Other than War: The Technological Dimension*, 2nd printing. Washington, DC: National Defense University, 1997.

Chapman, Anne W. *The Army's Training Revolution, 1973–1990: An Overview*. Fort Monroe, VA: Office of the Command Historian, Headquarters Training and Doctrine Command (TRADOC), 1991.

———. *The Origins and Development of the National Training Center, 1976–1984*. Fort Monroe, VA: Office of the Command Historian, Headquarters TRADOC, 1992.

———, and Carol J. Lilly, John L. Romjue, and Susan Canedy. *Prepare the Army for War: A Historical Overview of the Army Training and Doctrine Command, 1973–1998*. Fort Monroe, VA: Military History Office, Headquarters TRADOC, 1998.

Coats, Stephen D. *Gathering at the Golden Gate: Mobilizing for War in the Philippines, 1898*. Fort Leavenworth, KS: Combat Studies Institute Press, 2006.

Collins, John M. *Special Operations Forces: An Assessment*. Washington, DC: National Defense University Press, 1994.

Daso, Dik Alan. *Hap Arnold and the Evolution of American Airpower.* Washington, DC: Smithsonian Institution Press, 2000.

Department of Defense. *Department of Defense Directive 5100.46.* Washington, DC: Government Printing Office, 4 December 1975.

———. *Department of Defense Directive 3000.05.* Washington, DC: Government Printing Office, 28 November 2005.

Department of the Army. Field Manual 1, *The Army*, Prototype Draft version "K." Washington, DC: Government Printing Office, 14 June 2000.

———. Field Manual 25-101. *Training the Force: Battle Focused Training.* Washington, DC: Government Printing Office, September 1990.

———. Field Manual 3-0. *Operations.* Washington, DC: Department of the Army, 2001.

———. Field Manual 3-0. *Operations.* Washington, DC: Department of the Army, 2008.

———. Field Manual 3-24 and Marine Corps Warfighting Manual 3-33.5. *Counterinsurgency.* Washington, DC: Department of the Army, 2006.

———. Field Manual 100-8. *The Army in Multinational Operations.* Washington, DC: Government Printing Office, November 1997.

———. Field Service Regulations. Field Manual 100-5 *Operations, August 1949.* Washington, DC: Government Printing Office, 1949.

———. Field Service Regulations. Field Manual 100-5 *Operations, September 1954 and Change 1, December 1954.* Washington, DC: Government Printing Office, 1954, Change 1 (1956), Change 2 (1958).

———. Field Service Regulations. Field Manual 100-5 *Operations, February 1962.* Washington, DC: Government Printing Office, 1962.

———. Field Manual 100-5. *Operations of Army Forces in the Field, 1968.* Washington, DC: Government Printing Office, 1968, Change 1 (1971).

———. Field Manual 100-5. *Operations, 1 July 1976 with Change 1, 29 April 1977.* Washington, DC: Government Printing Office, 1976, 1977.

———. Field Manual 100-5. *Operations, 1982.* Washington, DC: Government Printing Office, 1982.

———. Field Manual 100-5. *Operations, 1986.* Washington, DC: Government Printing Office, 1986.

———. Field Manual 100-5. *Operations, June 1993.* Washington, DC: Government Printing Office, 1993.

———. Field Manual 100-5. *Operations, Final Draft, 5 August 1997.* Washington, DC: Department of the Army, 1997.

———. Field Manual 100-5. *Operations, Revised Final Draft, 19 June 1998.* Washington, DC: Department of the Army, 1998.

———. *FM 100-5 Content Summary, 1 February 2000.* Fort Monroe, VA: Headquarters, TRADOC, 2000.

———. Field Manual 100-19 and Fleet Marine Force Manual 7-10. *Domestic Support Operations*. Washington, DC: Government Printing Office, July 1993.

———. Field Manual 100-23. *Peace Operations*. Washington, DC: Government Printing Office, December 1994.

———. Field Manual 100-23-1. *HA: Multiservice Procedures for Humanitarian Assistance Operations*. Fort Monroe, VA: Headquarters, TRADOC, October 1994.

———. Field Manual 100-25. *Doctrine for Army Special Operations Forces*. Washington, DC: Government Printing Office, 1991.

———. Historical Division. *United States Army in the World War, 1917–1919*, 19 vols. Washington, DC: Government Printing Office, 1948.

———. *Vietnam Studies: U.S. Army Special Forces, 1961–1971*. Washington, DC: Government Printing Office, 1985.

Department of State. "Urgent National Needs: Special Message of the President to the Congress." *Department of State Bulletin*, 12 June 1961.

Dorn, Edward, Howard D. Graves, and Walter F. Ulmer Jr. *American Military Culture in the Twenty-First Century*. Washington, DC: Center for Strategic Studies Press, 2000.

Doughty, Robert A. *The Evolution of U.S. Army Tactical Doctrine, 1946–76*. Fort Leavenworth, KS: Combat Studies Institute, U.S. Army Command and General Staff College, 1979.

Drum, Hugh A. "Annual Report of the Director of the School of the Line, 1919–1920." In *Annual Report of the Commandant of the General Service Schools, 1920*. Fort Leavenworth, KS: General Service Schools Press, 1920.

Fishel, John T. *Liberation, Occupation, and Rescue: War Termination and Desert Storm*. Carlisle Barracks, PA: Strategic Studies Institute, 31 August 1992.

Fontenot, Gregory, E. J. Degen, and David Tohn. *On Point: The United States Army in Operation Iraqi Freedom*. Fort Leavenworth, KS: Combat Studies Institute Press, 2004.

Frisbee, John L., ed. *Makers of the United States Air Force*. Washington, DC: Office of Air Force History, 1987.

Gabel, Christopher R. *Seek, Strike and Destroy: U.S. Army Tank Destroyer Doctrine in World War II*. Fort Leavenworth, KS: Combat Studies Institute, U.S. Army Command and General Staff College, 1985.

———. *The U.S. Army GHQ Maneuvers of 1941*. Washington, DC: U.S. Army Center of Military History, 1991.

Headquarters 10th Mountain Division (Light). *Operations in Haiti: Planning/Preparation/Execution, August 1994 thru January 1995*. Fort Drum, NY: Headquarters 10th Mountain Division (Light), 1995.

Headquarters, U.S. Army Europe. *Operation Joint Endeavor After Action Report*, 2 vols. Heidelberg, Germany: Headquarters, U.S. Army Europe and Seventh Army, May 1997.

Herbert, Paul H. *Deciding What Has to Be Done: General William E. DePuy and the 1976 Edition of* FM 100-5 *Operations*. Fort Leavenworth, KS: Combat Studies Institute, U.S. Army Command and General Staff College, 1988.

Hermes, Walter G. *Truce Tent and Fighting Front*. Washington, DC: U.S. Army Center of Military History, 1992.

Hirsch, John L., and Robert B. Oakley. *Somalia and Operation Restore Hope: Reflections on Peacemaking and Peacekeeping*. Washington, DC: U.S. Institute for Peace Press, 1995.

House, Jonathan M. "Toward Combined Arms Warfare: A Survey of 20th-Century Tactics, Doctrine, and Organization." Fort Leavenworth, KS: Combat Studies Institute, U.S. Army Command and General Staff College, 1984.

Huston, James A. *The Sinews of War: Army Logistics, 1775–1953*. Washington, DC: Office of the Chief of Military History, 1966.

Ippolito, Dennis S. *Blunting the Sword: Budget Policy and the Future of Defense*. Washington, DC: Institute for National Strategic Studies, National Defense University Press, 1994.

Joint Chiefs of Staff. *Organizational Development of the Joint Chiefs of Staff, 1942–1989*. Washington, DC: Historical Division, Joint Secretariat, Joint Chiefs of Staff, 1989.

Joint Chiefs of Staff. *Department of Defense Dictionary of Military and Associated Terms*. Joint Pub 1-02. Washington, DC: Joint Chiefs of Staff, 23 March 1994.

Kretchik, Walter E., Robert F. Baumann, and John T. Fishel. *Invasion, Intervention, "Intervasion": A Concise History of the U.S. Army in Operation Uphold Democracy*. Fort Leavenworth, KS: U.S. Army Command and General Staff College Press, 1998.

Kriedberg, Marvin A., and Merton G. Henry. *History of Military Mobilization in the United States Army, 1775–1945*. Washington, DC: Department of the Army, 1955.

Laurie, Clayton D., and Ronald H. Cole. *The Role of Federal Military Forces in Domestic Disorders, 1877–1945*. Washington, DC: U.S. Army Center of Military History, 1997.

Libicki, Martin C. *What Is Information Warfare?* Washington, DC: Institute For National Strategic Studies, National Defense University, 1995.

MacGarrigle, George L. *Taking the Offensive: October 1966 to October 1967*. Washington, DC: U.S. Army Center of Military History, 1998.

MacIssac, David. "The Air Force and Strategic Thought, 1945–51." International Security Studies Program Working Paper no. 8. Washington, DC: Wilson Center, June 1979.

McClendon, R. Earl. *Autonomy of the Air Arm*. Washington, DC: Government Printing Office, Air University, 1993.

Mendel, William W., and David G. Bradford. *Interagency Cooperation: A Regional Model for Overseas Operations*. Washington, DC: Institute for National Strategic Studies, National Defense University, 1995.

Ney, Virgil. *Evolution of the United States Army Field Manual: Valley Forge to Vietnam.* Fort Belvoir, VA: Combat Operations Research Group, 1966.

North Atlantic Treaty Organization, Military Agency for Standardization. *Land Forces Tactical Doctrine,* Change 5. Allied Tactical Publication 35 (A). Mons, Belgium: Supreme Headquarters Allied Powers Europe, 5 December 1990; Department of Joint and Combined Operations, U.S. Army Command and General Staff College.

Oakley, Robert B., Michael J. Dziedzic, and Eliot M. Goldberg, eds. *Policing the New World Disorder: Peace Operations and Public Security.* Washington, DC: National Defense University Press, 1998.

Parton, James. *"Air Force Spoken Here": General Ira Eaker and the Command of the Air.* Maxwell, AL: Air University Press, 2000.

"Project Attack: Hypothetical Use of A-Bombs on Massed Armor Illustrated by Operation 'Totalize,' Phase I, 7-8 August 1944." Operations Research Office, Johns Hopkins University, Baltimore, MD, 1944.

Ramsey, Robert D., III. *Savage Wars of Peace: Case Studies of Pacification in the Philippines, 1900–1902.* Fort Leavenworth, KS: Combat Studies Institute Press, 2008.

Romjue, John L. *From Active Defense to AirLand Battle: The Development of Army Doctrine, 1973–1982.* Fort Monroe, VA: United States Army Training and Doctrine Command, 1984.

———. *American Army Doctrine for the Post Cold War.* Fort Monroe, VA: Military History Office, United States Army Training and Doctrine Command, 1996.

Scales, Robert H. *Certain Victory: The U.S. Army in the Gulf War.* Washington, DC: Office of the Chief of Staff, U.S. Army, 1993.

Schubert, Frank N., and Theresa L. Kraus, eds. *The Whirlwind War: The United States Army in Operations Desert Shield and Desert Storm.* Washington, DC: U.S. Army Center of Military History, 1995.

Sorley, Lewis, ed., *Press On! Selected Works of General Donn A. Starry,* 2 vols. Fort Leavenworth, KS: Combat Studies Institute, U.S. Army Combined Arms Center, 2009.

Stewart, Richard W., gen. ed. *American Military History: The United States Army and the Forging of a Nation, 1775–1917,* vol. 1. Washington, DC: U.S. Army Center of Military History, 2005.

Swain, Richard M., comp. *Selected Papers of General William E. DePuy.* Fort Leavenworth, KS: Combat Studies Institute, U.S. Army Command and General Staff College, 1994.

———. *"Lucky War": Third Army in Desert Storm.* Fort Leavenworth, KS: U.S. Army Command and General Staff College Press, 1994.

Thibault, George Edward, ed. *The Art and Practice of Military Strategy.* Washington, DC: National Defense University, 1984.

Thomas, Robert S., and Inez V. Allen. "The Mexican Punitive Expedition under

Brigadier General John J. Pershing, United States Army, 1916–1917, Chapters
I–V." Washington, DC: War Histories Division, Office of the Chief of Military
History, 1954.

Tunnell, Major Harry D., IV. *To Compel with Force: A Staff Ride Handbook for the
Battle of Tippecanoe*. Fort Leavenworth, KS: Combat Studies Institute Press,
2000.

UN Security Council. *United Nations Security Council Resolution 688*. New York:
United Nations, 5 April 1991.

U.S. Army Command and General Staff College. *Student Text 3-0, Operations, 1
October 2000*. Fort Leavenworth, KS: U.S. Army Command and General Staff
College, 2000.

U.S. Army Peacekeeping Institute. *Success in Peacekeeping: United Nations Mission
in Haiti: The Military Perspective*. Carlisle Barracks, PA: U.S. Army Peacekeeping
Institute, 1996.

U.S. Code Annotated. *Title 10 Armed Forces*, Sections 3001–6480. West Group
Press, 1998.

U.S. Military Academy. "History of West Point." West Point, NY: U.S. Military
Academy Catalog, 1965.

Upton, Emory. *The Military Policy of the United States*, Joseph P. Sanger, William
D. Beach, and Charles D. Rhodes, eds. Washington, DC: Government Printing
Office, 1904.

———. *Military Policy of the United States*. Washington, DC: Government Printing
Office, 1917.

Vetock, Dennis J. "Lessons Learned: A History of U.S. Army Lesson Learning."
Carlisle Barracks, PA: U.S. Army Military History Institute, 1988.

War Department. *Abstract of Infantry Tactics; including exercises and manoeuvers of
light-infantry and riflemen; for use of the militia of the United States. Published by
the Department of War; under the Authority of an Act of Congress of the 2d of March,
1829*. Boston: Hillard, Gray, Little, & Wilkins, 1830.

———. *Annual Reports of the Chief of Staff, 1920–1941*. Washington, DC:
Government Printing Office, 1920–1941.

———. *By Authority Infantry Tactics; or Rules for the Exercise and Manoeuvers of the
Infantry of the U.S. Army*. Washington, DC: Davis & Force, 1825.

———. *By Authority Infantry Tactics; or Rules for the exercise and manoeuvers of the
United States Infantry. By Major-General Scott, U.S. Army*. New York:
G. Dearborn, 1835.

———. *By Authority Infantry Tactics; or Rules for the exercise and manoeuvers of the
United States Infantry. By Major-General Scott, U.S. Army*, 3 vols. New York:
Harper & Bros., 1840.

———. *By Authority Infantry Tactics; or Rules for the exercise and manoeuvers of the
United States Infantry. New Edition*, 3 vols. *By Major-General Scott, U.S. Army*.
New York: Harper & Brothers, 1846.

————. *Cavalry Tactics or Regulations for the Instruction, Formations, and Movements of the Cavalry of the Army and Volunteers of the United States*. Philadelphia, PA: J. B. Lippincott, 1862.

————. *Extracts from the Report of Major General Adna R. Chaffee, Commanding United States Troops in China on Military Operations in China*. Washington, DC: Government Printing Office, 1900.

————. *Infantry Drill Regulations, United States Army*. Washington, DC: Government Printing Office, 1891. New York: D. Appleton, 1891, 1892, 1893, 1895, and 1898.

————. *Regulations of the Army of the United States, 1889*. Washington, DC: Government Printing Office, 1889.

————. *Report of the Commission Appointed by the President to Investigate the Conduct of the War Department in the War with Spain*, 8 vols. Washington, DC: Government Printing Office, 1900.

————. *Reports of Military Observers Attached to the Armies in Manchuria during the Russo-Japanese War*. Washington, DC: Government Printing Office, 1907.

————. *Rules and Regulations For the Field Exercise and Manoeuvers of Infantry Compiled and adapted to the Organization of the Army of the United States, Agreeably to A resolve of Congress, Dated December 1814. Published by order of the War Department*. New York: T. & W. Mercein, 1815.

————. *Troops in Campaign: Regulations for the Army of the United States*. Washington, DC: Government Printing Office, 1892.

————. General Staff, United States Army. *Field Service Regulations, 1905 with amendments to 1908*. Washington, DC: Government Printing Office, 1905, 1908.

————. *Field Service Regulations, U.S. Army, 1910*. Washington, DC: Government Printing Office, 1910.

————. *Field Service Regulations, U.S. Army, 1913, with Corrections to May 21, 1913*. Washington, DC: Government Printing Office, 1913.

————. *Field Service Regulations, U.S. Army, 1914*. Washington, DC: Government Printing Office, 1914.

————. *Field Service Regulations, United States Army (Including Questions), 1914 with Text Corrections to August 18, 1917 (Changes No. 7)*. Banta's Special Edition. Menasha, WI: Collegiate Press, George Banta Publishing, 1918.

————. *Field Service Regulations, United States Army, 1914, Corrected to July 31, 1918 (Changes Nos. 1 to 11)*. Washington, DC: Government Printing Office, 1918.

————. *Field Service Regulations, United States Army, 1923*. Washington, DC: Government Printing Office, 1924.

————. *Tentative Field Service Regulations FM 100-5 Operations*. Washington, DC: Government Printing Office, 1939.

————. *Field Service Regulations Field Manual 100-5 Operations, 1941*. Washington, DC: Government Printing Office, 1941.

———. *Field Service Regulations* Field Manual 100-5 *Operations*, 15 June 1944. Washington, DC: Government Printing Office, 1944.

———. *The Infantry Exercise of the United States Army, abridged, for the use of the militia of the United States*. Poughkeepsie, NY: Printed and published by P. Potter, for himself, and for S. Potter & Co., Philadelphia, PA, 1817.

Washington, George. *The Writings of George Washington, 1745–1799*, ed. John C. Fitzpatrick, 39 vols. Washington, DC: U.S. Government Printing Office, 1931–1944.

Wentz, Larry, ed. *Lessons from Bosnia: The IFOR Experience*. Washington, DC: Command and Control Research Program by National Defense University Press, 1997.

West, Togo D., Jr., and General Gordon R. Sullivan. *Decisive Victory: America's Power Projection Army*. Washington, DC: U.S. Government Printing Office, 1994.

———. *Force XXI: Meeting the 21st Century Challenge*. Washington, DC: Government Printing Office, January 1995.

The White House. *National Security Strategy of the United States*. Washington, DC: Government Printing Office, March 1990.

———. *The National Security Strategy of the United States, 1991*. Washington, DC: Government Printing Office, August 1991.

———. *National Security Strategy of the United States, January 1993*. Washington, DC: Government Printing Office, 1993.

———. *A National Security Strategy of Enlargement and Engagement, February 1996*. Washington, DC: Government Printing Office, 1996.

———. *A National Security Strategy for a New Century, May 1997*. Washington, DC: Government Printing Office, 1997.

Wilson, John B. *Maneuver and Firepower: The Evolution of the Division and Separate Brigade*. Washington, DC: U.S. Army Center of Military History, 1997.

Wolk, Herman S. *The Struggle for Air Force Independence, 1943–1947*, rev. ed. Washington, DC: Government Printing Office, Air University, 1997.

World War Records, First Division, A.E.F., Regular. Washington, DC: Society of the First Division, 1928–1930.

Wright, Donald P., Timothy R. Reese, with the Contemporary Operations Study Team. *On Point II, Transition to the New Campaign: The United States Army in Operation Iraqi Freedom, May 2003–January 2005*. Fort Leavenworth, KS: Combat Studies Institute Press, 2008.

Wright, Robert K., Jr. *The Continental Army*. Washington, DC: U.S. Army Center of Military History, 1983.

Yates, Lawrence A. *Power Pack: U.S. Intervention in the Dominican Republic, 1965–1966*. Fort Leavenworth, KS: Combat Studies Institute, U.S. Army Command and General Staff College, 1988.

BOOKS

Abbot, W. W., et al., eds. *The Papers of George Washington*. Presidential series. 2 vols. Charlottesville: University of Virginia Press, 1983.

Adams, Thomas K. *U.S. Special Operations Forces in Action: The Challenge of Unconventional Warfare*. Portland: Frank Cass, 1998.

Addington, Larry H. *The Patterns of War since the Eighteenth Century*, 2nd ed. Bloomington: Indiana University Press, 1994.

Adelman, Jonathan R. *Prelude to the Cold War: The Tsarist, Soviet, and U.S. Armies in the Two World Wars*. Boulder, CO: Lynne Reimer Publishers, 1988.

Adkins, Mark. *Urgent Fury: The Battle for Grenada*. Lexington, MA: D. C. Heath, 1989.

Alger, R. A. *The American-Spanish War*. Norwich, CT, 1899.

Allison, William T., Jeffrey Grey, and Janet G. Valentine. *American Military History: A Survey from Colonial Times to the Present*. Upper Saddle River, NJ: Pearson Prentice Hall, 2007.

Ambrose, Stephen. *Upton and the Army*. 1964. Reprint, Baton Rouge: Louisiana State University Press, 1992.

Ardant du Picq, Charles J. "Battle Studies." In *Roots of Strategy*, book 2. Harrisburg, PA: Stackpole Books, 1987.

Badger, Anthony J. *The New Deal: The Depression Years, 1933–1940*. New York: Hill & Wang, 1989.

Bailyn, Bernard, ed. *Pamphlets of the American Revolution*. Cambridge, MA: Harvard University Press, 1965.

———. *The Ideological Origins of the American Revolution*. Cambridge, MA: Belknap Press of Harvard University Press, 1967.

———. *Voyages to the West: A Passage in the Peopling of America on the Eve of Revolution*. New York: Vintage Books, 1986.

Ballard, John R. *Upholding Democracy: The United States Military Campaign in Haiti, 1994–1997*. Westport, CT: Praeger, 1998.

Bank, Aaron. *From OSS to Green Berets*. New York: Pocket Books, 1986.

Barbuto, Richard V. *Niagara 1814: America Invades Canada*. Lawrence: University Press of Kansas, 2000.

Barnett, Correlli. *Britain and Her Army, 1509–1970: A Military, Political, and Social Survey*. London: Allen Lane; Penguin Press, 1970.

Bauer, K. Jack. *The Mexican War, 1846–1848*. Lincoln: University of Nebraska Press, 1974.

Baxter, D. W. C. *Baxter's Volunteer's Manual: Containing Full Instructions for the Recruit, in the Schools of the Soldier and Squad, with One Hundred Illustrations of the Different Positions in the Facings and Manual of Arms and the Loadings and Firings: Arranged According to Scott's System of Infantry Tactics*. Philadelphia: King & Baird, 1861.

Baxter, William P. *Soviet AirLand Battle Tactics*. Novato, CA: Presidio Press, 1986.

Bernardo, C. Joseph, and Eugene H. Bacon. *American Military Policy: Its Development since 1775*. Harrisburg, PA: Stackpole Co., 1957.

Bigler, Philip. *Hostile Fire: The Life and Death of First Lieutenant Sharon Lane*. Arlington, VA: Vandamere Press, 1996.

Blair, Clay. *The Forgotten War: America in Korea, 1950–1953*. New York: Anchor Books, 1987.

Black, Jeremy. *America as a Military Power: From the American Revolution to the Civil War*. Westport, CT: Praeger Publishers, 2002.

Bland, Humphrey. *Treatise of Military Discipline: in which is laid down and explained the duty of the officer and soldier, through the several branches of the service*. London: Printed for R. Baldwin, 1759.

Blaufarb, Douglas S. *The Counterinsurgency Era: U.S. Doctrine and Performance 1950 to the Present*. New York: Free Press, 1977.

Bolger, Daniel P. *Americans at War, 1975–1986: An Era of Violent Peace*. Novato, CA: Presidio Press, 1988.

———. *Savage Peace: Americans at War in the 1990s*. Novato, CA: Presidio Press, 1995.

Bowden, Mark. *Blackhawk Down: A Story of Modern War*. New York: Atlantic Monthly Press, 1999.

Boyne, Walter J. *Beyond the Wild Blue: A History of the U.S. Air Force*. New York: St. Martin's Griffon, 1998.

Bradford, Zeb B., and Frederick J. Brown. *The United States Army in Transition*. Beverly Hills, CA: Sage, 1973.

Brereton, Todd R. *Educating the Army: Arthur L. Wagner and Reform, 1875–1905*. Lincoln: University of Nebraska Press, 2000.

Bridges, Eldridge S. *The Story of Our War with Spain*. Boston: Lothrop Publishing, 1899.

Brodie, Bernard, ed. *The Absolute Weapon*. New York: Harcourt Brace, 1946.

———. *Strategy in the Missile Age*. Princeton, NJ: Princeton University Press, 1959.

———, and Fawn M. Brodie. *From Crossbow to H-Bomb: The Evolution of the Weapons and Tactics of Warfare*, Midland ed, Bloomington: Indiana University Press, 1973.

Brown, Jerold E., ed. *Historical Dictionary of the U.S. Army*. Westport, CT: Greenwood Press, 2001.

Budiansky, Stephen. *Air Power: The Men, Machines, and Ideas That Revolutionized War, from Kitty Hawk to Gulf War II*. New York: Viking, 2004.

Canfield, Bruce N. *U.S. Infantry Weapons of World War II*. Lincoln, RI: Andrew Mowbray Publishers, 1996.

Carp, E. Wayne. *To Starve the Army at Pleasure: Continental Army Administration and American Political Culture, 1775–1783*. Chapel Hill: University of North Carolina Press, 1984.

Carver, Michael, ed. *The War Lords: Military Commanders of the Twentieth Century*. Boston: Little, Brown, 1976.

Casey, Silas. *Infantry Tactics for the Infantry, Exercise, and Maneuvers of the Soldier, a Company, Line of Skirmishers, Battalion, Brigade, or Corps d'Armee.* New York: D. Van Nostrand, 1862.

Catton, Bruce. *Never Call Retreat.* New York: Doubleday, 1961.

Chambers, John Whiteclay, II, ed. *The Oxford Companion to American Military History.* New York: Oxford University Press, 1999.

Chambers, John Whiteclay, II, and G. Kurt Piehler, eds. *Major Problems in American Military History.* Boston: Houghton Mifflin, 1999.

Chaliand, Gerard, ed. *A People without a Country,* trans. Michael Pallis. Brooklyn, NY: Olive Branch Press, 1993.

de Chanal, Francois V. A. *The American Army in the War of Secession.* Leavenworth, KS: G. A. Spooner, 1894.

Chandler, David, ed. *The Dictionary of Battles: The World's Key Battles from 405 BC to Today.* New York: Henry Holt, 1987.

Chapman, Bert. *Military Doctrine: A Reference Handbook.* Santa Barbara, CA: Praeger Security International, 2009.

Chartrand, René. *The Mexican Adventure, 1861–1867.* Oxford: Osprey Publishing, 1994.

Cheng, Christopher C. S. *Air Mobility: The Development of Tactical Doctrine.* Westport, CT: Praeger, 1994.

Church, Benjamin. *Diary of King Philip's War, 1675–76,* ed. Alan and Mary Simpson. Chester, CT: Pequot Press, 1975.

Church, Thomas. *The History of the Great Indian War of 1675 and 1676, Commonly Called Philip's War,* ed. Samuel G. Drake. Reprint, Whitefish, MT: Kessinger Publishing, 2007.

Citino, Robert M. *Blitzkrieg to Desert Storm: The Evolution of Operational Warfare.* Lawrence: University Press of Kansas, 2004.

Clausewitz, Carl von. *On War,* ed. and trans. Michael Howard and Peter Paret. Princeton, NJ: Princeton University Press, 1984.

Clark, Wesley. *Waging Modern War.* New York: Public Affairs, 2001.

Closen, Ludwig von. *The Revolutionary Journal of Baron Ludwig von Closen, 1780–1783,* ed. and trans. from the French by Evelyn M. Acomb. Chapel Hill: University of North Carolina Press, 1958.

Cooper, Samuel. *A Concise System of Instructions and Regulations for the militia and volunteers of the United States: comprehending the exercises and movements of the infantry, light infantry, and riflemen, cavalry and artillery: together with the manner of doing duty in garrison and in camp, and the forms of parades, reviews, and inspections, as established by authority for the government of the regular army.* Philadelphia: Robert P. Desilver, 1836.

Coffman, Edward M. *The Regulars: The American Army, 1898–1941.* Cambridge, MA: Belknap Press of Harvard University Press, 2004.

————. *The Old Army: A Portrait of the American Army in Peacetime, 1784–1898.* New York: Oxford University Press, 1986.

————. *The War to End All Wars: The American Military Experience in World War I.* Madison: University of Wisconsin Press, 1986.

Cohen, Stan. *Images of the Spanish-American War: April–August 1898.* Missoula, MT: Pictorial Histories Publishing Co., 1997.

Col. George Rogers Clark's Sketch of His Campaign in the Illinois in 1778–9 with an Introduction By Hon. Henry Pirtle, of Louisville and an Appendix containing The Public and Private Instructions to Col. Clark and Major Bowman's Journal of the Taking of Post St. Vincent. Reprint ed. Bedford, MA: Applewood Books, 2001.

Commager, Henry Steele, and Richard B. Morris, eds. *The Spirit of "Seventy-Six": The Story of the American Revolution as Told by Participants.* 2 vols. Indianapolis: Bobbs-Merrill, 1958.

Cooper, Jerry. *The Rise of the National Guard: The Evolution of the American Militia, 1865–1920.* Lincoln: Bison Books, 2002.

Corum, James S. *The Roots of Blitzkrieg: Hans von Seeckt and German Military Reform.* Lawrence: University Press of Kansas, 1992.

Cosmas, Graham A. *An Army for Empire: The United States Army in the Spanish American War.* College Station: Texas A&M University Press, 1994.

Cozzens, Peter. *Eyewitnesses to the Indian Wars, 1865–1890.* 3 vols. Mechanicsburg, PA: Stackpole Books, 2002.

Craig, Gordon A., and Alexander L. George. *Force and Statecraft: Diplomatic Problems of Our Time.* New York: Oxford University Press, 1983.

Davis, Richard Harding. *The Cuban and Porto Rican Campaigns.* New York: Privately published, 1898.

————. *Notes of a War Correspondent.* New York: Charles Scribner's Sons, 1914.

Dawisha, Karen, and Bruce Parrott. *Russia and the New States of Eurasia: The Politics of Upheaval.* Cambridge: Cambridge University Press, 1994.

Devlin, Gerard M. *Paratrooper! The Saga of Army and Marine Parachute and Glider Combat Troops during World War II.* New York: St. Martin's Press, 1979.

Donnelly, Thomas, Margaret Roth, and Caleb Baker. *Operation Just Cause: The Storming of Panama.* New York: Lexington Books, 1991.

Doughty, Robert A. *The Seeds of Disaster: The Development of French Army Doctrine, 1919–1939.* Hamden, CT: Archon Books, 1985.

————, and Ira D. Gruber. *American Military History and the Evolution of Western Warfare.* Lexington, MA: D. C. Heath, 1996.

Duane, William. *Military Dictionary, or, Explanation of the Several Systems of Discipline of Different Kinds of Troops, Infantry, Artillery, and Cavalry; The Principles of Fortification, and All the Modern Improvements in the Science of Tactics; Comprising the Pocket Gunner, or Little Bombardier; The Military Regulations of the United States; The Weights, Measures, and Monies of all Nations; The technical*

Terms and Phrases of the Art of War in the French Language, Particularly Adapted to the Use of the Military Institutions of the United States. Philadelphia: William Duane Printer and Publisher, 1810.

———. *Regulations to Be Received and Observed for the Discipline of Infantry in the Army of the United States.* Philadelphia: Printed for the Author, 1814.

Duffy, Christopher. *Frederick the Great: A Military Life.* New York: Routledge, 1993.

Dunnigan, James F., and Raymond M. Macedonia. *Getting It Right: American Military Reforms after Vietnam to the Gulf War and Beyond.* New York: William Morrow, 1993.

Dupuy, R. Ernest. *The Compact History of the United States Army.* New York: Hawthorn, 1956.

———, and William H. Baumer. *The Little Wars of the United States.* New York: Hawthorn Books, 1968.

Durch, William J., ed. *UN Peacekeeping, American Policy, and the Uncivil Wars of the 1990s.* New York: Henry L. Stimson Center; St. Martin's Press, 1996.

Ebert, James R. *A Life in a Year: The American Infantryman in Vietnam, 1965–1972.* Novato, CA: Presidio Press, 1993.

Elderkin, James D. *Biographical Sketches and Anecdotes of a Soldier of Three Wars.* Detroit: Record Printing, 1899.

Ellis, Joseph J. *His Excellency George Washington.* New York: Vintage Books, 2004.

Esposito, Vincent J., general ed. *The West Point Atlas of American Wars.* 2 vols. New York: Henry Holt, 1995.

von Ewald, Johann. *Abhandlung Über den Kleinen Krieg* (Treatise on Partisan Warfare). Cassel: J. J. Cramer, 1785.

Fehrenbach, T. R. *This Kind of War: A Study in Unpreparedness.* New York: Macmillan, 1963.

Finlayson, Kenneth. *An Uncertain Trumpet: The Evolution of U.S. Army Infantry Doctrine, 1919–1941.* Westport, CT: Greenwood Press, 2001.

Fishel, John T., ed. *"The Savage Wars of Peace": Toward a New Paradigm of Peace Operations.* Boulder: Westview Press, 1998.

Foster, John W. *American Diplomacy in the Orient.* Boston: Houghton Mifflin, 1903.

Frey, Sylvia R. *The British Soldier in America: A Social History of Military Life in the Revolutionary Period.* Austin: University of Texas Press, 1981.

Friedman, Norman. *Desert Victory: The War For Kuwait,* updated ed. Annapolis, MD: Naval Institute Press, 1992.

Gaddis, John Lewis. *We Now Know: Rethinking Cold War History.* Oxford: Oxford University Press, 1997.

Gaff, Alan D. *Bayonets in the Wilderness: Anthony Wayne's Legion in the Old Northwest.* Norman: University of Oklahoma Press, 2004.

Gailey, Harry. *MacArthur Strikes Back: Decision at Buna, New Guinea, 1942–1943.* Novato, CA: Presidio, 2000.

Galula, David. *Counterinsurgency Warfare: Theory and Practice.* New York: Praeger, 1964.

Ganoe, William A. *The History of the United States Army.* New York: D. Appleton & Co., 1924.

Gardner, Charles K. *A Dictionary of All Officers Who Have Been Commissioned, or Have Been Appointed and Served in the Army of the United States, Since the Inauguration of Their First President, In 1789, To The First January, 1853—With Every Commission of each;—Including The Distinguished Officers of the Volunteers and Militia of the States, Who Have Served In Any Campaign, or Conflict With an Enemy, Since That Date; and of the Navy and Marine Corps, Who Have Served With the Land Forces: Indicating The Battle, In Which Every Such Officer Has Been Killed, or Wounded,—and the Special Words of Every Brevet Commission,* 2nd ed. New York: D. Van Nostrand, 1860.

Gates, John Morgan. *Schoolbooks and Krags: The United States Army in the Philippines, 1898–1902.* Westport, CT: Greenwood Press, 1973.

Gavin, James M. *War and Peace in the Space Age.* New York: Harper & Row, 1958.

Gawrych, George W. *The Albatross of Decisive Victory: War and Policy between Egypt and Israel in the 1967 and 1973 Arab-Israeli Wars.* Westport, CT: Greenwood Press, 2000.

Geertz, Clifford. *The Interpretation of Cultures.* New York: Basic Books, 1973.

Gimbel, John. *The American Occupation of Germany: Politics and the Military, 1945–1949.* Stanford, CA: Stanford University Press, 1968.

Ginsburgh, Robert N. *U.S. Military Strategy in the Sixties.* New York: Norton, 1965.

Gilpin, Robert. *War and Change in World Politics,* reprint ed. New York: Cambridge University Press, 1985.

Gordon, Michael R., and Bernard E. Trainor. *The General's War.* New York: Little, Brown, 1995.

Grant, Ulysses S. *Personal Memoirs of U.S. Grant.* 2 vols. New York: Charles L. Webster & Co., 1886.

Greene, Jack P., and J. R. Pole, eds. *Colonial British America: Essays in the New History of the Early Modern Era.* Baltimore: Johns Hopkins University Press, 1984.

Grenier, John. *The First Way of War: American War Making on the Frontier.* New York: Cambridge University Press, 2005.

Griffith, Paddy. *Forward into Battle: Fighting Tactics from Waterloo to the Near Future.* Novato, CA: Presidio Press, 1992.

———. *Battle Tactics of the Civil War.* New Haven, CT: Yale University Press, 2001.

Guderian, Heinz. *Panzer Leader,* trans. Constantine Fitzgibbon. Washington, DC: Zenger Publishing, 1979.

Hagerman, Edward. *The American Civil War and the Origins of Modern Warfare: Ideas, Organization, and Field Command.* Bloomington: Indiana University Press, 1992.

Halleck, Henry Wagner. *Elements of Military Art and Science: or, Course of Instruction in Strategy, Fortification, Tactics of Battles &c. Embracing the distance of staff,* infantry, cavalry, artillery and engineers adapted to the use of volunteers and militia. New York: B. Appleton & Co., 1846, 1862.

Halliday, E. M. *The Ignorant Armies.* New York: Harper Bros., 1960.

Hammel, Eric. *The Root: The Marines in Lebanon, August 1982–February 1984.* San Diego: Harcourt Brace Jovanovich, 1985.

Hardee, William J. *Rifle and Light Infantry Tactics for the Exercise of Maneuvers of troops when acting as light infantrymen or riflemen.* 2 vols. Philadelphia: Lippincott, Grambo, & Co., 1855.

Harper's Pictorial History of the War with Spain with an Introduction by Maj-Gen Nelson A. Miles, Commanding United States Army, in Two Volumes. New York: Harper & Brothers, 1899.

Hassler, Warren W., Jr. *General George B. McClellan, Shield of the Union.* Baton Rouge: Louisiana State University Press, 1957.

Hastings, Max. *The Korean War.* New York: Touchstone Books, 1987.

Hattaway, Hermann, and Archer Jones. *How the North Won: A Military History of the Civil War.* Champaign: University of Illinois Press, 1991.

Hayashi, Saburo. *KoGun: The Japanese Army in the Pacific War.* Reprint ed. Quantico, VA: Marine Corps Association, 1959.

Heidler, David S., and Jeanne T. Heidler. *The War of 1812.* Westport, CT: Greenwood Press, 2002.

Heller, Charles E., and William A. Stofft, eds. *America's First Battles, 1776–1965.* Lawrence: University Press of Kansas, 1986.

Herring, George C. *America's Longest War: The United States and Vietnam, 1950–1975,* 2nd ed. New York: Alfred A. Knopf, 1986.

Hersey, Paul, Kenneth H. Blanchard, and Dewey E. Johnson. *Management of Organizational Behavior: Utilizing Human Resources.* Upper Saddle River, NJ: Prentice Hall, 1996.

Herzog, Chaim. *The War of Atonement, October, 1973: The Fateful Implications of the Arab-Israeli Conflict.* Boston: Little, Brown, 1975.

Hess, Earl J. *The Union Soldier in Battle: Enduring the Ordeal of Combat.* Lawrence: University Press of Kansas, 1997.

Hickey, Donald R. *The War of 1812, a Forgotten Conflict.* Urbana: University of Illinois Press, 1990.

Higginbotham, Don. *The War of American Independence: Military Attitudes, Policies, and Practice, 1763–1789.* Boston: Northeastern University Press, 1983.

Hilsman, Roger. *To Move a Nation.* New York: Doubleday, 1967.

Hixson, Walter L. *Parting the Curtain: Propaganda, Culture and the Cold War, 1945–1961.* New York: St. Martin's Griffin, 1998.

House, Jonathan. *Combined Arms Warfare in the Twentieth Century.* Lawrence: University Press of Kansas, 2001.

Howard, Michael. *War in European History.* Oxford: Oxford University Press, 1976.

----. *The Franco-Prussian War: The German Invasion of France, 1870–1871.* London: Methuen, 1981.

Huntington, Samuel P. *The Soldier and the State: The Theory and Politics of Civil-Military Relations.* New York: Vintage Books, 1957.

Hurley, Alfred F. *Billy Mitchell, Crusader for Air Power.* New ed. Bloomington: Indiana University Press, 1975.

Hurst, James W. *Pancho Villa and Black Jack Pershing: The Punitive Expedition in Mexico.* Westport, CT: Praeger, 2008.

Huse, Edgar. *Organizational Development and Change.* San Francisco: West Publishing, 1988.

Jacobs, James Ripley. *The Beginning of the U.S. Army, 1783–1812.* Princeton, NJ: Princeton University Press, 1947.

Jamieson, Milton. *Journal and Notes of a Campaign in Mexico: Containing a History of Company C, of the Second regiment of Ohio Volunteers.* Cincinnati: 1849.

Jamieson, Perry D. *Crossing the Deadly Ground, United States Army Tactics, 1865–1899.* Tuscaloosa: University of Alabama Press, 1994.

Jefferson, Thomas. *The Jefferson Cyclopedia: A Comprehensive Collection of the Views of Thomas Jefferson*, ed. John P. Foley. New York: Funk & Wagnalls, 1900.

Jessup, Philip C. *Elihu Root.* New York: Dodd, Mead, 1938. Reprint, New York: Archon Books, 1964.

Jomini, Baron Antoine Henri de. *The Art of War*, Greenhill ed. Mechanicsburg, PA: Stackpole Books, 1996.

Jordan, Amos A., and William J. Taylor Jr. *American National Security: Policy and Process*, rev. ed. Baltimore: Johns Hopkins University Press, 1984.

Johnson, Lawrence H., III. *Winged Sabers: The Air Cavalry in Vietnam, 1965–1973.* Harrisburg, PA: Stackpole Books, 1990.

Johnson, Robert Underwood, and Clarence Clough Buell. *Battles and Leaders of the Civil War: Being for the Part Contributions by Union and Confederate Officers.* New York: Castle Books, 1956.

Jones, Archer. *The Art of War in the Western World.* New York: Oxford University Press, 1987.

----. *Civil War Command and Strategy: The Process of Victory and Defeat.* New York: Free Press, 1992.

Jones, Kenneth J. *The Enemy Within.* El Dorado, Panama: Focus Publications, 1990.

Kane, Richard. *Campaigns of King William and Queen Anne From 1689 to 1712, Also A New System of Military Discipline for a Battalion of Foot on Action With the Most Essential Exercise of the Cavalry.* London: Printed for J. Millan, 1745.

Kaplan, Fred M. *The Wizards of Armageddon.* New York: Simon & Schuster, 1983.

Karnow, Stanley. *Vietnam, A History.* Revised and updated ed. New York: Viking, 1991.

Karsten, Peter. *Soldiers and Society: The Effects of Military Service and War on American Life*. Westport, CT: Greenwood Press, 1978.

Keegan, John. *The Face of Battle: A Study of Agincourt, Waterloo & the Somme*. New York: Vintage Press, 1977.

Kennan, George F. *Soviet-American Relations: 1917–1920: The Decision to Intervene*. 2 vols. Princeton, NJ: Princeton University Press, 1958.

Kennedy, David M. *Over Here: The First World War and American Society*. New York: Oxford University Press, 1980.

Kier, Elizabeth. *Imagining War: French and British Military Doctrine between the Wars*. Princeton, NJ: Princeton University Press, 1997.

Kissinger, Henry. *Nuclear Weapons and Foreign Policy*. New York: Harper, 1957.

———. *Diplomacy*. New York: Touchstone, 1994.

Kitson, Frank. *Low-Intensity Operations: Subversion, Insurgency, Peacekeeping*. London: Stackpole Books, 1971.

Krepinevich, Andrew F., Jr. *The Army and Vietnam*. Baltimore: Johns Hopkins University Press, 1986.

LaFeber, Walter. *America, Russia, and the Cold War, 1945–1971*. New York: John Wiley & Sons, 1972.

Lambeth, Benjamin S. *NATO's Air War for Kosovo: A Strategic and Operational Assessment*. Santa Monica, CA: Rand, 2001.

Landau, Saul. *The Dangerous Doctrine: National Security and U.S. Foreign Policy*. Boulder, CO: Westview Press, 1988.

Laurens, John. *The Army Correspondence of Colonel John Laurens in the Years 1777–78, Now First Printed from Original Letters of His father, Henry Laurens, President of Congress: With a Memoir by William Gilmore Simms*. New York: Bradford Club, 1867.

Light, Margot, and A. J. R. Groom, eds. *International Relations: A Handbook of Current Theory*. Boulder, CO: Lynne Rienner Publishers, 1985.

Linderman, Gerald F. *Embattled Courage: The Experience of Combat in the American Civil War*. New York: Free Press, 1987.

Linn, Brian McAllister. *The Philippine War, 1899–1902*. Lawrence: University Press of Kansas, 2000.

———. *The Echo of Battle: The Army's Way of War*. Cambridge, MA: Harvard University Press, 2007.

Leckie, Robert. *The Wars of America*. New York: Harper & Row, 1968.

Leopold, Richard W. *Elihu Root and the Conservative Tradition*, ed. Oscar Handlin. Boston: Little, Brown, 1954.

Lewis, Adrian R. *The American Culture of War: The History of U.S. Military Force from World War II to Operation Iraqi Freedom*. New York: Routledge, 2007.

Luvass, Jay. *Frederick the Great on the Art of War*. New York: Free Press, 1966.

MacDonald, Charles B. *Company Commander*. Reprint ed. New York: Bantam Books, 1987.

McCaffrey, James M. *Army of Manifest Destiny: The American Soldier in the Mexican War, 1846–1848*. New York: New York University Press, 1992.

McClellan, George B. *Manual of Bayonet Exercise, Prepared for use of the army of the United States*. Philadelphia: Lippincott, Grambo, & Co., 1852.

McConnell, Malcolm. *Just Cause: The Real Story of America's High-Tech Invasion of Panama*. New York: St. Martin's Press, 1991.

McCullough, David. *Truman*. New York: Simon & Schuster, 1992.

McDonald, Forrest. *E Pluribus Unum: The Formation of the American Republic*. Boston: Houghton & Mifflin, 1965; 2nd ed., Indianapolis, IN: Liberty Press, 1979.

———. *Novus Ordo Seclorum: The Intellectual Origins of the Constitution*. Lawrence: University of Kansas Press, 1985.

McDonough, James R. *Platoon Leader*. New York: Bantam Books, 1986.

McDougall, Walter A. *Promised Land, Crusader State: The American Encounter with the World since 1776*. Boston: Mariner Books, 1997.

McNamara, Robert S. *In Retrospect: The Tragedy and Lessons of Vietnam*. New York: Vintage Books, 1996.

McNeill, William H. *The Pursuit of Power: Technology, Armed Force, and Society since A.D. 1000*. Chicago: University of Chicago Press, 1982.

McWhitney, Grady, and Perry D. Jamieson. *Attack and Die: Civil War Military Tactics and the Southern Heritage*. Tuscaloosa: University of Alabama Press, 1984.

Mahan, Alfred T. *Lessons of the War with Spain and Other Articles*. Boston: Little, Brown, 1899.

Mahan, Dennis Hart. *An Elementary Treatise on an Advanced Guard, Out-post and Detachment Service of Troops and Handling Them in the Presence of an Enemy, with a Historical Sketch of the Rise and Progress of Tactics, &c., &c.* New York: Wiley & Long, 1847, 1853, 1863.

Mahon, John K. *History of the Second Seminole War, 1835–1842*. Rev. ed. Gainesville: University of Florida Press, 1985.

Malone, Patrick M. *The Skulking Way of War: Technology and Tactics among the New England Indians*. New York: Madison Books, 1991.

Mansoor, Peter. *The G.I. Offensive in Europe: The Triumph of American Infantry Divisions, 1941–1945*. Lawrence: University Press of Kansas, 1999.

Marston, Daniel, and Carter Malkasian, eds. *Counterinsurgency in Modern Warfare*. London: Osprey Publishing, 2008.

Martin, Joseph Plumb. *Private Yankee Doodle: Being a Narrative of Some of the Adventures, Dangers, and Sufferings of a Revolutionary Soldier*, ed. George F. Scheer. Boston: Little, Brown, 1962.

Mataxis, Theodore C., and Seymore L. Goldberg. *Nuclear Tactics*. Harrisburg, PA: Military Service Publishing Co., 1958.

Millett, Allan R. *Semper Fidelis: The History of the United States Marine Corps*, revised and expanded ed. New York: Free Press, 1991.

————, and Peter Maslowski. *For the Common Defense: A Military History of the United States of America*. New York: Free Press, 1994.

Millis, Walter, ed. *American Military Thought*. Indianapolis: Bobbs-Merrill, Inc., 1966.

Moltke, Helmuth Karl Bernhard von. *Moltke on the Art of War: Selected Writings*, ed. and trans. Daniel J. Hughes and Harry Bell. Novato, CA: Presidio Press, 1993.

Moore, Harold G., and Joseph L. Galloway. *We Were Soldiers Once...and Young: Ia Drang: The Battle That Changed the War in Vietnam*. New York: Random House, 1992.

Nation, R. Craig. *Black Earth, Red Star: A History of Soviet Security Policy, 1917–1991*. Ithaca, NY: Cornell University Press, 1992.

Neimeyer, Charles Patrick. *America Goes to War: A Social History of the Continental Army*. New York: New York University Press, 1996.

Nenninger, Timothy K. *The Leavenworth Schools and the Old Army: Education, Professionalism, and the Officer Corps of the United States Army, 1881–1918*. Westport, CT: Greenwood Press, 1978.

Nesmith, James H. *The Soldier's Manual For Cavalry, Artillery, Light Infantry, and Infantry. Embellished with Twelve Plates, Representing Different Volunteer Corps, in the First Division Pennsylvania Militia*. Philadelphia: Published by James H. Nesmith, 1824; reprint, Washington, DC: Company of Military Collectors and Historians, 1963.

Nosworthy, Brent. *The Anatomy of Victory: Battle Tactics, 1689–1763*. New York: Hippocrene Books, 1990.

Novick, Peter. *That Noble Dream: The "Objectivity Question" and the American Historical Profession*. Cambridge: Cambridge University Press, 1988.

Odom, William O. *After the Trenches: The Transformation of U.S. Army Doctrine, 1918–1939*. College Station: Texas A&M University Press, 1999.

O'Toole, G. J. A. *The Spanish War: An American Epic, 1898*. Norton ed. 1984; reprint, New York: W. W. Norton, 1986.

Palmer, Scott W. *Dictatorship of the Air: Aviation Culture and the Fate of Modern Russia*. New York: Cambridge University Press, 2009.

Paret, Peter, ed. *Makers of Modern Strategy: From Machiavelli to the Nuclear Age*. Princeton, NJ: Princeton University Press, 1986.

Parkes, Henry Bamford. *A History of Mexico*. Boston: Houghton Mifflin, 1938.

Parks, David. *G.I. Diary*. New York: Harper & Row, Publishers, 1968.

Pearlman, Michael D. *Warmaking and American Democracy: The Struggle over Military Strategy, 1700 to the Present*. Lawrence: University of Kansas Press, 1999.

Pershing, John J. *My Experiences in the World War*. 2 vols. New York: Frederick A. Stokes Co., 1931.

Pickering, Timothy. *An Easy Plan of Discipline for a Militia*. Salem, MA: Printed by Samuel and Ebenezer Hall, 1775.

Pike, Douglas. *PAVN: People's Army of Vietnam*. Novato, CA: Presidio Press, 1986.

Polmar, Norman. *Strategic Weapons: An Introduction*. Rev. ed. New York: Crane Russack, National Strategic Information Center, Inc., 1982.

Porch, Douglas. *The Path to Victory: The Mediterranean Theater in World War II*. New York: Farrar, Straus & Giroux, 2004.

Porter, Bruce D. *War and the Rise of the State: The Military Foundations of Modern Politics*. New York: Free Press, 2002.

Posen, Barry R. *The Sources of Military Doctrine: France, Britain, and Germany between the World Wars*. Ithaca, NY: Cornell University Press, 1984.

Post, Charles Johnson. *The Little War of Private Post*. Boston: Little, Brown, 1960.

Prange, Gordon W. *At Dawn We Slept: The Untold Story of Pearl Harbor*. New York: McGraw-Hill, 1981.

Preston, Diana. *The Boxer Rebellion: The Dramatic Story of China's War on Foreigners That Shook the World in the Summer of 1900*. New York: Berkley Books, 2000.

Preston, Richard A., and Sydney F. Wise. *Men in Arms: A History of Warfare and Its Interrelationships with Western Society*, 4th ed. New York: Holt, Rinehart, & Winston, 1979.

Prucha, Francis Paul. *The Sword of the Republic: The United States Army on the Frontier, 1783–1846*. Bloomington: Indiana University Press, 1977.

Reardon, Carol. *Soldiers and Scholars: The U.S. Army and the Uses of Military History, 1865–1920*. Lawrence: University Press of Kansas, 1990.

Reichart, John F., and Steven R. Sturm, eds. *American Defense Policy*, 5th ed. Baltimore: Johns Hopkins University Press, 1982.

Remini, Robert V. *Andrew Jackson and His Indian Wars*. New York: Viking, 2001.

Riasanovsky, Nicholas V. *A History of Russia*, 4th ed. New York: Oxford University Press, 1984.

Ridgway, General Matthew B. *The Korean War*. New York: Doubleday, Inc., 1967.

Rogers, Alan. *Empire and Liberty: American Resistance to British Authority, 1755–1763*. Berkeley: University of California Press, 1974.

Roosevelt, Theodore. *The Rough Riders*. Dallas: Taylor Publishing, 1997.

Root, Elihu. *The Military and Colonial Policy of the United States; Addresses and Reports by Elihu Root*. Collected and edited by Robert Bacon and James Brown Scott. Cambridge, MA: Harvard University Press, 1916.

Roots of Strategy. 3 vols. Harrisburg, PA: Stackpole Books, 1987.

Rosinski, Herbert. *The German Army*, 2nd ed. London: Hogarth Press, 1940.

Ross, Steven T. *From Flintlock to Rifle: Infantry Tactics, 1740–1866*, 2nd ed. London: Frank Cass, 1996.

Rothenberg, Gunther E. *The Art of Warfare in the Age of Napoleon*. Bloomington: Indiana University Press, 1980.

Royster, Charles. *A Revolutionary People at War: The Continental Army and American Character, 1775–1783*. Chapel Hill: The University of North Carolina Press, 1979.

Ryan, Paul B. *The Iranian Rescue Mission*. Annapolis: U.S. Naval Institute Press, 1985.

Sarkesian, Sam C. *The New Battlefield*. New York: Greenwood Press, 1986.

———. *U.S. National Security, Policymakers, Processes, and Politics*. Boulder, CO: Lynne Rienner Publishers, 1989.

Saul, Norman. *War and Revolution: The United States and Russia, 1914–1921*. Lawrence: University Press of Kansas, 2001.

Savas, Theodore P., and J. David Dameron. *A Guide to the Battles of the American Revolution*. El Dorado Hills, CA: Savas Beatie, 2006.

Schneider, Benjamin, ed. *Organizational Climate and Culture*. San Francisco: Jossey-Bass Publishers, 1990.

Schlesinger, Arthur M., Jr. *The Almanac of American History*. New York: Barnes & Noble Books, 1993.

Scott, Winfield. *Infantry Tactics*. 3 vols., new ed. New York: Hagen & Brothers, 1840.

———. *Memoirs of Lieutenant General Scott*. 2 vols. New York: Sheldon, 1864.

Smyth, Alexander. *Regulations for the Field Exercise, Maneuvers, and Conduct of the Infantry of the United States Drawn Up and Adapted to the Organization of the Militia and Regular Troops*. Philadelphia: Anthony Finley, 1812.

Sherman, William Tecumseh. *Memoirs of General W. T. Sherman*. New York: Literary Classics of the United States, 1990.

Shy, John. *Toward Lexington: The Role of the British Army in the Coming of the American Revolution*. New York: Oxford University Press, 1976.

———. *A People Numerous and Armed: Reflections on the Military Struggle for American Independence*. Rev. ed. Ann Arbor: University of Michigan Press, 1990.

Simpson, Charles M. *Inside the Green Berets: The Story of the U.S. Army Special Forces*. New York: Berkley Books, 1984.

Skelton, William B. *An American Profession of Arms: The Army Officer Corps, 1784–1861*. Lawrence: University Press of Kansas, 1992.

Smith, John. "The Proceedings of the English Colonies in Virginia." *The Complete Works of John Smith*, ed. Philip L. Barbour. 3 vols. Chapel Hill: University of North Carolina Press, 1986.

Smith, Sherry L. *The View from Officers' Row: Army Perceptions of Western Indians*. 3rd printing. Tucson: University of Arizona Press, 1995.

Spaulding, Oliver L. *The United States Army in War and Peace*. New York: G. P. Putnam's Sons, 1937.

Spiller, Roger J., Joseph G. Dawson III, and T. Harry Williams, eds. *Dictionary of American Military Biography*, 3 vols. Westport, CT: Greenwood Press, 1984.

Stanton, Shelby L. *Vietnam: Order of Battle*. Washington, DC: U.S. News Books, 1981.

———. *The Rise and Fall of an American Army: U.S. Ground Forces in Vietnam, 1965–1973*. Novato, CA: Presidio Press, 1985.

Stephenson, Roger. *Military Instruction for Officers in the Field: Containing a Scheme for forming a corps of a partisan, 1775*. Philadelphia: R. Aitken, 1775.

Steuben, Baron Friedrich Wilhelm Ludolf Gerhard Augustin. *Regulations for the Order and Discipline of the Troops of the United States, Part 1*. Albany, NY: Backus & Whiting, 1897.

Sullivan, John. *The Letters and Papers of Major General John Sullivan*, ed. Otis G. Hammond. Collections of the New Hampshire Historical Society. 3 vols. Concord, New Hampshire, 1930–1939.

Thatcher, James. *A Military Journal during the American Revolutionary War from 1775–1783*. Boston: Cottons & Barnard, 1827.

Thompson, Sir Robert. *Defeating Communist Insurgency: The Lessons of Malaya and Vietnam*. New York: Praeger, 1966.

Tierney, John J., Jr. *Chasing Ghosts: Unconventional Warfare in American History*. Washington, DC: Potomac Books, 2007.

Titus, James. *The Old Dominion at War: Society, Politics, and Warfare in Late Colonial Virginia*. Columbia: University of South Carolina Press, 1991.

Toffler, Alvin, and Heidi Toffler. *War and Anti-War, Survival at the Dawn of the 21st Century*. Boston: Little, Brown, 1993.

Toland, John. *In Mortal Combat: Korea, 1950–1953*. New York: Quill, William Morrow, 1991.

Trask, David F. *The War with Spain in 1898*. New York: Macmillan, 1981.

Trauschweizer, Ingo. *The Cold War U.S. Army: Building Deterrence for Limited War*. Lawrence: University Press of Kansas, 2008.

Truman, Harry S. *Memoirs by Harry S. Truman*. 2 vols. Garden City, NY: Doubleday, 1956.

Tsouras, Peter G., ed. *The Greenhill Dictionary of Military Quotations*. London: Greenhill Books, 2000.

Turpin, Lancelot de Crisse. *An Essay on the Art of War*, trans. Joseph Otway. London: Printed by A. Hamilton for W. Johnson, 1761.

Unterberger, Betty Miller. *America's Siberian Expedition, 1918–1920*. Durham: Duke University Press, 1956.

Upton, Emory. *A New System of Infantry Tactics, Double and Single Rank, Adapted to American Topography and Improved Fire-Arms*. New York: D. Appleton & Co., 1867.

Utley, Robert M. *Frontiersmen in Blue: The United States Army and the Indian, 1848–1865*. Lincoln: University of Nebraska Press, 1967.

———. *Frontier Regulars: The United States Army and the Indian, 1866–1891*. Reprint ed. Lincoln: University of Nebraska Press, 1984.

Wagner, Arthur L. *Organization and Tactics*, 6th ed. Kansas City: Hudson & Kimberly Publishing, 1905.

Waller, Douglas C. *The Commandos: The Inside Story of America's Secret Soldiers*. New York: Simon & Schuster, 1994.

Webb, Thomas. *A Military Treatise on the Appointments of the Army Containing Many Useful Hints, Not touched upon by any Author, and proposing some New regulations* in the Army, which will be particularly Useful in Carrying out the War in North-America, A Short Treatise with Military Honors. Philadelphia: W. Dunlop, 1759.

Weigley, Russell F. History of the United States Army. New York: Macmillan, Inc., 1967.

———. The American Way of War: A History of United States Military Strategy and Policy. Bloomington: Indiana University Press, 1977.

———. Eisenhower's Lieutenants: The Campaigns of France and Germany, 1944–1945. Bloomington: Indiana University Press, 1981.

White, N. D. Keeping the Peace: The United Nations and the Maintenance of International Peace and Security. Reprint ed. Manchester, NH: Manchester University Press, 1995.

Wilson, Frazer Ells. Arthur St. Clair, Rugged Ruler of the Old Northwest: An Epic of the American Frontier. Richmond, VA: Garrett & Massie, Publishers, 1944.

Wood, Gordon S. The Creation of the American Republic, 1776–1787. Chapel Hill: University of North Carolina Press, 1969.

Wood, W. J. Civil War Generalship: The Art of Command. New York: Da Capo Press, 2000.

Wooster, Robert. The American Military Frontiers; The United States Army in the West, 1783–1900. Albuquerque: University of New Mexico Press, 2009.

Wright, William C., ed. New Jersey in the American Revolution. 3 vols. New Jersey Historical Commission, 1976.

Zinn, Howard. The Twentieth Century: A People's History. New York: Harper Perennial, 1998.

UNPUBLISHED MANUSCRIPTS, DISSERTATIONS, AND THESES

Adams, Thomas K. "Military Doctrine and the Organization Culture of the United States Army." Ph.D. diss., Syracuse University, 1990.

Akard, Bruce E. "Strategic Deployment: An Analysis of How U.S. Army Europe Deployed VIIth Corps to Desert Storm and the 1st Armored Division to Bosnia." Master's thesis, U.S. Army Command and General Staff College, 1997.

Bowman, Stephen L. "The Evolution of United States Army Doctrine for Counterinsurgency Warfare: From World War II to the Commitment of Combat Units in Vietnam." Ph.D. diss., Duke University, 1985.

Chase, Philander Dean. "Baron von Steuben in the War of Independence." Ph.D. diss., Duke University, 1972.

DeMontravel, Peter. "The Career of Lieutenant General Nelson A. Miles from the Civil War through the Indian Wars." Ph.D. diss., Ohio State University, 1986.

Dederer, John Morgan. "The Origins and Development of American Conceptions of War to 1775." Ph.D. diss., University of Alabama, 1988.

Driggers, Dennis S. "The United States Army's Long March from Saigon to Baghdad: The Development of War-Fighting Doctrine in the Post-Vietnam Era." Ph.D. diss., Syracuse University, 1995.

Edwards, Britt Lynn. "Reforming the Army: The Formulation and Implementation of AirLand Battle 2000." Ph.D. diss., University of California Santa Barbara, 1985.

Haynes, Peter D. "American Culture, Military Services' Cultures, and Military Strategy." Master's thesis, Naval Post Graduate School, Monterey, California, 1998.

Lowe, Alan C. "Foreign Devils and Boxers: A Concise History of Combined Interoperability during the Boxer Rebellion." Master's thesis, U.S. Army Command and General Staff College, Fort Leavenworth, KS, 2000.

Price, Barrye L. "Against All Enemies: A Study of Civil Unrest and Federal Intervention within the United States from 1794 to 1968." Unpublished research paper (photocopy), n.d. Combined Arms Research Library, Fort Leavenworth, KS.

Rudd, Gordon W. "Operation Provide Comfort: Humanitarian Intervention in Northern Iraq, 1991." Ph.D. diss., Duke University, 1993.

Schrader, Charles R. "'Such History as Every Young Gentleman Should Be Presumed to Know': Officer Education and Military History—An Opinion, 1999 (?)." Photocopy provided to author by Roger J. Spiller, George C. Marshall Chair of Military History, U.S. Army Command and General Staff College.

Schneider, James J. "Vulcan's Anvil: The American Civil War and the Emergence of Operational Art." Fort Leavenworth, KS: School of Advanced Military Studies, Theoretical Paper no. 4, 16 June 1991.

Ulrich, William. "The Northern Military Mind in Regard to Reconstruction, 1865–72: The Attitude of Ten Leading Union Generals." Ph.D. diss., Ohio State University, 1959.

ARTICLES AND MONOGRAPHS

Abizaid, John P. "Lessons for Peacekeepers." *Military Review* 73, no. 3 (March 1993): 11–19.

Adams, Thomas K. "Intervention in Haiti: Lessons Relearned." *Military Review* 76, no. 5 (September–October 1996): 45–56.

Arnold, S. L. "Somalia: An Operation Other than War." *Military Review* 73, no. 12 (December 1993): 26–35.

Atkeson, Edward B. "FM 3-0: An Assessment." *Army* 58, no. 8 (August 2008): 20–21.

Auguste le Grandmaison, Thomas. *La Petite Guerre, ou traite du service des troupes légères en campagne.* Not Published, 1756. Special Collections, United States Military Academy Library at West Point, West Point, NY.

Baratto, David J. "Special Forces in the 1980s: A Strategic Reorientation." *Military Review* 63, no. 3 (March 1983): 2–14.

Barrell, Marjorie. "Cleanup on Kauai." *Soldiers* 47, no. 12 (December 1992): 6–8.

Baum, Dan. "Battle Lessons, What the Generals Don't Know." *New Yorker* 80, no. 43 (17 January 2005): 42–48.

Benson, Kevin C., and Christopher Thrash. "Declaring Victory: Planning Exit Strategies for Peace Operations." *Parameters* 26, no. 3 (Autumn 1996): 69–80.

Birtle, Andrew J. "The Origins of the Legion of the United States." *Journal of Military History* 67, no. 4 (October 2003): 1249–1262.

Bradley, Francis X. "The Fallacy of Dual Capability." *Army* 10, no. 3 (October 1959): 18–19.

Bradley, Omar N. "Creating a Sound Military Force." *Military Review* 29, no. 2 (May 1949): 3–6.

Brown, Frederic J. "AirLand Battle Future, The Other Side of the Coin." *Military Review* 71, no. 2 (February 1991): 14–24.

Bush, George. "The Possibility of a New World Order." April 13, 1991. *Vital Speeches of the Day* 57, no. 15 (15 May 1991).

Bush, Larry E., Jr. "Building Special Forces to Last: Redesigning the Organizational Culture." *Special Warfare* 12, no. 3 (Summer 1999): 16–20.

Carleton, Linwood A., and Frank A. Farnsworth. "A Philosophy for Tactics." *Military Review* 40, no. 4 (July 1960): 12–22.

Carlin, Thomas M., and Mike Sanders. "Soldier of the Future: Assessment and Selection of Force XXI." *Special Warfare* 9, no. 2 (May 1996): 16–21.

Cherrie, Stanley F. "Task Force Eagle." *Military Review* 77, no. 4 (July–August 1997): 63–72.

Click, Robert L. "Combat Jump." *Army* 42, no. 8 (August 1992): 24–30.

Collins, Arthur S. "The Other Side of the Atom." *Army* 10, no. 4 (November 1959): 18–19.

Collins, Steven N. "*Just Cause* Up Close: A Light Infantryman's View of LIC." *Parameters* 22, no. 2 (Summer 1992): 55–65.

Combat Studies Institute. "The History of Transformation." *Military Review* 80, no. 3 (May–June 2000): 17–29.

Cook, Steven E. "Field Manual 100-25: Updating Army SOF Doctrine." *Special Warfare* 9, no. 3 (August 1996): 36–37.

Creel, George. "Four Million Citizen Defenders." *Everybody's Magazine*, no. 36 (1917): 553.

Crider, James R. "A View from inside the Surge." *Military Review* 89, no. 2 (March–April 2009): 81–88.

Currie, James T. "The First Congressional Investigation: St. Clair's Military Disaster of 1791." *Parameters* 20, no. 4 (December 1990): 95–102.

Cushman, John H. "What Is the Army's Story." *Army Combat Forces Journal* 5, no. 3 (October 1955): 49–51.

DePuy, William E. "The Case for a Dual Capability." *Army* 10, no. 6 (January 1960): 32–38.

Eid, Leroy V. "American Indian Military Leadership: St. Clair's 1791 Defeat." *Journal of Military History* 57, no. 1 (January 1993): 71–88.

Elmo, David S. "Food Distribution during Operation PROVIDE COMFORT."
    Special Warfare 5, no. 1 (March 1992): 8–9.

Ephron, Dan. "Conrad Crane: The Military: With Iraq in Flames, a Historian
    Rethinks the Way We Fight the Enemy." *Newsweek* 148, no. 26 (25 December
    2006): 63.

Fastabend, David. "The Categorization of Conflict." *Parameters* 27, no. 2 (Summer
    1997): 75–85.

———. "FM 100-5, 1998: Endless Evolution." *Army* 47, no. 5 (May 1997): 45–50.

Finch, Frank R. "Piercing Winds Hit Hawaii." *Military Engineer* 85, no. 54
    (January–February 1993): 4–6.

Franks, Frederick M., Jr. "Full-Dimensional Operations: A Doctrine for an Era of
    Change." *Military Review* 73, no. 12 (December 1993): 5–10.

Fridovich, David P., and Fred T. Krawchuk. "The Special Operations Forces Indirect
    Approach." *Joint Forces Quarterly*, Winter 2007, no. 44 (1st Quarter 2007): 24–27.

Ferling, John. "The New England Soldier: A Study in Changing Perceptions."
    American Quarterly 33, no. 1 (Spring 1981): 26–45.

Furlong, Patrick J. "The Investigation of General Arthur St. Clair, 1792–1793."
    Capitol Studies no. 5 (Fall 1977): 65–86.

Garabedian, John. "Terrorism Kills over 500,000 in Rwanda: Violence and
    Combating Terrorism Update." *Lamp* (Fort Leavenworth, KS), September 1994.

Gates, John M. "The Pacification of the Philippines, 1898–1902." *American Military
    and the Far East: Proceedings of the Ninth Military History Symposium, United
    States Air Force Academy, 1–3 October 1980*, Joe C. Dixon, ed. Washington, DC:
    United States Air Force Academy and Office of Air Force History, Headquarters
    USAF, 1980.

Gavin, James M. "The Tactical Use of the Atomic Bomb." *Combat Forces Journal* 1,
    no. 4 (November 1950): 9–11.

Gildrie, Richard P. "Defiance, Diversion, and the Exercise of Arms: The Several
    Meanings of Colonial Training Days in Colonial Massachusetts." *Military Affairs*
    52, no. 2 (April 1988): 52–54.

Graham, Bradley. "Pentagon Officials Worry Aid Missions Will Sap Military
    Strength." *Washington Post* (Washington, DC), 29 July 1994.

Graves, Donald E. "Dry Books of Tactics: U.S. Infantry Manuals of the War of 1812
    and After, Part I." *Military Collector and Historian* 38, no. 2 (Summer 1986): 51–61.

———. "Dry Books of Tactics: U.S. Infantry Manuals of the War of 1812 and After,
    Part II." *Military Collector and Historian* 38, no. 4 (Winter 1986): 173–177.

Gray, Colin S. "Force Planning, Political Guidance and the Decision to Fight."
    Military Review 58, no. 4 (April 1978): 27–36.

Harmon, Robert E., Ramon A. Malave, Charles A. Miller III, and William K.
    Nadolski. "Counterdrug Assistance: The Number One Priority." *Military Review*
    73, no. 3 (March 1993): 26–35.

Holder, L. D. "Doctrinal Development, 1975–85." *Military Review* 55, no. 5 (May 1985): 50–52.

Honeyman, A. Van Doren, ed. "The Revolutionary War Record of Samuel Sutphin, Slave." *Somerset County Historical Quarterly*, no. 3 (1914): 186–190.

Howard, Michael. "Military Science in an Age of Peace." *Journal of the United Services Institute for Defence Studies* 119, no. 1 (March 1974): 3–9.

Hunt, John B. "OOTW: A Conflict in Flux." *Military Review* 76, no. 5 (September–October 1996): 3–9.

Jameson, Hugh. "Equipment for the Militia of the Middle States, 1775–1781." *Journal of the American Military Institute* 3, no. 1 (Spring 1939): 26–38.

Jones, Archer. "The New FM 100-5: A View from the Ivory Tower." *Military Review* 58, no. 2 (February 1978): 27–36.

Johnson, Douglas, V., II, ed. *Warriors in Peace Operations*. Carlisle Barracks, PA: Strategic Studies Institute, U.S. Army War College, 1999.

Karber, Philip A. "Dynamic Doctrine for Dynamic Defense." *Armed Forces Journal International* 114, no. 2 (October 1976): 28–29.

Kelly, Henry E. "Verbal Defense." *Military Review* 35, no. 7 (October 1955): 45–52.

Kennon, L. W. V. "Some Points on Tactics." *Army and Navy Register*, no. 9 (August 25, 1888): 540–541.

Kimball, Jeffrey. "The Battle of Chippawa: Infantry Tactics in the War of 1812." *Military Affairs* 31, no. 4 (Winter 1967–68): 169–186.

Kretchik, Walter E. "Multinational Staff Effectiveness in UN Peace Operations: The Case of the U.S. Army and UNMIH, 1994–1995." *Armed Forces and Society* 29, no. 3 (Spring 2003): 393–413.

Kuehl, Dale. "Testing Galula in Ameriyah: The People Are the Key." *Military Review* 89, no. 2 (March–April 2009): 72–80.

Lange, John E. "Civilian-Military Cooperation and Humanitarian Assistance: Lessons from Rwanda." *Parameters* 28, no. 2 (Summer 1998): 106–122.

Linn, Brian M., and Russell Weigley. "The American Way of War Revisited." *Journal of Military History* 66, no. 2 (April 2002): 501–533.

Marshall, S. L. A. "Arms in Wonderland." *Army*, no. 7 (June 1957): 18–19.

Meek, Basil. "General Harmar's Expedition." *Ohio History*, no. 20 (January 1911): 74–108.

Metz, Steven. "U.S. Strategy and the Changing LIC Threat." *Military Review* 71, no. 6 (June 1991): 22–29.

Miller, John. "The Militia and the Army in the Reign of James II." *Historical Journal* 16, no. 4 (1973): 659–679.

McDonough, James. "Building the New FM 100-5, Process and Product." *Military Review* 71, no. 10 (October 1991): 2–12.

McGrane, R. C. "William Clark's Journal of General Wayne's Campaign." *Mississippi Valley Historical Review* 1, no. 3 (December 1914): 418–444.

Naylor, Sean. "The Invasion That Never Was." *Army Times* (Springfield, VA), 26 (February 1996): 13.

Neighbour, M. R., et al. "Providing Operational Analysis to a Peace Support Operation: The Kosovo Experience." *Journal of the Operational Research Society* 53, no. 5 (2002): 523–543.

Nenninger, Timothy K. "Tactical Dysfunction in the AEF, 1917–1918." *Military Affairs* 51, no. 4 (October 1987): 77–81.

Negrier, General de. "Some Lessons of the Russo-Japanese War." *Journal of the Royal United Service Institution* 50, no. 339 (July–December 1906): 910–919.

Norman, Lloyd, and John B. Spore. "Big Push in Guerrilla Warfare." *Army* 12, no. 9 (March 1962): 28–44.

Odom, William O. "Destined for Defeat: An Analysis of the St. Clair Expedition of 1791." *Northwest Ohio Quarterly* 65, no. 2 (Spring 1993): 68–93.

Ord, Robert L., III, and Ed Mornston. "Light Forces in the Force-Projection Army." *Military Review* 74, no. 1 (January 1994): 22–33.

Packer, George. "Letter from Iraq, the Lesson of Tal Afar: Is It Too Late for the Administration to Correct Its Course in Iraq?" *New Yorker* 82, no. 8 (10 April 2006): 48–65.

Peckham, Howard K. "Josiah Harmar and His Indian Expedition." Speech given at the Ohio State Archeological and Historical Society, 11 April 1946.

Percy, George. "A Trewe Relacyon." *Tyler's Quarterly Historical and Genealogical Magazine* 3, no. 4 (April 1922): 270–273.

"Perspective on Rwanda Support: Commander of the 21st Theater Army Area Support Command Assesses His Command's Support for Operation Support Hope." *Army Logistician* 27, no. 3 (May–June 1995): 4–6.

Phillips, Ed. "Army SOF: Right Tool for OOTW." *Special Warfare* 10, no. 3 (Summer 1997): 2–13.

Pitcavage, Mark. "Ropes of Sand: Territorial Militias, 1801–1812." *Journal of the Early Republic* 13, no. 4 (Winter 1993): 481–500.

Powers, Sandra L. "Studying the Art of War: Military Books Known to American Officers and Their French Counterparts during the Second Half of the Eighteenth Century." *Journal of Military History* 70, no. 3 (July 2006): 781–814.

Pratt, Julius W. "Fur Trader Strategy and the American Left Flank in the War of 1812." *American Historical Review*, no. 40 (January 1935): 246–273.

Prucha, Francis Paul. "The United States Army as Viewed by British Travelers, 1825–1860." *Military Affairs* 17, no. 3 (Fall 1953): 113–124.

Reeve, Frank D., ed. "Puritan and Apache: A Diary [of Lieutenant H. M. Lazelle]." *New Mexico Historical Review* 23, no. 4 (October 1948): 269–301.

Richardson, William R. "FM 100-5: The AirLand Battle in 1986." *Military Review* 66, no. 2 (March 1986): 4–11.

Roberts, William E. "Keeping Pace with the Future—Training Officers to Fight on Atomic Battlefields." *Military Review* 37, no. 7 (October 1957): 22–29.

Romjue, John L. "AirLand Battle: The Historical Background." *Military Review* 66, no. 2 (March 1986): 52–55.

Russell, Peter E. "Redcoats in the Wilderness: British Officers and Irregular Warfare in Europe and America, 1740 to 1760." *William and Mary Quarterly*, 3rd ser., 35, no. 4 (October 1978): 629–652.

Ruston, Jay C. "Anthony Wayne and the Indian Campaign, 1792–1794." *Indiana Military History Journal*, no. 7 (May 1982): 21–25.

Sanger, David E. "Plans for Postwar Iraq Are Re-evaluated as Fast Military Exit Looks Less Likely." *New York Times*, 2 April 2003.

Shacochis, Bob. "Our Two Armies in Haiti." *New York Times*, 8 January 1995.

Schaffel, Kenneth. "The American Board of War." *Military Affairs* 50, no. 4 (October 1986): 185–189.

Shipton, Clifford K. "Literary Leaven in Provincial New England." *New England Quarterly* 9, no. 2 (June 1936): 203–217.

Shea, William L. "The First American Militia." *Military Affairs* 46, no. 1 (February 1982): 15–18.

Shy, John. "A New Look at Colonial Militia." *William and Mary Quarterly*, 3rd ser., 20, no. 2 (April 1963): 175–185.

Silvasy, Stephen, Jr. "AirLand Battle Future, The Tactical Battlefield." *Military Review* 71, no. 2 (February 1991): 3–4.

Siochru, Michael O. "Atrocity, Codes of Conduct and the Irish in the British Civil Wars, 1641–1653." *Past and Present* no. 195 (May 2007): 55–86.

Snyder, James M. "ROAD Division Command Staff Relationships." *Military Review* 43, no. 1 (January 1963): 57–62.

Spiller, Roger J. "War History and the History Wars: Establishing the Combat Studies Institute." *Public Historian* 10, no. 4 (Fall 1988): 65–81.

———. "The Tenth Imperative." *Military Review* 69, no. 4 (April 1989): 2–13.

Steele, Dennis. "Digging Out from Despair." *Army* 42, no. 11 (November 1992): 22–25.

Stiner, Carl W. "The Strategic Employment of Special Operations Forces." *Military Review* 71, no. 6 (June 1991): 3–13.

Sullivan, Brian R. "Special Operations and LIC in the 21st Century: The Joint Strategic Perspective." *Special Warfare* 9, no. 2 (May 1996): 2–7.

Sullivan, Gordon R. "Doctrine: A Guide to the Future." *Military Review* 72, no. 2 (February 1992): 2–9.

———. "Hurricane Andrew: An After-Action Report." *Army* 43, no. 1 (January 1993): 16–22.

———. *America's Army into the Twenty First Century*. National Security Paper no. 14. Cambridge: Institute for Foreign Policy Analysis, 1993.

———. "A Vision for the Future." *Military Review* 75, no. 3 (May–June 1995): 5–14.

Thurman, Edward E. "Shaping an Army for Peace, Crisis and War: The Continuum of Military Operations." *Military Review* 72, no. 4 (April 1992): 27–35.

Thurman, Maxwell R., and William Hartzog. "Simultaneity: The Panama Case." *Army* 43, no. 11 (November 1993): 16–24.

Tooke, Lamar. "Blending Maneuver and Attrition." *Military Review* 90, no. 2 (March–April 2000): 7–13.

Vestal, Samuel C. "Frederick William von Steuben." *Coast Artillery Journal* 75, no. 4 (July–August 1932): 289–293.

Vigman, Fred K. "A 1745 Plan for . . . a National Militia in Great Britain . . . and America." *Military Affairs* 9, no. 4 (Winter 1945): 355–360.

Washington Post. "U.S. Flies Iraqis to Guam." 7 December 1996.

Wallace, Stephen O. "Joint Task Force Support Hope: The Role of the CMOC in Humanitarian Operations." *Special Warfare* 9, no. 1 (January 1996): 36–41.

Wallace, William S. "FM 3-0 *Operations*: The Army's Blueprint." *Military Review* 88, no. 2 (March–April 2008): 2–7.

Warner, Michael S. "General Josiah Harmar's Campaign Reconsidered: How the Americans Lost the Battle of Kekionga." *Indiana Magazine of History* 83, no. 1 (March 1987): 45–63.

Wells, Samuel F., Jr. "The Origins of Massive Retaliation." *Political Science Quarterly* 96, no. 1 (Spring 1981): 31–52.

West, Bing. "Counterinsurgency Lessons from Iraq." *Military Review* 89, no. 2 (March–April 2009): 2–12.

Wright, John W. "Some Notes on the Continental Army." First Installment. *William and Mary College Quarterly Historical Magazine*, 2nd ser., 11, no. 2 (April 1931): 81–105.

Willcox, William B. "Too Many Cooks: British Planning before Saratoga." *Journal of British Studies* 2, no. 1 (November 1962): 56–90.

Zierdt, William H., Jr. "The Structure of the New Army Divisions." *Army* 11, no. 7 (July 1961): 49–52, 62.

ORAL HISTORIES AND INTERVIEWS

Lieutenant Colonel Kevin C. Benson. By author, Fort Leavenworth, KS, 20 January 1998, and by author [telephone], Fort Lewis, Washington, June 2000.

Lieutenant Colonel (Ret.) Glen Boney. By author, Leavenworth, KS, March 2000.

Lieutenant Colonel Gordon C. Bonham. By author, School of Advanced Military Studies (SAMS), Fort Leavenworth, KS, January 1999.

Colonel Mark C. Boyatt. By author, Fort Leavenworth, KS, November 1997.

Major Stewart Braden. By author, Fort Leavenworth, KS, July 2000.

Lieutenant Colonel (Ret.) Michael Burke. By author, Fort Leavenworth, KS, 29 January 2001.

Lieutenant Colonel John Carmichael. By author, Fort Leavenworth, KS, September 1997.

Lieutenant Colonel Edward Donnelly. By author, Headquarters U.S. Atlantic Command, Norfolk, VA, December 1995.

Major Joseph Doyle. By author, Headquarters U.S. Atlantic Command, Norfolk, VA, December 1995.

Major Len Gaddis. By author, Fort Leavenworth, KS, March 1997.

Lieutenant Colonel (Ret.) Russell Glenn. By author [telephone], Santa Monica, CA, 1 February 2001.

Lieutenant Colonel (Ret.) John B. Hunt. By author, Leavenworth, KS, February 2001.

Lieutenant Colonel Phil Idiart. By author, Headquarters U.S. Atlantic Command, Norfolk, VA, December 1995.

Lieutenant General Joseph Kinzer. By author, San Antonio, TX, 2 June 1998.

Major Berthony Ladouceur. By author, Fort Leavenworth, KS, May 1997.

Lieutenant Colonel Alan C. Lowe. By author, Fort Leavenworth, KS, May 1999.

Major General (Ret.) David C. Meade. By author, Stafford, VA, 29 April 2000.

Major Drew Meyerowich. By author, Fort Leavenworth, KS, March 1998.

Lieutenant Colonel John J. Moore. By author, U.S. Army Command and General Staff College, Fort Leavenworth, KS, February 1991.

Major James Mulvenna. By author, U.S. Army Command and General Staff College, Fort Leavenworth, KS, May 1991.

William G. O'Neill. By author, Port-au-Prince, Haiti, October 1996.

Colonel Michael Parker. By author, Fort Leavenworth, KS, June 1998.

Major Daniel A. Pinnell. By Robert F. Baumann, Combat Studies Institute, U.S. Army Command and General Staff College, Leavenworth, KS, 20 April 1999.

Lieutenant Colonel Michael Rampy. By author, School of Advanced Military Science (SAMS), Fort Leavenworth, KS, October 1991.

Lieutenant Colonel James Rentz. By author, Fort Leavenworth, KS, May 2000.

Lieutenant Colonel (Ret.) Richard Rinaldo. By author [telephone], Fort Monroe, VA, 11 March 1999.

Lieutenant Colonel Victor Robertson. By author, Fort Leavenworth, KS, March 1999.

Lieutenant General Henry H. Shelton. By Steve Dietrich, Port-au-Prince, Haiti, 24 October 1994. In Cynthia Hayden, ed., *JTF-180 Uphold Democracy: Oral History Interviews*. Fort Bragg, NC: XVIIIth Airborne Corps History Office, 1996.

Lieutenant Colonel George Steuber. By author, Fort Leavenworth, KS, April 1998.

Roger J. Spiller, George C. Marshall Chair of Military History, conversation with author, U.S. Army Command and General Staff College, Fort Leavenworth, KS, April 1995.

General (Ret.) Donn Starry. By author, U.S. Army Command and General Staff College, Fort Leavenworth, KS, May 2000.

Major John Valledor. By author, Fort Leavenworth, KS, December 1998.

Major Kristen Vlahos-Schafer. By author, Fort Leavenworth, KS, March 1997.

"Alexander Smyth." *War of 1812: People and Stories.* Companion website to documentary by Galafilm. http://www.galafilm.com/1812/e/people/smyth.html.

Appleton's *Cyclopedia of American Biography.* Edited by James Grant Wilson, John Fiske, and Stanley L. Klos. Six vols. New York: D. Appleton & Co., 1887–1889. StanKlos.com, 1999. http://www.famousamericans.net/williamduane/.

Association of the United States Army. "Revolution in Army Doctrine: The 2008 Field Manual 2008 3-0 *Operations.*" February 2008. http://www.ausa.org/ publications/torchbearercampaign/torchbearerissuepapers/Documents/ TBIP_022508.pdf

Author unknown. "A Brief History of U.S. Infantry Tactics (1855–1865)." http:// www.usregulars.com/drill_history.html.

Center for Army Lessons Learned. "After Action Report: Joint Task Force Los Angeles, 1992." http://call.army.mil/call/newsltrs/93-7/937ch2.htm.

"General Framework Agreement for Peace in Bosnia and Herzegovina." http:// www.state/gov/www/current/bosnia/dayframe.html.

Headquarters, Department of the Army. Field Manual 3-24 / Headquarters, U.S. Marine Corps, Marine Corps Warfighting Publication 3-33.5 *Counterinsurgency.* Washington, DC: Government Printing Office, December 2006. http://www .fas.org/irp/doddir/army/fm3-24.pdf.

Hussain, Hamid. "'Courageous Colonels'—Current History Recap." *Small Wars Journal.* December 23, 2008. http://smallwarsjournal.com/blog/2008/12/ courageous-colonels-current-hi/.

Kahl, Colin. "The Four Phases of the U.S. COIN Effort in Iraq." E-mail posted online. *Small Wars Journal.* March 18, 2007. http://smallwarsjournal.com/ blog/2007/03/the-four-phases-of-the-us-coin/.

Kretchik, Walter E., producer and director. *Rhythm of the Street,* VHS film. 69 min. Fort Leavenworth, KS: Combat Studies Institute, 1998.

Lovelace, Douglas C., Jr. "Unification of the United States Armed Forces: Implementing the 1986 Department of Defense Reorganization Act." Carlisle, PA: Strategic Studies Institute, 1996, foreword. http://carlisle-www.army.mil/ usassi/ssipubs/pubs96/dodact/dodact.htm.

Military.com. "New Counterinsurgency Manual" [Comments on FM 3-24]. December 18, 2006. http://www.military.com/features/0,15240,120810,00 .html.

Nagl, John. "The Evolution and Importance of Army/Marine Corps Field Manual 3-24, Counterinsurgency." *Small Wars Journal.* 27 June 2007. http:// smallwarsjournal.com/blog/2007/06/the-evolution-and-importance-0/.

Power, Samantha. "Our War on Terror." 9 July 2007. Posted at *Small Wars Journal,* 26 July 2007. http://smallwarsjournal.com/blog/2007/07/ny-times-book -review-fm-324-1/.

Riley, Kevin. "Theorist, Instructors, and Practitioners: The Evolution of French

Doctrine in the Revolutionary and Napoleonic Wars, 1792–1815," *The Napoleon Series.* http://www.napoleon-series.org/military/organization/c_tactics.html.

Tunnel IV, Major Harry D. *To Compel with Armed Force: A Staff Ride Handbook for the Battle of Tippecanoe.* Fort Leavenworth, KS: Combat Studies Institute Press, 2000. http://www.cgsc.edu/carl/resources/csi/tunnell/tunnell.asp.

The U.S. Army/Marine Corps Counterinsurgency Field Manual. U.S. Army Field Manual no. 3-24/Marine Corps Warfighting Publication no. 3-33.5. Chicago: University of Chicago Press, 2007.

Walden, Geoff, and Dom Dal Bello. "Manual of Arms for Infantry: A Re-Examination, Part I." http://www.drillnet/manalarms_1.htm.

van Patten, R. E. "Arms Drill for Volunteer Militia in the Old Northwest." *Casebook—The War of 1812.* http://www.warof1812.casebook.org/articles/dissertation.html; Internet.

Office of the Secretary of Defense. "Civil War Casualty Figures." http://web1.whs.osd.mil/mmid/m01/sms223r.htm.

INDEX

Abrams, Creighton W., Jr., 189, 194
Abrams, John, 242
active defense, 197, 200, 226
advance guard, 113, 119, 170. *See also*
    tactics, offensive
"aeroplanes." *See* technology
Afghanistan, 256–258
Aguinaldo, Emilio, 99–100
airborne operations, 155–156, 214–216
Air Corps
    mission, 144–146
    1926 Air Corps Act, 141
AirLand Battle
    development, 196–197, 209–211
    discussion and evolution, 219, 222–226
    Operation Desert Storm, 216–217
    Operation Just Cause, 214–216
    Operation Provide Comfort, 219–220
    Operation Restore Hope, 232–234
    Operation Uphold Democracy, 238–239
    Operation Urgent Fury, 209
    tenets, 200, 206, 209
AirLand Battle III, 210
AirLand Battle 2000, 211
airmobile operations, 182–183, 192–193
airpower
    strategic reconnaissance, 117, 122
    tactical reconnaissance, 117
    World War I, 129–130, 150
    World War II, 153–157
Allen, Robert H., 133
American Mission to Aid Greece (AMAG),
    160
American Revolution
    Battle of Saratoga, 15
    lack of doctrine, 13–14
    Steuben and, 16–20
    Washington, George, 14–17, 20–22
amphibious operations, 61, 151–152, 155
Arab-Israeli War (1973), 195–196, 199,
    282
Archangel (Russia), 131
Ardant du Picq, Charles J., 80, 85
area of influence, 206, 274

area of interest, 206, 274
Arista, Mariano, 60–61
armor, 130, 156, 163, 170–172, 180,
    198–199
    Modern Mobile Army (MOMAR), 179
    Pentomic division, 175
    pre–WWII, 140, 145, 148
    Vietnam, 186, 188–189
Army School of the Line, 109
Army Training and Evaluation Program
    (ARTEP), 201
Army War College, 109, 138–140, 169,
    180, 200
Arnold, Henry Harley (Hap), 150, 159
Arnold, Steven L., 234
artillery
    Boxer Rebellion, 104
    FM 100-5 *Operations* (1949), 162–163
    FSR (1905), 113
    FSR (1914), 125
    FSR (1923), 137
    Mexican War, 60–61
    Napoleon, 50
    Tentative *FSR* FM 100-5 *Operations*
        (1939), 145, 146
    vs. infantry, 115
Artillery School of Practice (Fort Monroe,
    VA), 56
Artillery Tactics (1875), 85
Art of War (Jomini), 56, 71, 111
assault, 66, 119
    amphibious, 151
    frontal, 75–76, 113, 127, 174–175, 178
    Veracruz (Mexico), 61
    See also attacks
atomic weapons. *See* nuclear weapons
attacks
    concentrated, 149
    counterattack, 119, 171, 191, 230
    decisive, 149
    deep, 203–204, 208
    deliberate, 207
    enveloping, 119, 149
    extended order, 85